GCSE Chemistry

No doubt about it, GCSE Chemistry is a tough subject. Not to worry — this CGP book has it covered, from facts and theory to practical skills.

What's more, we've included exam-style practice in every topic, *and* put in a full set of practice papers to help you sharpen up your exam skills.

There are fully-worked answers at the back too, so it's easy to check how you're doing and find out how to pick up the best marks possible!

How to access your free Online Edition

This book includes a free Online Edition to read on your PC, Mac or tablet. You'll just need to go to **cgpbooks.co.uk/extras** and enter this code:

0467 3777 4643 9539

By the way, this code only works for one person. If somebody else has used this book before you, they might have already claimed the Online Edition.

Complete Revision & Practice

Everything you need to pass the exams!

Contents

Working Scientifically

Topic 1 — Atomic Structure and the Periodic Table

Topic 2 — Bonding, Structure and Properties of Matter

Topic 3 — Quantitative Chemistry

Throughout this book you'll see grade stamps like these: Grade 4-6, Grade 6-7, Grade 7-9

These grade stamps help to show how difficult the questions are.

Remember — to get a top grade you need to be able to answer **all** the questions, not just the hardest ones.

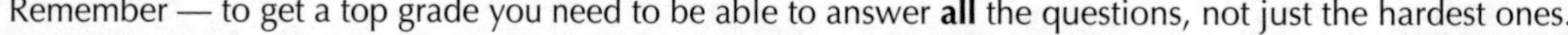

In the real exams, some questions test how well you can write (as well as your scientific knowledge). In this book, we've marked these questions with an asterisk (*).

Topic 4 — Chemical Changes

Topic 5 — Energy Changes

Topic 6 — The Rate and Extent of Chemical Change

Topic 7 — Organic Chemistry

Topic 8 — Chemical Analysis

Topic 9 — Chemistry of the Atmosphere

Topic 10 — Using Resources

Practical Skills

Practice Exams

Published by CGP

Editors: Mary Falkner, Robin Flello, Emily Howe, Paul Jordin, Christopher Lindle and Ciara McGlade.

Contributors: Mike Bossart and Paddy Gannon.

From original material by Richard Parsons.

With thanks to Glenn Rogers, Sophie Scott, Jamie Sinclair and Hayley Thompson for the proofreading.

With thanks to Ana Pungartnik for the copyright research.

Printed by Elanders Ltd, Newcastle upon Tyne.
Clipart from Corel®
Illustrations by: Sandy Gardner Artist, email sandy@sandygardner.co.uk

The Scientific Method

This section isn't about how to 'do' science — but it does show you the way most scientists work.

Scientists Come Up With **Hypotheses** — Then **Test** Them

1) Scientists try to explain things. They start by observing something they don't understand.
2) They then come up with a hypothesis — a possible explanation for what they've observed.
3) The next step is to test whether the hypothesis might be right or not. This involves making a prediction based on the hypothesis and testing it by gathering evidence (i.e. data) from investigations. If evidence from experiments backs up a prediction, you're a step closer to figuring out if the hypothesis is true.

About 100 years ago, scientists hypothesised that atoms looked like this.

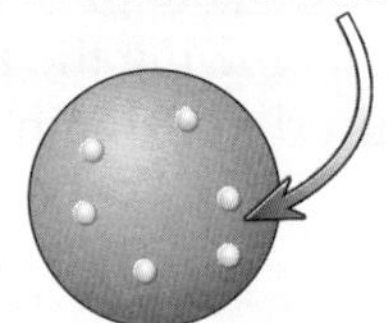

Several Scientists Will **Test** a Hypothesis

1) Normally, scientists share their findings in peer-reviewed journals, or at conferences.
2) Peer-review is where other scientists check results and scientific explanations to make sure they're 'scientific' (e.g. that experiments have been done in a sensible way) before they're published. It helps to detect false claims, but it doesn't mean that findings are correct — just that they're not wrong in any obvious way.
3) Once other scientists have found out about a hypothesis, they'll start basing their own predictions on it and carry out their own experiments. They'll also try to reproduce the original experiments to check the results — and if all the experiments in the world back up the hypothesis, then scientists start to think the hypothesis is true.
4) However, if a scientist does an experiment that doesn't fit with the hypothesis (and other scientists can reproduce the results) then the hypothesis may need to be modified or scrapped altogether.

After more evidence was gathered, scientists changed their hypothesis to this.

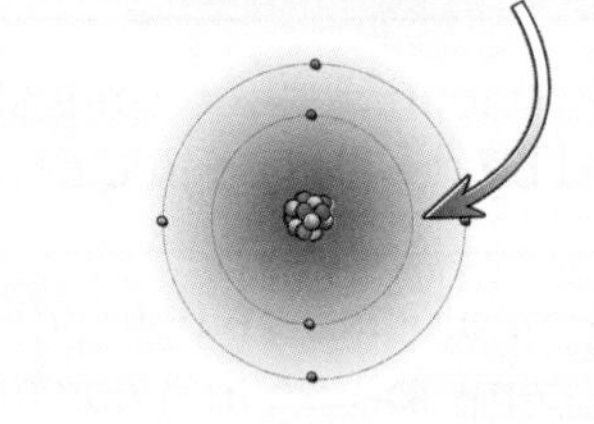

If **All** the **Evidence** Supports a Hypothesis, It's **Accepted** — For Now

1) Accepted hypotheses are often referred to as theories. Our currently accepted theories are the ones that have survived this 'trial by evidence' — they've been tested many times over the years and survived.
2) However, theories never become totally indisputable fact. If new evidence comes along that can't be explained using the existing theory, then the hypothesising and testing is likely to start all over again.

Now we think it's more like this.

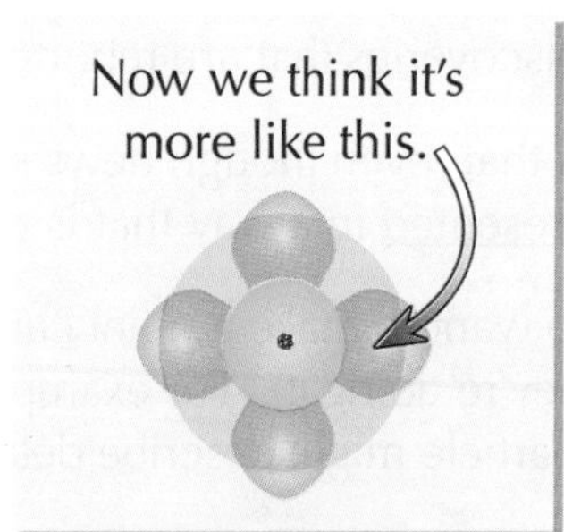

Scientific models are constantly being refined...

The scientific method has been developed over time. Aristotle (a Greek philosopher) was the first person to realise that theories need to be based on observations. Muslim scholars then introduced the ideas of creating a hypothesis, testing it, and repeating work to check results.

Models and Communication

Once scientists have made a new discovery, they don't just keep it to themselves. Oh no. Time to learn about how scientific discoveries are communicated, and the models that are used to represent theories.

Theories Can Involve Different Types of Models

1) A representational model is a simplified description or picture of what's going on in real life. Like all models, it can be used to explain observations and make predictions. E.g. the Bohr model of an atom is a simplified way of showing the arrangement of electrons in an atom (see p.32). It can be used to explain trends down groups in the periodic table.

Scientists test models by carrying out experiments to check that the predictions made by the model happen as expected.

2) Computational models use computers to make simulations of complex real-life processes, such as climate change. They're used when there are a lot of different variables (factors that change) to consider, and because you can easily change their design to take into account new data.

3) All models have limitations on what they can explain or predict. E.g. ball and stick models (a type of spatial model) can be used to show how ions are arranged in an ionic compound. One of their limitations is that they don't show the relative sizes of the ions (see p.50).

Scientific Discoveries are Communicated to the General Public

Some scientific discoveries show that people should change their habits, or they might provide ideas that could be developed into new technology. So scientists need to tell the world about their discoveries.

Technologies are being developed that make use of fullerenes (see p.60). These include drug delivery systems for use in medicine. Information about these systems needs to be communicated to doctors so they can use them, and to patients, so they can make informed decisions about their treatment.

Scientific Evidence can be Presented in a Biased Way

1) Scientific discoveries that are reported in the media (e.g. newspapers or television) aren't peer-reviewed.
2) This means that, even though news stories are often based on data that has been peer-reviewed, the data might be presented in a way that is over-simplified or inaccurate, making it open to misinterpretation.
3) People who want to make a point can sometimes present data in a biased way (sometimes without knowing they're doing it). For example, a scientist might overemphasise a relationship in the data, or a newspaper article might describe details of data supporting an idea without giving any evidence against it.

Companies can present biased data to help sell products...

Sometimes a company may only want you to see half of the story so they present the data in a biased way. For example, a pharmaceutical company may want to encourage you to buy their drugs by telling you about all the positives, but not report the results of any unfavourable studies.

Issues Created by Science

Science has helped us make progress in loads of areas, from medicine to space travel. But science still has its issues. And it can't answer everything, as you're about to find out.

Scientific Developments are Great, but they can Raise Issues

Scientific knowledge is increased by doing experiments. And this knowledge leads to scientific developments, e.g. new technologies or new advice. These developments can create issues though. For example:

Economic issues: Society can't always afford to do things scientists recommend (e.g. investing in alternative energy sources) without cutting back elsewhere.

Social issues: Decisions based on scientific evidence affect people — e.g. should fossil fuels be taxed more highly? Would the effect on people's lifestyles be acceptable?

Personal issues: Some decisions will affect individuals. For example, someone might support alternative energy, but object if a wind farm is built next to their house.

Environmental issues: Human activity often affects the natural environment. For example, building a dam to produce electricity will change the local habitat so some species might be displaced. But it will also reduce our need for fossil fuels, so will help to reduce climate change.

Science Can't Answer Every Question — Especially Ethical Ones

1) We don't understand everything. We're always finding out more, but we'll never know all the answers.
2) In order to answer scientific questions, scientists need data to provide evidence for their hypotheses.
3) Some questions can't be answered yet because the data can't currently be collected, or because there's not enough data to support a theory.
4) Eventually, as we get more evidence, we'll answer some of the questions that currently can't be answered, e.g. what the impact of global warming on sea levels will be. But there will always be the "Should we be doing this at all?"-type questions that experiments can't help us to answer...

Think about new drugs which can be taken to boost your 'brain power'.

- Some people think they're good as they could improve concentration or memory. New drugs could let people think in ways beyond the powers of normal brains.
- Other people say they're bad — they could give some people an unfair advantage in exams. And people might be pressured into taking them so that they could work more effectively, and for longer hours.

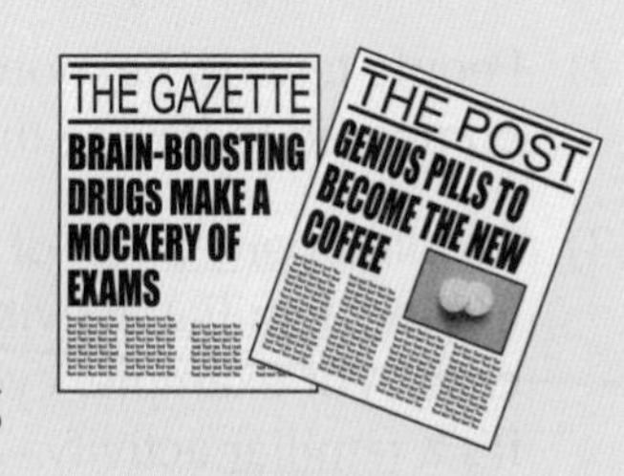

There are often issues with new scientific developments...

The trouble is, there's often no clear right answer where these issues are concerned. Different people have different views, depending on their priorities. These issues are full of grey areas.

Risk

Scientific discoveries are often great, but they can prove risky. With dangers all around, you've got to be aware of hazards — this includes how likely they are to cause harm and how serious the effects may be.

Nothing is Completely **Risk-Free**

1) A hazard is something that could potentially cause harm.

2) All hazards have a risk attached to them — this is the chance that the hazard will cause harm.

3) The risks of some things seem pretty obvious, or we've known about them for a while, like the risk of causing acid rain by polluting the atmosphere, or of having a car accident when you're travelling in a car.

4) New technology arising from scientific advances can bring new risks, e.g. scientists are unsure whether nanoparticles that are being used in cosmetics and suncream might be harming the cells in our bodies. These risks need to be considered alongside the benefits of the technology, e.g. improved sun protection.

5) You can estimate the size of a risk based on how many times something happens in a big sample (e.g. 100 000 people) over a given period (e.g. a year). For example, you could assess the risk of a driver crashing by recording how many people in a group of 100 000 drivers crashed their cars over a year.

6) To make decisions about activities that involve hazards, we need to take into account the chance of the hazard causing harm, and how serious the consequences would be if it did. If an activity involves a hazard that's very likely to cause harm, with serious consequences if it does, that activity is considered high risk.

People Make Their **Own Decisions** About Risk

1) Not all risks have the same consequences, e.g. if you chop veg with a sharp knife you risk cutting your finger, but if you go scuba-diving you risk death. You're much more likely to cut your finger during half an hour of chopping than to die during half an hour of scuba-diving. But most people are happier to accept a higher probability of an accident if the consequences are short-lived and fairly minor.
2) People tend to be more willing to accept a risk if they choose to do something (e.g. go scuba diving), compared to having the risk imposed on them (e.g. having a nuclear power station built next door).
3) People's perception of risk (how risky they think something is) isn't always accurate. They tend to view familiar activities as low-risk and unfamiliar activities as high-risk — even if that's not the case. For example, cycling on roads is often high-risk, but many people are happy to do it because it's a familiar activity. Air travel is actually pretty safe, but a lot of people perceive it as high-risk.
4) People may over-estimate the risk of things with long-term or invisible effects, e.g. ionising radiation.

The pros and cons of new technology must be weighed up...

The world's a dangerous place and it's impossible to rule out the chance of an accident altogether. But if you can recognise hazards and take steps to reduce the risks, you're more likely to stay safe.

Designing Investigations

Dig out your lab coat and dust off your badly-scratched safety goggles... it's investigation time.

Evidence Can Support or Disprove a Hypothesis

1) Scientists observe things and come up with hypotheses to test them (see p.1). You need to be able to do the same. For example:

> Observation: People with big feet have spots. Hypothesis: Having big feet causes spots.

2) To determine whether or not a hypothesis is right, you need to do an investigation to gather evidence. To do this, you need to use your hypothesis to make a prediction — something you think will happen that you can test. E.g. people who have bigger feet will have more spots.

Investigations include experiments and studies.

3) Investigations are used to see if there are patterns or relationships between two variables, e.g. to see if there's a pattern or relationship between the variables 'number of spots' and 'size of feet'.

Evidence Needs to be Repeatable, Reproducible and Valid

1) Repeatable means that if the same person does an experiment again using the same methods and equipment, they'll get similar results.
2) Reproducible means that if someone else does the experiment, or a different method or piece of equipment is used, the results will still be similar.
3) If data is repeatable and reproducible, it's reliable and scientists are more likely to have confidence in it.
4) Valid results are both repeatable and reproducible AND they answer the original question. They come from experiments that were designed to be a fair test...

Make an Investigation a Fair Test By Controlling the Variables

1) In a lab experiment you usually change one variable and measure how it affects another variable.
2) To make it a fair test, everything else that could affect the results should stay the same — otherwise you can't tell if the thing you're changing is causing the results or not.
3) The variable you CHANGE is called the INDEPENDENT variable.
4) The variable you MEASURE when you change the independent variable is the DEPENDENT variable.
5) The variables that you KEEP THE SAME are called CONTROL variables.

> You could find how temperature affects reaction rate by measuring the volume of gas formed over time. The independent variable is the temperature.
> The dependent variable is the volume of gas produced. Control variables include the concentration and amounts of reactants, the time period you measure, etc.

6) Because you can't always control all the variables, you often need to use a control experiment. This is an experiment that's kept under the same conditions as the rest of the investigation, but doesn't have anything done to it. This is so that you can see what happens when you don't change anything at all.

Designing Investigations

The **Bigger** the **Sample Size** the **Better**

1) Data based on small samples isn't as good as data based on large samples. A sample should represent the whole population (i.e. it should share as many of the characteristics in the population as possible) — a small sample can't do that as well. It's also harder to spot anomalies if your sample size is too small.

2) The bigger the sample size the better, but scientists have to be realistic when choosing how big. For example, if you were studying the effects of a chemical used to sterilise water on the people drinking it, it'd be great to study everyone who was drinking the water (a huge sample), but it'd take ages and cost a bomb. It's more realistic to study a thousand people, with a mixture of ages, gender, and race.

Your **Equipment** has to be **Right for the Job**

1) The measuring equipment you use has to be sensitive enough to measure the changes you're looking for. For example, if you need to measure changes of 1 cm^3 you need to use a measuring cylinder that can measure in 1 cm^3 steps — it'd be no good trying with one that only measures 10 cm^3 steps.
2) The smallest change a measuring instrument can detect is called its resolution. E.g. some mass balances have a resolution of 1 g, some have a resolution of 0.1 g, and some are even more sensitive.
3) Also, equipment needs to be calibrated by measuring a known value. If there's a difference between the measured and known value, you can use this to correct the inaccuracy of the equipment.

Data Should be **Repeatable, Reproducible, Accurate** and **Precise**

1) To check repeatability you need to repeat the readings and check that the results are similar. You need to repeat each reading at least three times.
2) To make sure your results are reproducible you can cross check them by taking a second set of readings with another instrument (or a different observer).
3) Your data also needs to be accurate. Really accurate results are those that are really close to the true answer. The accuracy of your results usually depends on your method — you need to make sure you're measuring the right thing and that you don't miss anything that should be included in the measurements. E.g. estimating the amount of gas released from a reaction by counting the bubbles isn't very accurate because you might miss some of the bubbles and they might have different volumes. It's more accurate to measure the volume of gas released using a gas syringe (see p.121).
4) Your data also needs to be precise. Precise results are ones where the data is all really close to the mean (average) of your repeated results (i.e. not spread out).

Repeat	Data set 1	Data set 2
1	12	11
2	14	17
3	13	14
Mean	13	14

Data set 1 is more precise than data set 2.

Designing Investigations

You Need to Look out for **Errors** and **Anomalous Results**

1) The results of your experiment will always vary a bit because of random errors — unpredictable differences caused by things like human errors in measuring. E.g. the errors you make when you take a reading from a burette are random. You have to estimate or round the level when it's between two marks — so sometimes your figure will be a bit above the real one, and sometimes it will be a bit below.
2) You can reduce the effect of random errors by taking repeat readings and finding the mean. This will make your results more precise.
3) If a measurement is wrong by the same amount every time, it's called a systematic error. For example, if you measured from the very end of your ruler instead of from the 0 cm mark every time, all your measurements would be a bit small. Repeating the experiment in the exact same way and calculating a mean won't correct a systematic error.

If there's no systematic error, then doing repeats and calculating a mean could make your results more accurate.

4) Just to make things more complicated, if a systematic error is caused by using equipment that isn't zeroed properly, it's called a zero error. For example, if a mass balance always reads 1 gram before you put anything on it, all your measurements will be 1 gram too heavy.
5) You can compensate for some systematic errors if you know about them, e.g. if a mass balance always reads 1 gram before you put anything on it, you can subtract 1 gram from all your results.
6) Sometimes you get a result that doesn't fit in with the rest at all. This is called an anomalous result. You should investigate it and try to work out what happened. If you can work out what happened (e.g. you measured something wrong) you can ignore it when processing your results.

Investigations Can be **Hazardous**

1) Hazards from science experiments might include:

- Microorganisms, e.g. some bacteria can make you ill.
- Chemicals, e.g. sulfuric acid can burn your skin and alcohols catch fire easily.
- Fire, e.g. an unattended Bunsen burner is a fire hazard.
- Electricity, e.g. faulty electrical equipment could give you a shock.

You can find out about potential hazards by looking in textbooks, doing some internet research, or asking your teacher.

2) Part of planning an investigation is making sure that it's safe.
3) You should always make sure that you identify all the hazards that you might encounter. Then you should think of ways of reducing the risks from the hazards you've identified. For example:

- If you're working with sulfuric acid, always wear gloves and safety goggles. This will reduce the risk of the acid coming into contact with your skin and eyes.
- If you're using a Bunsen burner, stand it on a heat proof mat. This will reduce the risk of starting a fire.

Designing an investigation is an involved process...

Collecting data is what investigations are all about. Designing a good investigation is really important to make sure that any data collected is accurate, precise, repeatable and reproducible.

Processing Data

Processing your data means doing some calculations with it to make it more useful.

Data Needs to be Organised

1) Tables are dead useful for organising data.
2) When you draw a table use a ruler and make sure each column has a heading (including the units).

There are Different Ways of Processing Your Data

1) When you've done repeats of an experiment you should always calculate the mean (average). To do this add together all the data values and divide by the total number of values in the sample.
2) You might also need to calculate the range (how spread out the data is). To do this find the largest number and subtract the smallest number from it.

Ignore anomalous results when calculating the mean and the range.

The results of an experiment to find the mass of gas lost from two reactions are shown below. Calculate the mean and the range for the mass of gas lost in each reaction.

Test tube	Repeat 1 (g)	Repeat 2 (g)	Repeat 3 (g)	Mean (g)	Range (g)
A	28	37	32	(28 + 37 + 32) ÷ 3 = 32	37 – 28 = 9
B	47	51	60	(47 + 51 + 60) ÷ 3 = 53	60 – 47 = 13

Round to the Lowest Number of Significant Figures

The first significant figure of a number is the first digit that's not zero. The second and third significant figures come straight after (even if they're zeros). You should be aware of significant figures in calculations.

1) In any calculation where you need to round, you should round the answer to the lowest number of significant figures (s.f.) given.
2) Remember to write down how many significant figures you've rounded to after your answer.
3) If your calculation has multiple steps, only round the final answer, or it won't be as accurate.

The volume of one mole of gas is 24.0 dm^3 at room temperature and pressure. How many moles are there in 4.6 dm^3 of gas under the same conditions?

No. of moles of gas = 4.6 dm^3 ÷ 24.0 dm^3 = 0.19166... = 0.19 mol (2 s.f.)

2 s.f. (4.6) — 3 s.f. (24.0) — Final answer should be rounded to 2 s.f.

Don't forget your calculator...

In the exam you could be given some data and be expected to process it in some way. Make sure you keep an eye on significant figures in your answers and always write down your working.

Presenting Data

Once you've processed your data, e.g. by calculating the mean, you can present your results in a nice chart or graph. This will help you to spot any patterns in your data.

If Your Data Comes in **Categories**, Present It in a **Bar Chart**

1) If the independent variable is categoric (comes in distinct categories, e.g. alkane chain length, metals) you should use a bar chart to display the data.
2) You also use them if the independent variable is discrete (the data can be counted in chunks, where there's no in-between value, e.g. number of protons is discrete because you can't have half a proton).
3) There are some golden rules you need to follow for drawing bar charts:

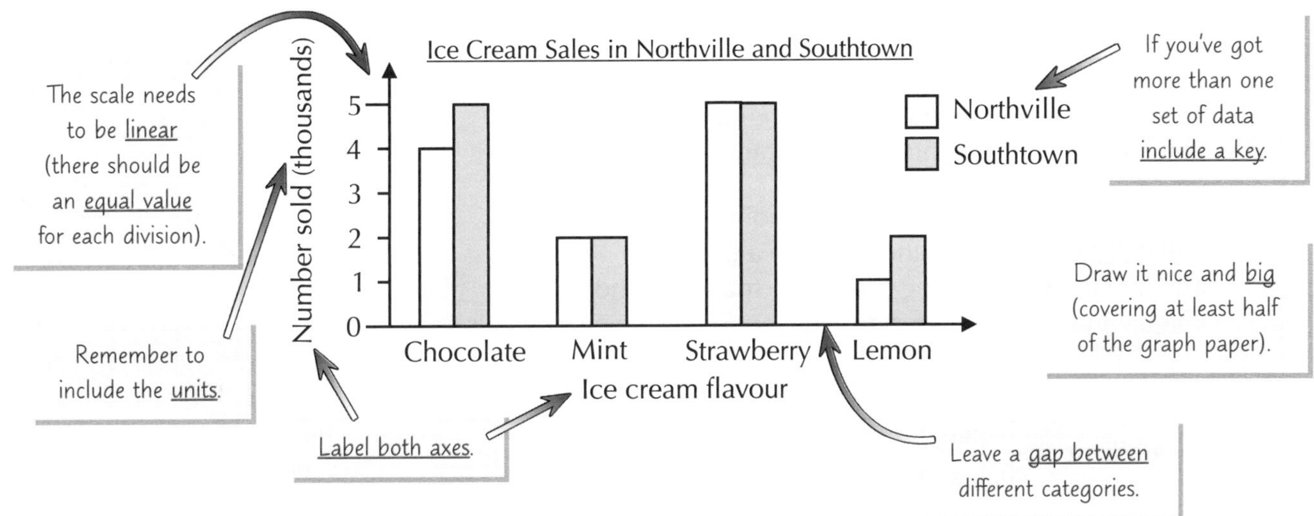

If Your Data is **Continuous**, Plot a **Graph**

1) If both variables are continuous (numerical data that can have any value within a range, e.g. length, volume, temperature) you should use a graph to display the data.
2) Here are the rules for plotting points on a graph:

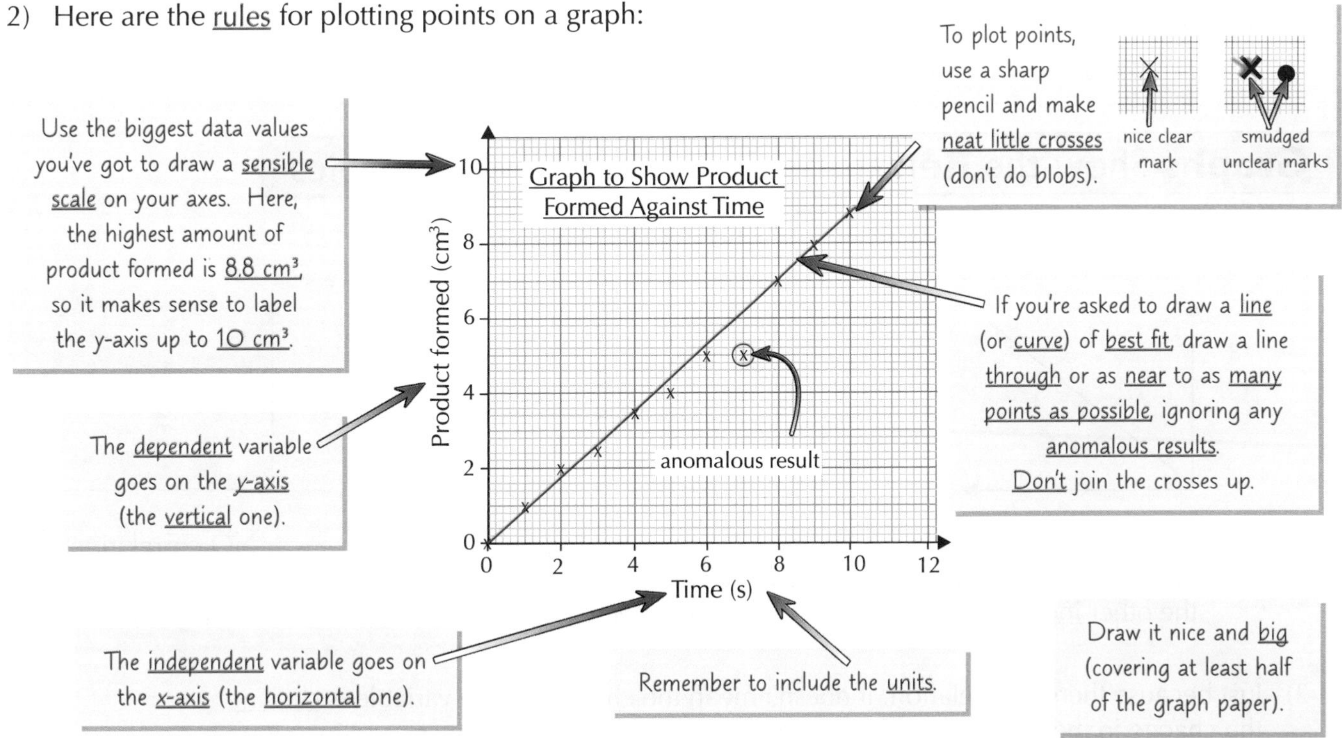

Use the biggest data values you've got to draw a sensible scale on your axes. Here, the highest amount of product formed is 8.8 cm³, so it makes sense to label the y-axis up to 10 cm³.
The dependent variable goes on the y-axis (the vertical one).
The independent variable goes on the x-axis (the horizontal one).
Graph to Show Product Formed Against Time
Product formed (cm³)
0
2
4
6
8
10
12
Time (s)
anomalous result
Remember to include the units.
To plot points, use a sharp pencil and make neat little crosses (don't do blobs).
nice clear mark
smudged unclear marks
If you're asked to draw a line (or curve) of best fit, draw a line through or as near to as many points as possible, ignoring any anomalous results. Don't join the crosses up.
Draw it nice and big (covering at least half of the graph paper).

More on Graphs

Graphs aren't just fun to plot, they're also really useful for showing trends in your data.

Graphs Can Give You a Lot of Information About Your Data

1) The gradient (slope) of a graph tells you how quickly the dependent variable changes if you change the independent variable.

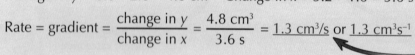

$$\text{gradient} = \frac{\text{change in } y}{\text{change in } x}$$

You can use this method to calculate any rates from a graph, not just the rate of a reaction. Just remember that a rate is how much something changes over time, so x needs to be the time.

The graph below shows the volume of gas produced in a reaction against time. The graph is linear (it's a straight line graph), so you can simply calculate the gradient of the line to find out the rate of reaction.

1) To calculate the gradient, pick two points on the line that are easy to read and a good distance apart.
2) Draw a line down from one of the points and a line across from the other to make a triangle. The line drawn down the side of the triangle is the change in y and the line across the bottom is the change in x.

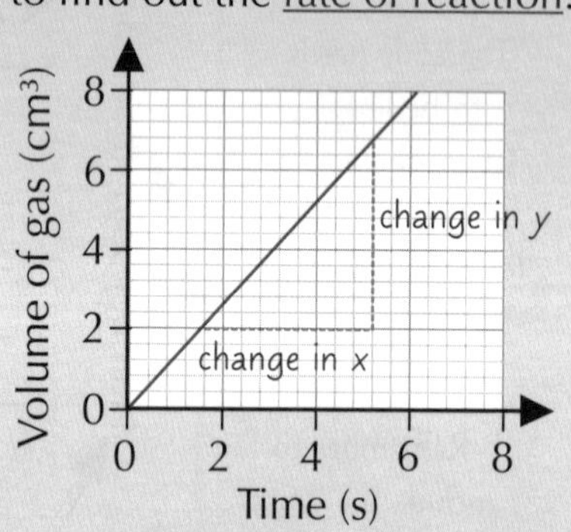

Change in $y = 6.8 - 2.0 = 4.8$ cm³ Change in $x = 5.2 - 1.6 = 3.6$ s

$$\text{Rate} = \text{gradient} = \frac{\text{change in } y}{\text{change in } x} = \frac{4.8 \text{ cm}^3}{3.6 \text{ s}} = 1.3 \text{ cm}^3/\text{s or } 1.3 \text{ cm}^3\text{s}^{-1}$$

The units of the gradient are (units of y)/(units of x). cm³/s can also be written as cm³s⁻¹.

2) If you've got a curved graph, you can find the rate at any point by drawing a tangent — a straight line that touches a single point on a curve. You can then find the gradient of the tangent in the usual way, to give you the rate at that point (see page 124 for more on calculating gradients).

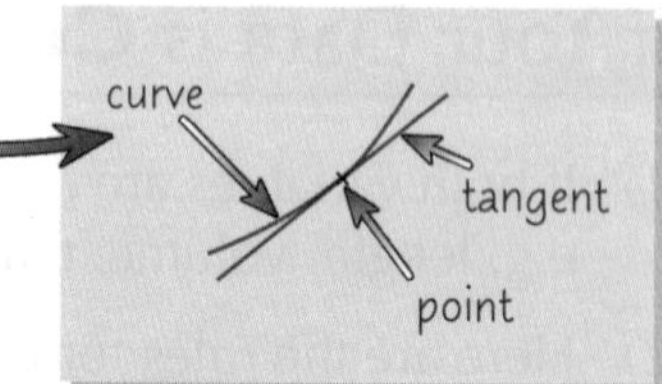

3) The intercept of a graph is where the line of best fit crosses one of the axes. The x-intercept is where the line of best fit crosses the x-axis and the y-intercept is where it crosses the y-axis.

Graphs Show the Relationship Between Two Variables

1) You can get three types of correlation (relationship) between variables:

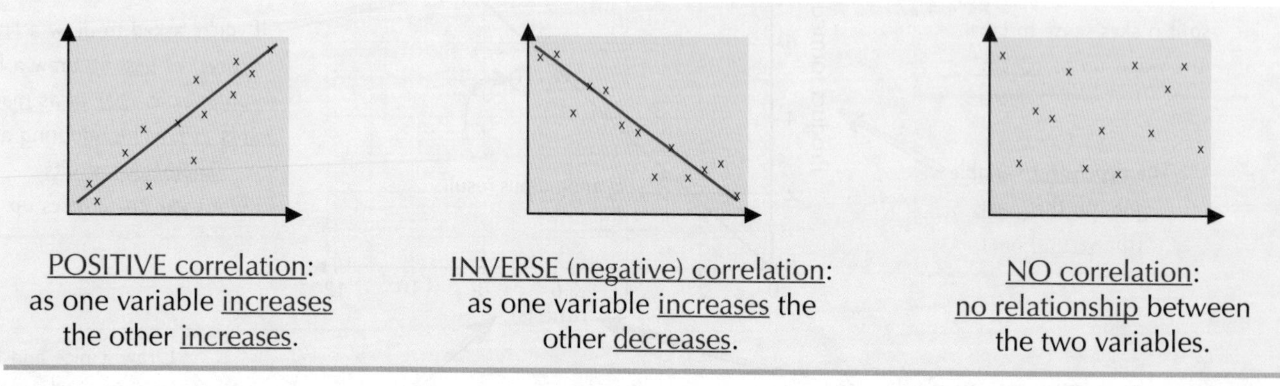

POSITIVE correlation: as one variable increases the other increases.

INVERSE (negative) correlation: as one variable increases the other decreases.

NO correlation: no relationship between the two variables.

2) Just because there's correlation, it doesn't mean the change in one variable is causing the change in the other — there might be other factors involved (see page 13).

Units

Graphs and maths skills are all very well, but the numbers don't mean much if you don't get the units right.

S.I. Units Are Used All Round the World

1) It wouldn't be all that useful if I defined volume in terms of bath tubs, you defined it in terms of egg-cups and my pal Fred defined it in terms of balloons — we'd never be able to compare our data.
2) To stop this happening, scientists have come up with a set of standard units, called S.I. units, that all scientists use to measure their data. Here are some S.I. units you'll see in chemistry:

Quantity	S.I. Base Unit
mass	kilogram, kg
length	metre, m
time	second, s
amount of substance	mole, mol

Always Check The Values Used in Equations Have the Right Units

1) Equations (sometimes called formulas) show relationships between variables.
2) To rearrange an equation, make sure that whatever you do to one side of the equation you also do to the other side.

For example, you can find the number of moles of something using the equation: moles = mass ÷ molar mass.
You can rearrange this equation to find the mass by multiplying each side by molar mass to give: mass = moles × molar mass.

3) To use an equation, you need to know the values of all but one of the variables. Substitute the values you do know into the formula, and do the calculation to work out the final variable.
4) Always make sure the values you put into an equation have the right units. For example, you might have done a titration experiment to work out the concentration of a solution. The volume of the solution will probably have been measured in cm^3, but the equation to find concentration uses volume in dm^3. So you'll have to convert your volume from cm^3 to dm^3 before you put it into the equation.
5) To make sure your units are correct, it can help to write down the units on each line of your calculation.

S.I. units help scientists to compare data...

You can only really compare things if they're in the same units. E.g. if the concentration of an acid was measured in mol dm^{-3} and another acid in g cm^{-3}, it'd be hard to compare them.

Converting Units

You can convert units using scaling prefixes. This can save you from having to write a lot of 0's...

Scaling Prefixes Can Be Used for **Large** and **Small** Quantities

1) Quantities come in a huge range of sizes. For example, the volume of a swimming pool might be around 2 000 000 000 cm^3, while the volume of a cup is around 250 cm^3.
2) To make the size of numbers more manageable, larger or smaller units are used. These are the S.I. base units (e.g. metres) with a prefix in front:

Prefix	tera (T)	giga (G)	mega (M)	kilo (k)	deci (d)	centi (c)	milli (m)	micro (μ)	nano (n)
Multiple of Unit	10^{12}	10^{9}	1 000 000 (10^{6})	1000	0.1	0.01	0.001	0.000001 (10^{-6})	10^{-9}

3) These prefixes tell you how much bigger or smaller a unit is than the base unit. So one kilometre is one thousand metres.
4) To swap from one unit to another, all you need to know is what number you have to divide or multiply by to get from the original unit to the new unit — this is called the conversion factor.

The conversion factor is the number of times the smaller unit goes into the larger unit.

- To go from a bigger unit (like m) to a smaller unit (like cm), you multiply by the conversion factor.
- To go from a smaller unit (like g) to a bigger unit (like kg), you divide by the conversion factor.

5) Here are some conversions that'll be useful for GCSE chemistry:

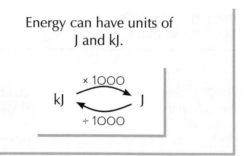

Energy can have units of J and kJ.

kJ → × 1000 → J
J → ÷ 1000 → kJ

Mass can have units of kg and g.

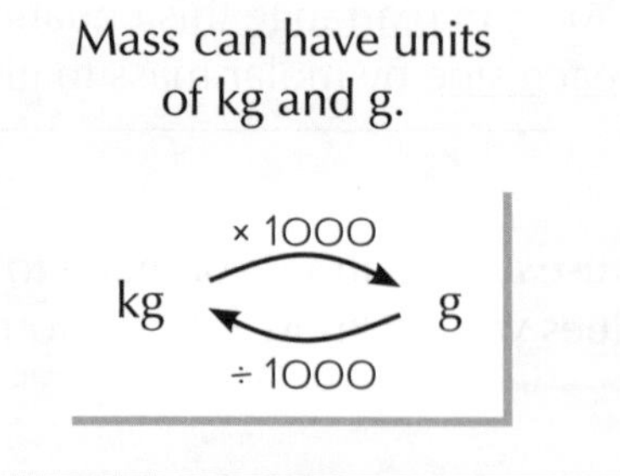

Volume can have units of m^3, dm^3 and cm^3.

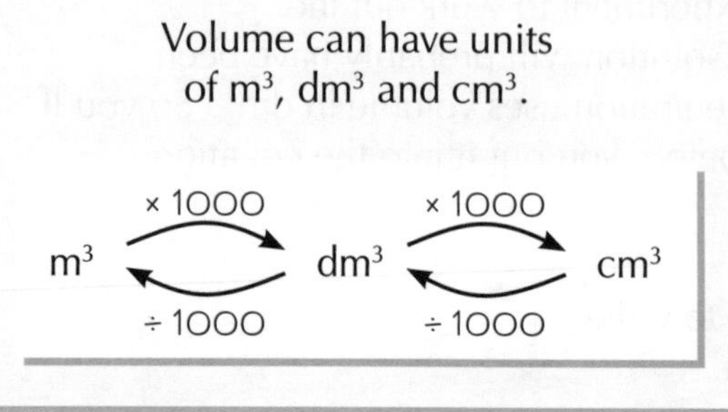

Concentration can have units of mol/dm^3 and mol/cm^3.

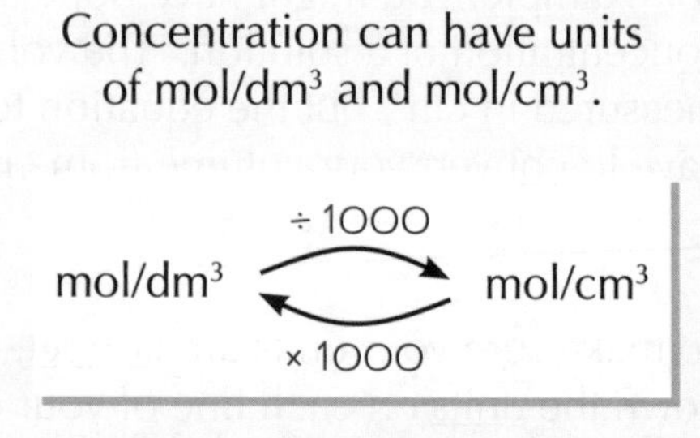

MATHS TIP

To convert from bigger units to smaller units...

...multiply by the conversion factor, and to convert from smaller units to bigger units, divide by the conversion factor. Don't go getting this the wrong way round or you'll get some odd answers.

Drawing Conclusions

Once you've carried out an experiment and processed your data, it's time to work out what your data shows.

You Can **Only Conclude** What the Data Shows and **No More**

1) Drawing conclusions might seem pretty straightforward — you just look at your data and say what pattern or relationship you see between the dependent and independent variables.

The table on the right shows the rate of a reaction in the presence of two different catalysts:

Catalyst	Rate of Reaction / cm^3/s
A	13.5
B	19.5
No catalyst	5.5

CONCLUSION: Catalyst B makes this reaction go faster than catalyst A.

2) But you've got to be really careful that your conclusion matches the data you've got and doesn't go any further.

You can't conclude that catalyst B increases the rate of any other reaction more than catalyst A — the results might be completely different.

3) You also need to be able to use your results to justify your conclusion (i.e. back up your conclusion with some specific data).

The rate of this reaction was 6 cm^3/s faster using catalyst B compared with catalyst A.

4) When writing a conclusion you need to refer back to the original hypothesis and say whether the data supports it or not:

The hypothesis for this experiment might have been that catalyst B would make the reaction go quicker than catalyst A. If so, the data supports the hypothesis.

Correlation DOES NOT Mean **Cause**

If two things are correlated (i.e. there's a relationship between them) it doesn't necessarily mean a change in one variable is causing the change in the other — this is REALLY IMPORTANT — DON'T FORGET IT. There are three possible reasons for a correlation:

1) CHANCE: It might seem strange, but two things can show a correlation purely due to chance.

For example, one study might find a correlation between people's hair colour and how good they are at frisbee. But other scientists don't get a correlation when they investigate it — the results of the first study are just a fluke.

2) LINKED BY A 3RD VARIABLE: A lot of the time it may look as if a change in one variable is causing a change in the other, but it isn't — a third variable links the two things.

For example, there's a correlation between water temperature and shark attacks. This isn't because warmer water makes sharks crazy. Instead, they're linked by a third variable — the number of people swimming (more people swim when the water's hotter, and with more people in the water you get more shark attacks).

3) CAUSE: Sometimes a change in one variable does cause a change in the other. You can only conclude that a correlation is due to cause when you've controlled all the variables that could affect the result.

For example, there's a correlation between smoking and lung cancer. This is because chemicals in tobacco smoke cause lung cancer. This conclusion was only made once other variables (such as age and exposure to other things that cause cancer) had been controlled.

Uncertainty

Uncertainty is how sure you can really be about your data. There's a little bit of maths to do, and also a formula to learn. But don't worry too much — it's no more than a simple bit of subtraction and division.

Uncertainty is the Amount of **Error** Your Measurements Might Have

1) When you repeat a measurement, you often get a slightly different figure each time you do it due to random error (see page 7). This means that each result has some uncertainty to it.

2) The measurements you make will also have some uncertainty in them due to limits in the resolution of the equipment you use (see page 6).

3) This all means that the mean of a set of results will also have some uncertainty to it. You can calculate the uncertainty of a mean result using the equation:

$$\text{uncertainty} = \frac{\text{range}}{2}$$

The range is the largest value minus the smallest value (see p.8).

4) The larger the range, the less precise your results are and the more uncertainty there will be in your results. Uncertainties are shown using the '±' symbol.

The table below shows the results of a titration experiment to determine the volume of 0.5 mol/dm³ sodium hydroxide solution needed to neutralise 25 cm³ of a solution of hydrochloric acid with unknown concentration. Calculate the uncertainty of the mean.

1) First work out the range:
Range = 20.10 – 19.80
= 0.300 cm³

Repeat	1	2	3	mean
Volume of sodium hydroxide (cm³)	20.10	19.80	20.00	19.97

2) Use the range to find the uncertainty:
Uncertainty = range ÷ 2 = 0.300 ÷ 2 = 0.150 cm³
So the uncertainty of the mean = 19.97 ± 0.15 cm³

5) Measuring a greater amount of something helps to reduce uncertainty.

For example, in a rate of reaction experiment, measuring the amount of product formed over a longer period compared to a shorter period will reduce the uncertainty in your results.

The smaller the uncertainty, the more precise your results...

Remember that equation for uncertainty. You never know when you might need it — you could be expected to use it in the exams. You need to make sure all the data is in the same units though. For example, if you had some measurements in metres, and some in centimetres, you'd need to convert them all into either metres or centimetres before you set about calculating uncertainty.

Evaluations

Hurrah! The end of another investigation. Well, now you have to work out all the things you did wrong. That's what evaluations are all about I'm afraid. Best get cracking with this page...

Evaluations — Describe **How** Investigations Could be **Improved**

An evaluation is a critical analysis of the whole investigation.

1) You should comment on the method — was it valid? Did you control all the other variables to make it a fair test?

2) Comment on the quality of the results — was there enough evidence to reach a valid conclusion? Were the results repeatable, reproducible, accurate and precise?

3) Were there any anomalous results? If there were none then say so. If there were any, try to explain them — were they caused by errors in measurement? Were there any other variables that could have affected the results? You should comment on the level of uncertainty in your results too.

4) All this analysis will allow you to say how confident you are that your conclusion is right.

5) Then you can suggest any changes to the method that would improve the quality of the results, so that you could have more confidence in your conclusion. For example, you might suggest changing the way you controlled a variable, or increasing the number of measurements you took. Taking more measurements at narrower intervals could give you a more accurate result. For example:

> Enzymes have an optimum temperature (a temperature at which they work best). Say you do an experiment to find an enzyme's optimum temperature and take measurements at 10 °C, 20 °C, 30 °C, 40 °C and 50 °C. The results of this experiment tell you the optimum is 40 °C. You could then repeat the experiment, taking more measurements around 40 °C to a get a more accurate value for the optimum.

6) You could also make more predictions based on your conclusion, then further experiments could be carried out to test them.

When suggesting improvements to the investigation, always make sure that you say why you think this would make the results better.

Always look for ways to improve your investigations

So there you have it — Working Scientifically. Make sure you know this stuff like the back of your hand. It's not just in the lab, when you're carrying out your groundbreaking investigations, that you'll need to know how to work scientifically. You can be asked about it in the exams as well. So swot up...

Atoms

All substances are made of atoms. They're really tiny — too small to see, even with your microscope. Atoms are so tiny that a 50p piece contains about 77 400 000 000 000 000 000 000 of them.

Atoms Contain **Protons, Neutrons** and **Electrons**

Atoms have a radius of about 0.1 nanometers (that's 1×10^{-10} m). There are a few different (and equally useful) modern models of the atom — but chemists tend to like the model below best.

A nanometer (nm) is one billionth of a meter. Shown in standard form, that's 1×10^{-9} m. Standard form is used for showing really large or really small numbers.

The **Nucleus**

1) It's in the middle of the atom.
2) It contains protons and neutrons.
3) The nucleus has a radius of around 1×10^{-14} m (that's around 1/10 000 of the radius of an atom).
4) It has a positive charge because of the protons.
5) Almost the whole mass of the atom is concentrated in the nucleus.

The **Electrons**

1) Move around the nucleus in electron shells.
2) They're negatively charged and tiny, but they cover a lot of space.
3) The volume of their orbits determines the size of the atom.
4) Electrons have virtually no mass.

Particle	Relative Mass	Relative Charge
Proton	1	+1
Neutron	1	0
Electron	Very small	−1

Protons are heavy and positively charged.
Neutrons are heavy and neutral.
Electrons are tiny and negatively charged.

Number of Protons **Equals** Number of Electrons

1) Atoms are neutral — they have no charge overall (unlike ions).
2) This is because they have the same number of protons as electrons.
3) The charge on the electrons is the same size as the charge on the protons, but opposite — so the charges cancel out.
4) In an ion, the number of protons doesn't equal the number of electrons. This means it has an overall charge. For example, an ion with a 2– charge, has two more electrons than protons.

An ion is an atom or group of atoms that has lost or gained electrons.

Atomic Number and **Mass Number** Describe an Atom

1) The nuclear symbol of an atom tells you its atomic (proton) number and mass number.
2) The atomic number tells you how many protons there are.
3) The mass number tells you the total number of protons and neutrons in the atom.
4) To get the number of neutrons, just subtract the atomic number from the mass number.

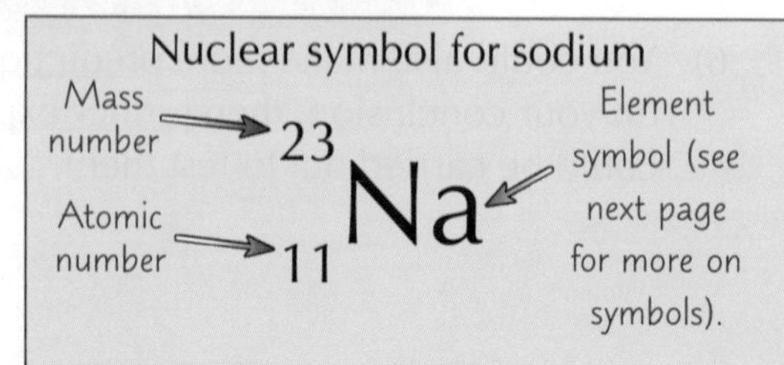

Atoms make up pretty much everything

So here we are — the very beginning of GCSE Chemistry. This stuff is super important — if you get to grips with the basic facts then you'll have a better chance understanding the rest of chemistry.

Elements

An element is a substance made up of atoms that all have the same number of protons in their nucleus.

Elements Consist of Atoms With the Same Atomic Number

Atoms can have different numbers of protons, neutrons and electrons. It's the number of protons in the nucleus that decides what type of atom it is.

For example, an atom with one proton in its nucleus is hydrogen and an atom with two protons is helium.

If a substance only contains atoms with the same number of protons it's called an element.

So all the atoms of a particular element (e.g. nitrogen) have the same number of protons and different elements have atoms with different numbers of protons.

There are about 100 different elements. Each element consists of only one type of atom — see diagram below.

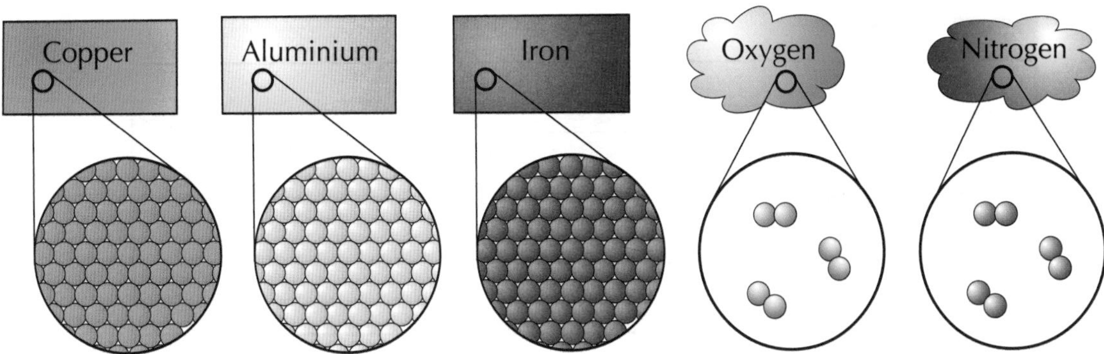

Atoms Can be Represented by Symbols

1) Atoms of each element can be represented by a one or two letter symbol — it's a type of shorthand that saves you the bother of having to write the full name of the element.
2) Some make perfect sense, e.g. C = carbon O = oxygen Mg = magnesium
3) Others less so, e.g. Na = sodium Fe = iron Pb = lead

Most of these odd symbols actually come from the Latin names of the elements.

4) You'll see these symbols on the periodic table (see page 35).

All atoms in an element have the same number of protons

Atoms and elements — make sure you know what they are and the differences between them. You don't need to learn all the symbols as you can use a periodic table but it's handy to know the common ones.

Isotopes

What's inside different atoms of the same element can vary. Read on to find out how...

Isotopes are the Same Except for Extra **Neutrons**

1) Isotopes are defined as:

> Different forms of the same element, which have the SAME number of PROTONS but a DIFFERENT number of NEUTRONS.

2) So isotopes have the same atomic number but different mass numbers.

3) A very popular example of a pair of isotopes are carbon-12 and carbon-13.

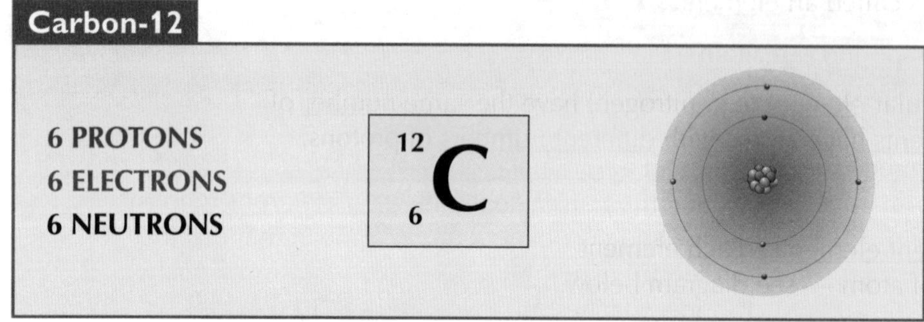

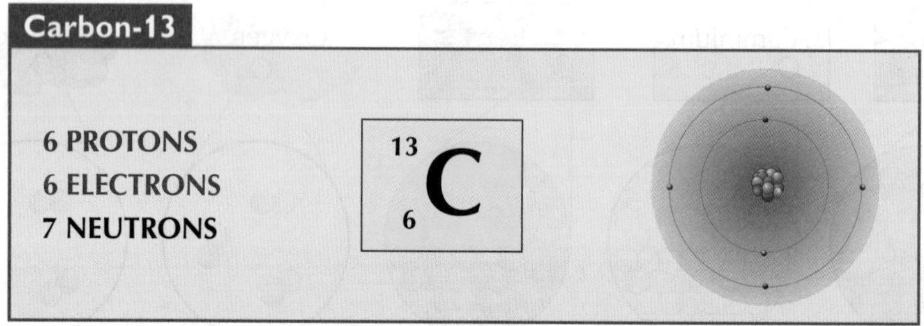

Carbon-13
6 PROTONS
6 ELECTRONS
7 NEUTRONS
$^{13}_{6}C$

4) Because many elements can exist as a number of different isotopes, relative atomic mass (A_r) is used instead of mass number when referring to the element as a whole.

> Relative atomic mass is an average mass taking into account the different masses and abundances (amounts) of all the isotopes that make up the element.

5) You can use this formula to work out the relative atomic mass of an element:

$$\text{relative atomic mass } (A_r) = \frac{\text{sum of (isotope abundance} \times \text{isotope mass number)}}{\text{sum of abundances of all the isotopes}}$$

Copper has two stable isotopes. Cu-63 has an abundance of 69.2% and Cu-65 has an abundance of 30.8%. Calculate the relative atomic mass of copper to 1 decimal place.

$$\text{Relative atomic mass} = \frac{(69.2 \times 63) + (30.8 \times 65)}{69.2 + 30.8} = \frac{4359.6 + 2002}{100} = \frac{6361.6}{100} = 63.616 = 63.6$$

Compounds

It would be great if we only had to deal with elements. But unluckily for you, elements can mix and match to make lots of new substances called compounds. And this makes things a little bit more complicated...

Atoms **Join Together** to Make **Compounds**

1) When elements react, atoms combine with other atoms to form compounds.

2) Compounds are substances formed from two or more elements, the atoms of each are in fixed proportions throughout the compound and they're held together by chemical bonds.

3) Making bonds involves atoms giving away, taking or sharing electrons. Only the electrons are involved — the nuclei of the atoms aren't affected at all when a bond is made.

4) It's usually difficult to separate the original elements of a compound out again — a chemical reaction is needed to do this.

During a chemical reaction, at least one new substance is made. You can usually measure a change in energy, such as a temperature change, as well.

5) A compound which is formed from a metal and a non-metal consists of ions. The metal atoms lose electrons to form positive ions and the non-metal atoms gain electrons to form negative ions. The opposite charges (positive and negative) of the ions mean that they're strongly attracted to each other. This is called ionic bonding. Examples of compounds which are bonded ionically include sodium chloride, magnesium oxide and calcium oxide.

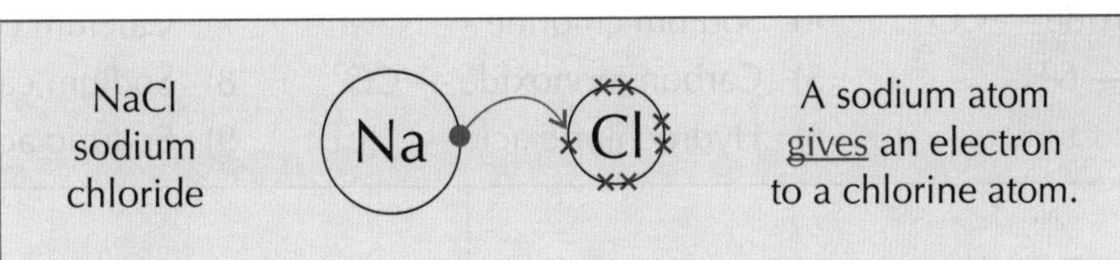

6) A compound formed from non-metals consists of molecules. Each atom shares an electron with another atom — this is called covalent bonding. Examples of compounds that are bonded covalently include hydrogen chloride gas, carbon monoxide, and water.

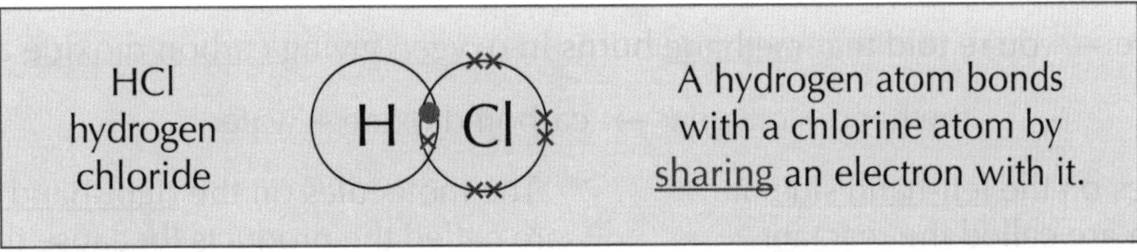

HCl
hydrogen
chloride
H
Cl
A hydrogen atom bonds with a chlorine atom by sharing an electron with it.

7) The properties of a compound are usually totally different from the properties of the original elements.

For example, if iron (a lustrous magnetic metal) and sulfur (a nice yellow powder) react, the compound formed (iron sulfide) is a dull grey solid lump, and doesn't behave anything like either iron or sulfur.

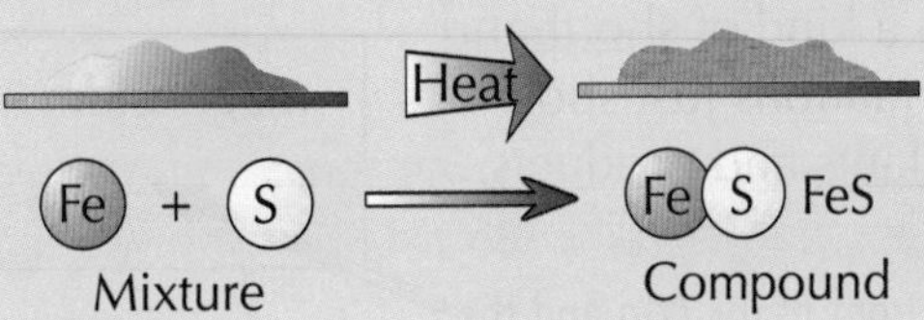

Formulas and Equations

Formulas and equations are used to show what is happening to substances involved in chemical reactions. They tell us what atoms are involved and how the substances change during a reaction.

A Formula Shows What Atoms are in a Compound

Just as elements can be represented by symbols, compounds can be represented by formulas. The formulas are made up of elemental symbols in the same proportions that the elements can be found in the compound.

1) For example, carbon dioxide, CO_2, is a compound formed from a chemical reaction between carbon and oxygen. It contains 1 carbon atom and 2 oxygen atoms.
2) Here's another example: the formula of sulfuric acid is H_2SO_4. So, each molecule contains 2 hydrogen atoms, 1 sulfur atom and 4 oxygen atoms.

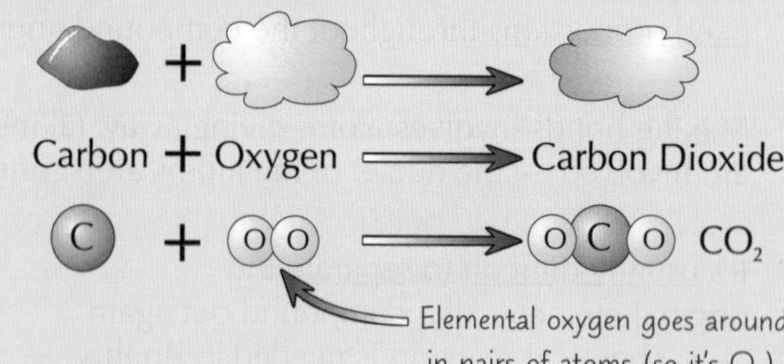

Carbon + Oxygen ⟶ Carbon Dioxide
C + O O ⟶ O C O CO_2
Elemental oxygen goes around in pairs of atoms (so it's O_2).

3) There might be brackets in a formula, e.g. calcium hydroxide is $Ca(OH)_2$. The little number outside the bracket applies to everything inside the brackets. So in $Ca(OH)_2$ there's 1 calcium atom, 2 oxygen atoms and 2 hydrogen atoms.

Here are some examples of formulas which might come in handy:

1) Carbon dioxide — CO_2
2) Ammonia — NH_3
3) Water — H_2O
4) Sodium chloride — NaCl
5) Carbon monoxide — CO
6) Hydrochloric acid — HCl
7) Calcium chloride — $CaCl_2$
8) Sodium carbonate — Na_2CO_3
9) Sulfuric acid — H_2SO_4

Chemical Changes are Shown Using Chemical Equations

One way to show a chemical reaction is to write a word equation. It's not as quick as using chemical symbols and you can't tell straight away what's happened to each of the atoms, but it's dead easy.

Here's an example — you're told that methane burns in oxygen giving carbon dioxide and water:

methane + oxygen → carbon dioxide + water

The molecules on the left-hand side of the equation are called the reactants (because they react with each other).

The molecules on the right-hand side are called the products (because they've been produced from the reactants).

Symbol Equations Show the Atoms on Both Sides

Chemical changes can be shown in a kind of shorthand using symbol equations. Symbol equations just show the symbols or formulas of the reactants and products...

magnesium + oxygen → magnesium oxide

$$2Mg + O_2 \rightarrow 2MgO$$

You'll have spotted that there's a '2' in front of the Mg and the MgO. The reason for this is explained on the next page.

Formulas and Equations

Symbol Equations Need to be **Balanced**

1) There must always be the same number of atoms on both sides — they can't just disappear.
2) You balance the equation by putting numbers in front of the formulas where needed.
3) Take this equation for reacting sulfuric acid (H_2SO_4) with sodium hydroxide (NaOH) to get sodium sulfate (Na_2SO_4) and water (H_2O):

$$H_2SO_4 + NaOH \rightarrow Na_2SO_4 + H_2O$$

The formulas for all of the compounds are correct but the numbers of some atoms don't match up on both sides. E.g. there are 3 Hs on the left, but only 2 on the right. You can't change formulas like H_2SO_4 to H_2SO_5. You can only put numbers in front of them.

4) This equation needs balancing — see below for how to do this.

Here's how to **Balance** an **Equation**

The more you practise, the quicker you get, but all you do is this:

1) Find an element that doesn't balance and pencil in a number to try and sort it out.
2) See where it gets you. It may create another imbalance, but if so, pencil in another number and see where that gets you.
3) Carry on chasing unbalanced elements and it'll sort itself out pretty quickly.

EXAMPLE:

In the equation above you soon notice we're short of H atoms on the RHS (Right-Hand Side).

1) The only thing you can do about that is make it $2H_2O$ instead of just H_2O:

$$H_2SO_4 + NaOH \rightarrow Na_2SO_4 + 2H_2O$$

2) But that now causes too many H atoms and O atoms on the RHS, so to balance that up you could try putting 2NaOH on the LHS (Left-Hand Side):

$$H_2SO_4 + 2NaOH \rightarrow Na_2SO_4 + 2H_2O$$

3) And suddenly there it is! Everything balances. And you'll notice the Na just sorted itself out.

Getting good at balancing equations takes patience and practice

Remember, a number in front of a formula applies to the entire formula — so, $3Na_2SO_4$ means three lots of Na_2SO_4. The little numbers within or at the end of a formula only apply to the atom or brackets immediately before. So the 4 in Na_2SO_4 means there are 4 Os, but there's just 1 S not 4.

Warm-Up & Exam Questions

So, you reckon you know your elements from your compounds?... Have a go at these questions and see how you do. If you get stuck on something just flick back and give it another read through.

Warm-Up Questions

1) What does the mass number tell you about an atom?
2) What is the definition of an element?
3) Describe what a compound is.
4) What does a compound formed from a metal and a non-metal consist of?
5) Balance this equation for the reaction of glucose ($C_6H_{12}O_6$) and oxygen:
   $C_6H_{12}O_6 + O_2 \rightarrow CO_2 + H_2O$

Exam Questions

1 This question is about atomic structure. 

1.1 Use your knowledge to complete **Table 1**.

Table 1

Name of particle	Relative charge
Proton	....
Neutron	....

[2 marks]

1.2 Where are protons and neutrons found in an atom?

[1 mark]

1.3 An atom has 8 electrons. How many protons does the atom have?

[1 mark]

1.4 What is the relative mass of a proton?

[1 mark]

2 **Table 2** gives some information about nitrogen. 

Table 2

Element	Number of protons	Mass number
nitrogen	7	14

2.1 How many neutrons does nitrogen have?

[1 mark]

2.2 Describe how the information in **Table 2** can be used to work out the atomic number of nitrogen.

[1 mark]

2.3 What is the chemical symbol for nitrogen?

[1 mark]

Exam Questions

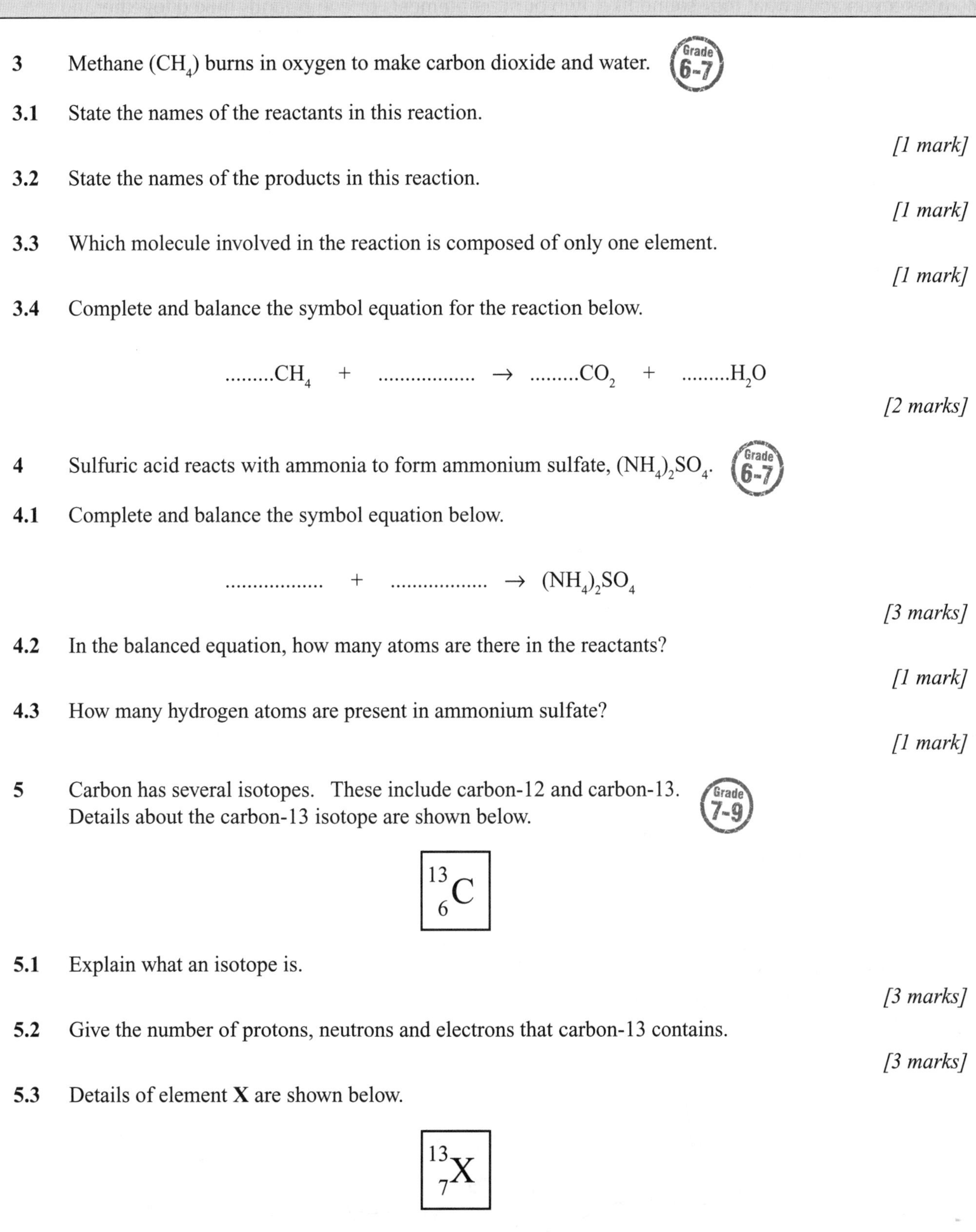

3 Methane (CH_4) burns in oxygen to make carbon dioxide and water. Grade 6-7

3.1 State the names of the reactants in this reaction.

[1 mark]

3.2 State the names of the products in this reaction.

[1 mark]

3.3 Which molecule involved in the reaction is composed of only one element.

[1 mark]

3.4 Complete and balance the symbol equation for the reaction below.

$$\ldots\ldots CH_4 + \ldots\ldots\ldots \rightarrow \ldots\ldots CO_2 + \ldots\ldots H_2O$$

[2 marks]

4 Sulfuric acid reacts with ammonia to form ammonium sulfate, $(NH_4)_2SO_4$. Grade 6-7

4.1 Complete and balance the symbol equation below.

$$\ldots\ldots\ldots + \ldots\ldots\ldots \rightarrow (NH_4)_2SO_4$$

[3 marks]

4.2 In the balanced equation, how many atoms are there in the reactants?

[1 mark]

4.3 How many hydrogen atoms are present in ammonium sulfate?

[1 mark]

5 Carbon has several isotopes. These include carbon-12 and carbon-13. Details about the carbon-13 isotope are shown below. Grade 7-9

$${}^{13}_{6}C$$

5.1 Explain what an isotope is.

[3 marks]

5.2 Give the number of protons, neutrons and electrons that carbon-13 contains.

[3 marks]

5.3 Details of element **X** are shown below.

$${}^{13}_{7}X$$

Explain how you can tell that element **X** is not an isotope of carbon.

[1 mark]

5.4 There are two isotopes of element **X**. One isotope has a mass number of 13 and a percentage abundance of 79%. The other isotope has a mass number of 14 and a percentage abundance of 21%.
Use this information to calculate the relative atomic mass of element **X**. Give your answer to 1 d.p.

[4 marks]

Mixtures

Mixtures are exactly what they sound like, two or more elements or compounds mixed together.

Mixtures are Easily Separated — Not Like Compounds

1) Unlike in a compound, there's no chemical bond between the different parts of a mixture.

2) The parts of a mixture can be either elements or compounds, and they can be separated out by physical methods such as filtration (p. 26), crystallisation (p.26), simple distillation (p.28), fractional distillation (p.29) and chromatography (see next page).

A physical method is one that doesn't involve a chemical reaction, so doesn't form any new substances.

Air is a mixture of gases, mainly nitrogen, oxygen, carbon dioxide and argon. The gases can all be separated out fairly easily.

Crude oil is a mixture of different length hydrocarbon molecules.

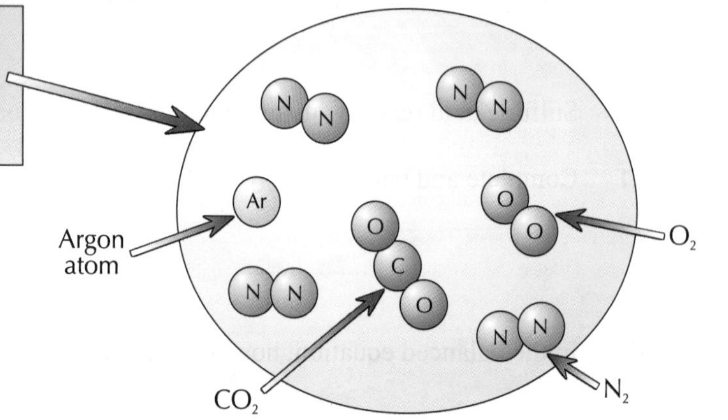

Argon atom

O_2

CO_2

N_2

3) The properties of a mixture are just a mixture of the properties of the separate parts — the chemical properties of a substance aren't affected by it being part of a mixture.

For example, a mixture of iron powder and sulfur powder will show the properties of both iron and sulfur. It will contain grey magnetic bits of iron and bright yellow bits of sulfur.

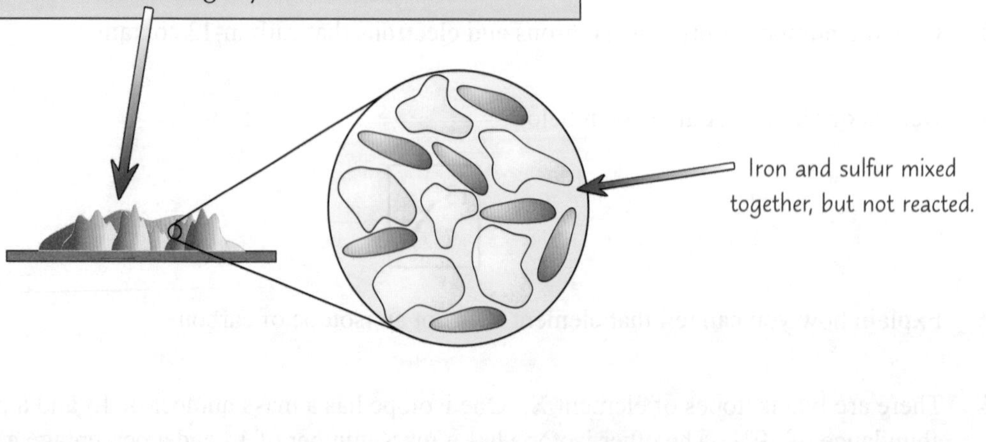

Iron and sulfur mixed together, but not reacted.

Mixtures can be separated without a chemical reaction

Remember that the different parts of mixtures aren't chemically combined, this means their chemical properties remain unchanged and the compounds or elements can be separated by physical methods.

Chromatography

Paper chromatography is a really useful technique to separate compounds out of a mixture.

You Need to Know How to Do **Paper Chromatography**

One method of separating substances in a mixture is through chromatography.
This technique can be used to separate different dyes in an ink. Here's how you can do it:

1) Draw a line near the bottom of a sheet of filter paper. (Use a pencil to do this — pencil marks are insoluble and won't dissolve in the solvent.)

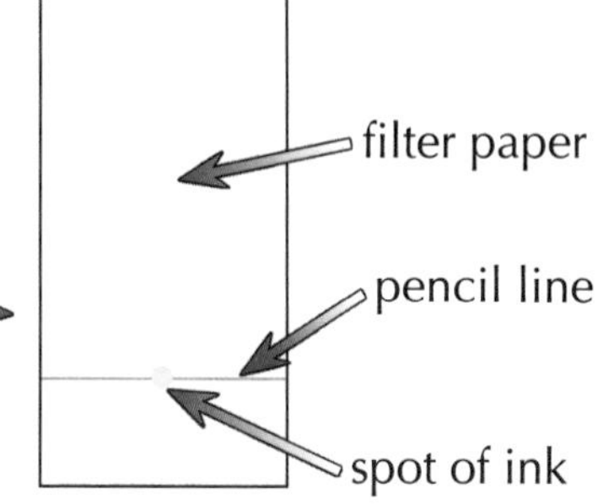

2) Add a spot of the ink to the line and place the sheet in a beaker of solvent, e.g. water.

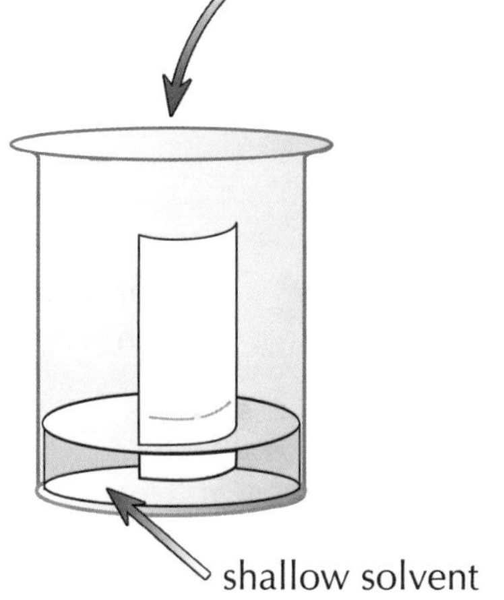

3) The solvent used depends on what's being tested. Some compounds dissolve well in water, but sometimes other solvents, like ethanol, are needed.

4) Make sure the ink isn't touching the solvent — you don't want it to dissolve into it.

5) Place a lid on top of the container to stop the solvent evaporating.

6) The solvent seeps up the paper, carrying the ink with it.

The point the solvent has reached as it moves up the paper is the solvent front.

7) Each different dye in the ink will move up the paper at a different rate so the dyes will separate out. Each dye will form a spot in a different place — 1 spot per dye in the ink.

8) If any of the dyes in the ink are insoluble (won't dissolve) in the solvent you've used, they'll stay on the baseline.

9) When the solvent has nearly reached the top of the paper, take the paper out of the beaker and leave it to dry.

10) The end result is a pattern of spots called a chromatogram.

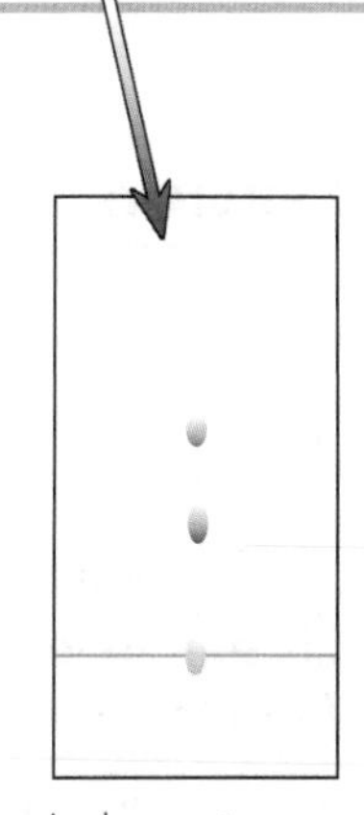

A chromatogram

Chromatography separates the different dyes in inks

Make sure you use a pencil to draw your baseline on the sheet of paper for your chromatogram. If you use a pen, all the components of the ink in the pen will get separated, along with the substance you're analysing, which will make your results very confusing.

Filtration and Crystallisation

Filtration and crystallisation are methods of separating mixtures. Chemists use these techniques all the time to separate solids from liquids, so it's worth making sure you know how to do them.

Filtration Separates Insoluble Solids from Liquids

1) Filtration can be used if your product is an insoluble solid that needs to be separated from a liquid reaction mixture.
2) It can be used in purification as well. For example, solid impurities in the reaction mixture can be separated out using filtration.

Insoluble means the solid can't be dissolved in the liquid.

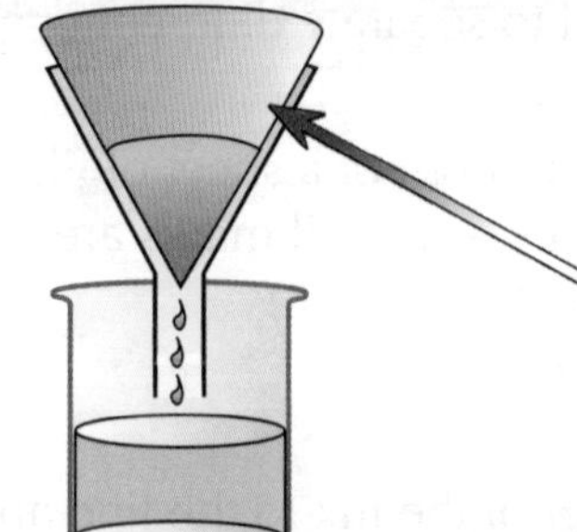

Filter paper folded into a cone shape — the solid is left in the filter paper.

Two Ways to Separate Soluble Solids from Solutions

If a solid can be dissolved it's described as being soluble. There are two methods you can use to separate a soluble salt from a solution — evaporation and crystallisation.

Evaporation

1) Pour the solution into an evaporating dish.

2) Slowly heat the solution. The solvent will evaporate and the solution will get more concentrated. Eventually, crystals will start to form.

3) Keep heating the evaporating dish until all you have left are dry crystals.

You don't have to use a Bunsen burner, you could use a water bath, or an electric heater.

Evaporation is a really quick way of separating a soluble salt from a solution, but you can only use it if the salt doesn't decompose (break down) when its heated. Otherwise, you'll have to use crystallisation.

Crystallisation

1) Pour the solution into an evaporating dish and gently heat the solution. Some of the solvent will evaporate and the solution will get more concentrated.
2) Once some of the solvent has evaporated, or when you see crystals start to form (the point of crystallisation), remove the dish from the heat and leave the solution to cool.
3) The salt should start to form crystals as it becomes insoluble in the cold, highly concentrated solution.
4) Filter the crystals out of the solution, and leave them in a warm place to dry. You could also use a drying oven or a dessicator.

You should also use crystallisation if you want to make nice big crystals of your salt.

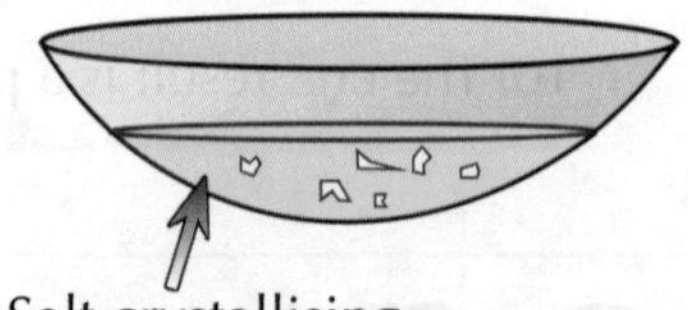

Salt crystallising out of solution.

Filtration and Crystallisation

Here's how you can put filtration and crystallisation to good use. Separating rock salt...

Filtration and Crystallisation can be Used to Separate Rock Salt

1) Rock salt is simply a mixture of salt and sand (they spread it on the roads in winter).
2) Salt and sand are both compounds — but salt dissolves in water and sand doesn't. This vital difference in their physical properties gives a great way to separate them. Here's what to do...

1. Grinding

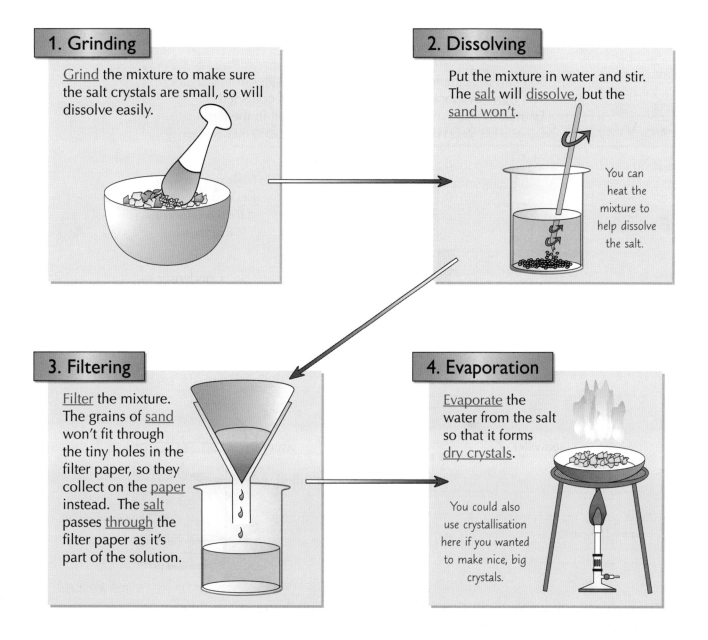

Grind the mixture to make sure the salt crystals are small, so will dissolve easily.

2. Dissolving

Put the mixture in water and stir. The salt will dissolve, but the sand won't.

You can heat the mixture to help dissolve the salt.

3. Filtering

Filter the mixture. The grains of sand won't fit through the tiny holes in the filter paper, so they collect on the paper instead. The salt passes through the filter paper as it's part of the solution.

4. Evaporation

Evaporate the water from the salt so that it forms dry crystals.

You could also use crystallisation here if you wanted to make nice, big crystals.

Separating rock salt requires filtration and evaporation

You may be asked how to separate another type of mixture containing insoluble and soluble solids — just apply the same method and think through what is happening in each stage.

Distillation

Distillation is used to separate mixtures which contain liquids. This first page looks at simple distillation.

Simple Distillation is Used to Separate Out Solutions

1) Simple distillation is used for separating out a liquid from a solution.
2) The solution is heated. The part of the solution that has the lowest boiling point evaporates first.
3) The vapour is then cooled, condenses (turns back into a liquid) and is collected.
4) The rest of the solution is left behind in the flask.

You can use simple distillation to get pure water from seawater. The water evaporates and is condensed and collected. Eventually you'll end up with just the salt left in the flask.

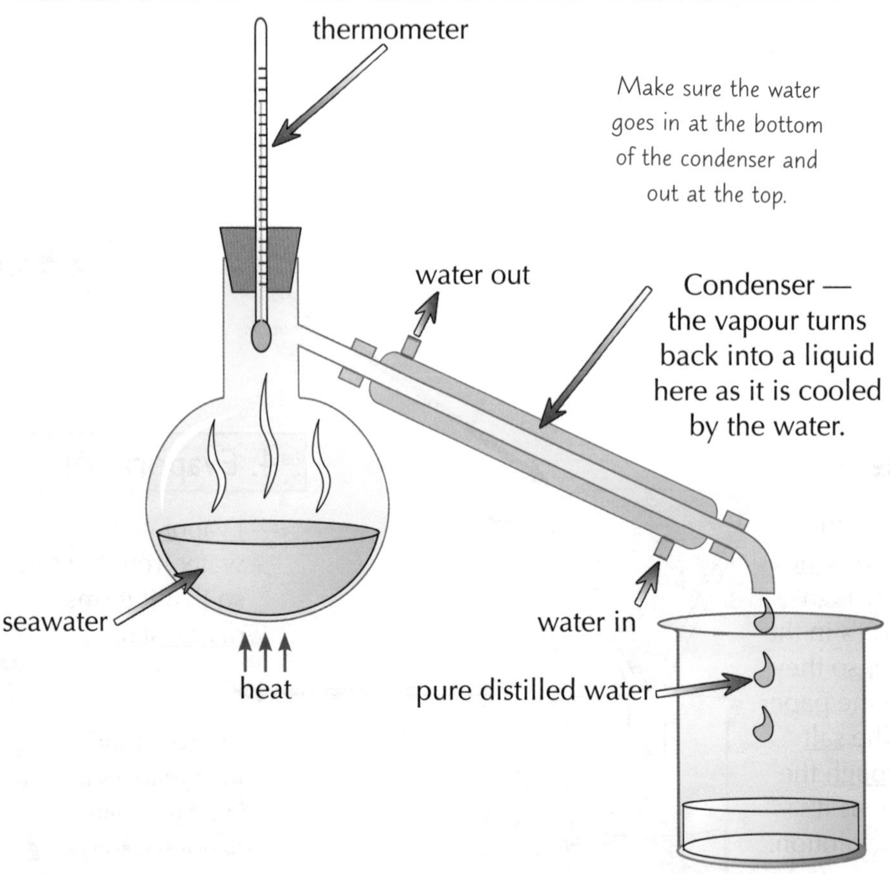

5) The problem with simple distillation is that you can only use it to separate things with very different boiling points — if the temperature goes higher than the boiling point of the substance with the higher boiling point, they will mix again.
6) If you have a mixture of liquids with similar boiling points you need another method to separate them — like fractional distillation (see next page).

Heating → Evaporating → Cooling → Condensing

You might have used distilled water in Chemistry lessons. Because it's been distilled, there aren't any impurities in it (like ions, see page 47) that might interfere with experimental results. Clever stuff.

Distillation

Another type of distillation is fractional distillation. This is more complicated to carry out than simple distillation but it can separate out mixtures of liquids even if their boiling points are close together.

Fractional Distillation is Used to Separate a Mixture of Liquids

Here is a lab demonstration that can be used to model the fractional distillation of crude oil at a refinery.

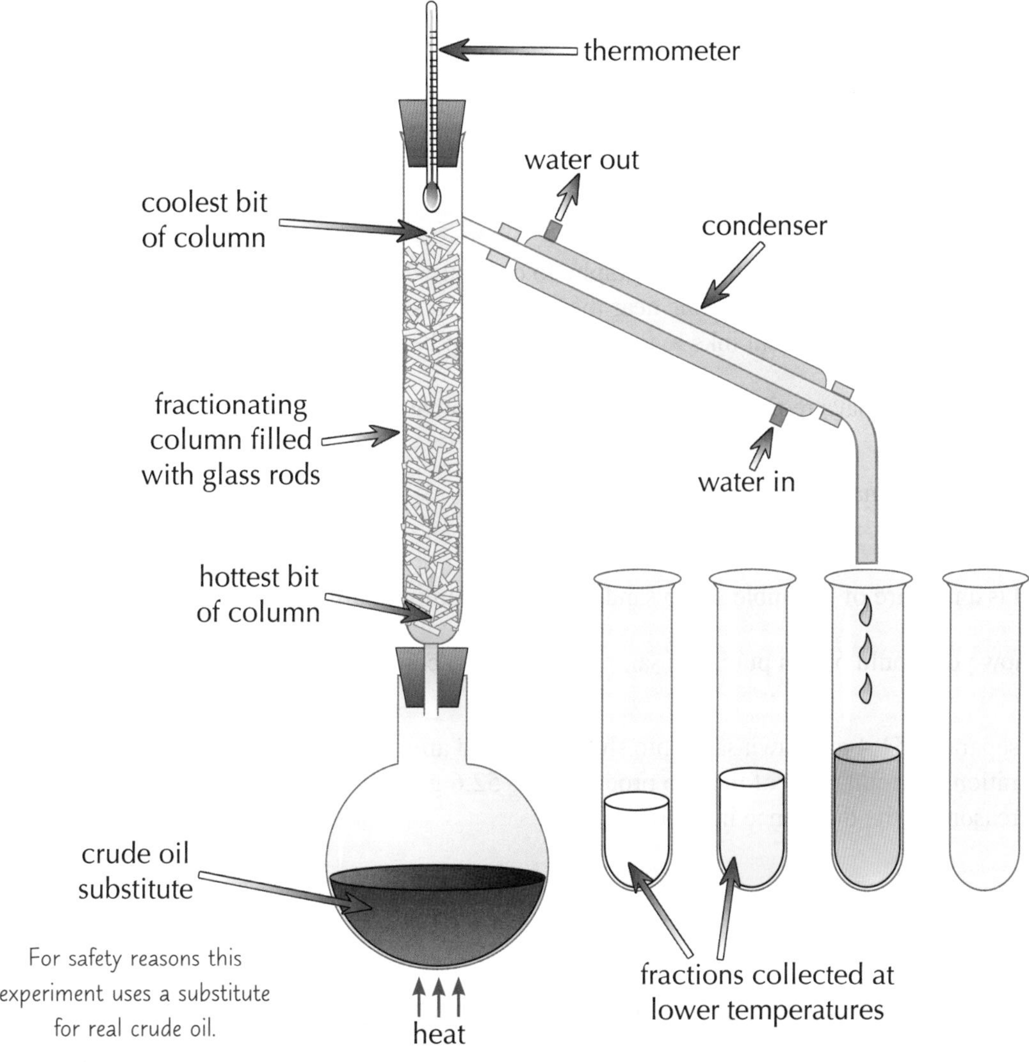

1) You put your mixture in a flask and stick a fractionating column on top. Then you heat it.
2) The different liquids will all have different boiling points — so they will evaporate at different temperatures.
3) The liquid with the lowest boiling point evaporates first. When the temperature on the thermometer matches the boiling point of this liquid, it will reach the top of the column.
4) Liquids with higher boiling points might also start to evaporate. But the column is cooler towards the top. So they will only get part of the way up before condensing and running back down towards the flask.
5) When the first liquid has been collected, you raise the temperature until the next one reaches the top.

Fractional distillation is used in the lab and industry

You've made it to the end of the pages on separation techniques, so make sure you understand what each of the methods can be used to separate and the apparatus set up for each technique.

PRACTICAL Warm-Up & Exam Questions

So the last few pages have all been about mixtures and how to separate them. Here are some questions to get stuck into and make sure you know your filtration from your distillation...

Warm-Up Questions

1) What is the definition of a mixture?
2) Name a separation technique that could be used to separate a soluble solid from a solution.
3) Which technique could you use to separate a mixture of liquids with similar boiling points?

Exam Questions

1 A forensic scientist is using paper chromatography to compare different inks which contain a mixture of dyes. He draws a pencil line near the bottom of a sheet of filter paper and adds a spot of each of the different inks to the line with a gap between each spot.

1.1 Describe the next step that's required to carry out paper chromatography.

[1 mark]

1.2 Why is pencil used to make the line on the filter paper?

[1 mark]

2 Lawn sand is a mixture of insoluble sharp sand and soluble ammonium sulfate fertiliser.

2.1 Describe how you would obtain pure, dry samples of the two components of lawn sand in the lab.

[3 marks]

2.2 A student separated 51.4 g of lawn sand into sharp sand and ammonium sulfate.
After separation, the total mass of the two products was 52.6 g.
Suggest a reason for the difference in mass.

[1 mark]

3 **Table 1** gives the boiling points of three liquids.

Table 1

Liquid	Boiling point (°C)
Methanoic acid	101
Propanone	56
Water	100

3.1 State why simple distillation cannot be used to separate water from a solution of water and methanoic acid.

[1 mark]

3.2 The apparatus in **Figure 1** was used to separate a mixture of propanone and water.
Complete the table using the options below.

no liquid **water** **propanone** **both liquids**

Temperature on thermometer	Contents of the flask	Contents of the beaker
30 °C	....	....
65 °C	....	....
110 °C	....	....

Figure 1

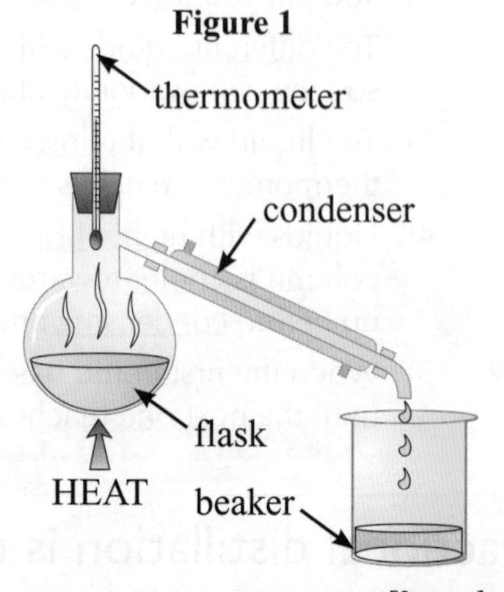

[3 marks]

The History of the Atom

You may have thought you were done with the atom after page 16. Unfortunately, you don't get away that easily. The next couple of pages are all about how scientists came to understand the atom as we do today.

The Theory of **Atomic Structure** Has **Changed** Over Time

1) At the start of the 19th century John Dalton described atoms as solid spheres, and said that different spheres made up the different elements.

2) In 1897 J J Thomson concluded from his experiments that atoms weren't solid spheres. His measurements of charge and mass showed that an atom must contain even smaller, negatively charged particles — electrons.

3) The 'solid sphere' idea of atomic structure had to be changed. The new theory was known as the 'plum pudding model'.

4) The plum pudding model showed the atom as a ball of positive charge with electrons stuck in it.

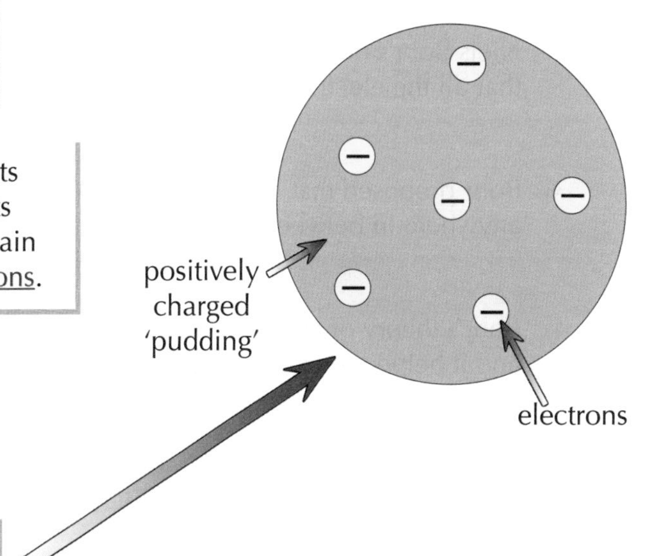
positively charged 'pudding'
electrons

Rutherford Showed that the **Plum Pudding Model** Was Wrong

1) In 1909 Ernest Rutherford and his student Ernest Marsden conducted the famous alpha particle scattering experiments. They fired positively charged alpha particles at an extremely thin sheet of gold.

2) From the plum pudding model, they were expecting the particles to pass straight through the sheet or be slightly deflected at most. This was because the positive charge of each atom was thought to be very spread out through the 'pudding' of the atom. But, whilst most of the particles did go straight through the gold sheet, some were deflected more than expected, and a small number were deflected backwards. So the plum pudding model couldn't be right.

3) Rutherford came up with an idea to explain this new evidence — the nuclear model of the atom. In this, there's a tiny, positively charged nucleus at the centre, where most of the mass is concentrated. A 'cloud' of negative electrons surrounds this nucleus — so most of the atom is empty space. When alpha particles came near the concentrated, positive charge of the nucleus, they were deflected. If they were fired directly at the nucleus, they were deflected backwards. Otherwise, they passed through the empty space.

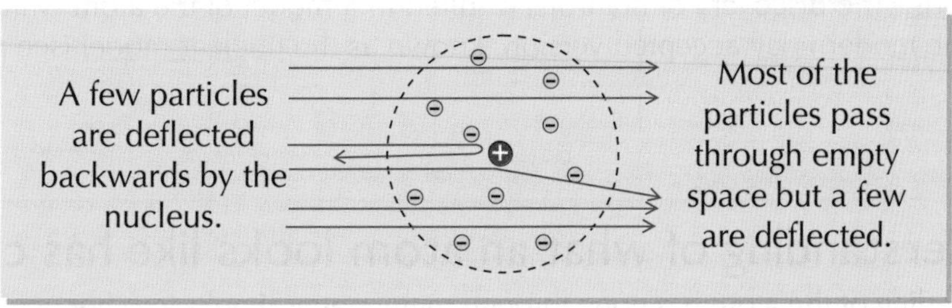
A few particles are deflected backwards by the nucleus.
Most of the particles pass through empty space but a few are deflected.

The History of the Atom

Bohr's **Nuclear Model** Explains a Lot

1) Scientists realised that electrons in a 'cloud' around the nucleus of an atom, as Rutherford described, would be attracted to the nucleus, causing the atom to collapse.

2) Niels Bohr's nuclear model of the atom suggested that all the electrons were contained in shells.

3) Bohr proposed that electrons orbit the nucleus in fixed shells and aren't anywhere in between. Each shell is a fixed distance from the nucleus.

4) Bohr's theory of atomic structure was supported by many experiments and it helped to explain lots of other scientists' observations at the time.

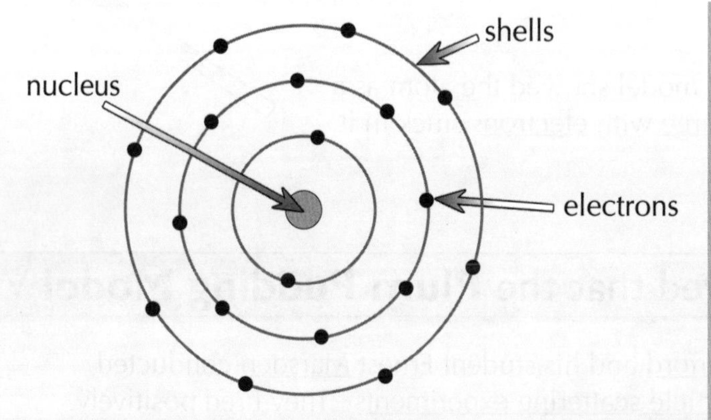

Further Experiments Showed the **Existence** of **Protons**

1) Further experimentation by Rutherford and others gave the conclusion that the nucleus could be divided into smaller particles, each of which has the same charge as a hydrogen nucleus.

2) About 20 years after scientists had accepted that atoms have nuclei, James Chadwick carried out an experiment which provided evidence for neutral particles in the nucleus. These became known as neutrons. The discovery of neutrons resulted in a model of the atom which was pretty close to the modern day accepted version, known as the nuclear model (see page 16).

Our understanding of what an atom looks like has changed

Our understanding of the atom has gone through many stages thanks to other people's work being built upon with new evidence and new predictions made. A fine example of the scientific method.

Electronic Structure

The fact that electrons occupy 'shells' around the nucleus is what causes the whole of chemistry. Remember that, and watch how it applies to each bit of it.

Electron Shell **Rules:**

1) Electrons always occupy shells (sometimes called energy levels).
2) The lowest energy levels are always filled first — these are the ones closest to the nucleus.
3) Only a certain number of electrons are allowed in each shell:

1st shell: 2 2nd shell: 8 3rd shell: 8

4) Atoms are much happier when they have full electron shells — like the noble gases in Group 0.
5) In most atoms, the outer shell is not full and this makes the atom want to react to fill it.

Electron configurations can be shown as diagrams like this...

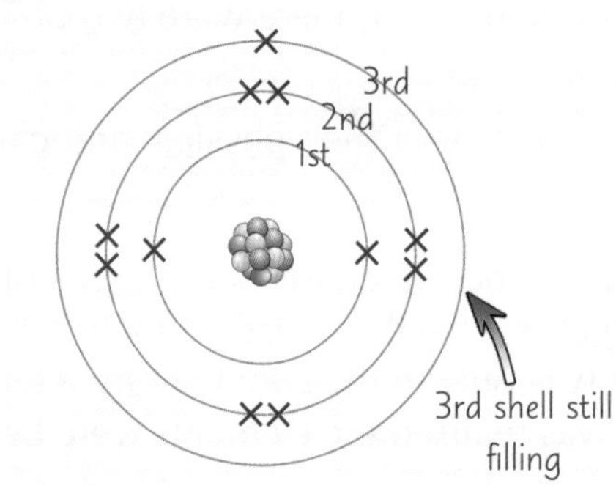

...or as numbers like this: 2, 8, 1

Both of the configurations above are for sodium.

Follow the Rules to **Work Out** Electronic Structures

You can easily work out the electronic structures for the first 20 elements of the periodic table (things get a bit more complicated after that).

EXAMPLE:

What is the electronic structure of nitrogen?

1) Nitrogen's atomic number is 7. This means it has 7 protons... so it must have 7 electrons.
2) Follow the 'Electron Shell Rules' above. The first shell can only take 2 electrons and the second shell can take a maximum of 8 electrons.

So the electronic structure for nitrogen must be 2, 5.

EXAMPLE:

What is the electronic structure of magnesium?

1) Magnesium's atomic number is 12. This means it has 12 protons... so it must have 12 electrons.
2) Follow the 'Electron Shell Rules' above. The first shell can only take 2 electrons and the second shell can take a maximum of 8 electrons, so the third shell must also be partially filled.

So the electronic structure for magnesium must be 2, 8, 2.

Here are some more examples of electronic structures:

H Hydrogen

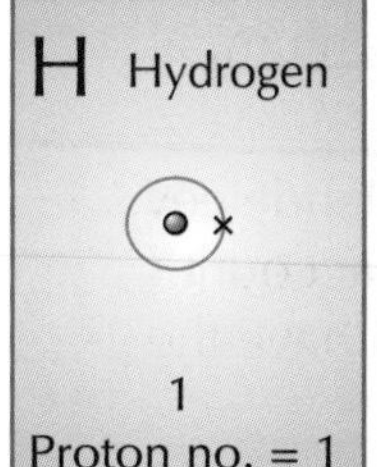

1
Proton no. = 1

He Helium

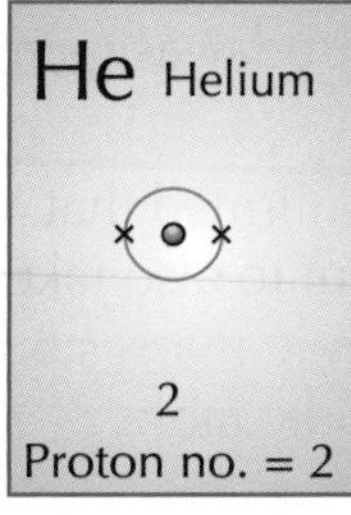

2
Proton no. = 2

Li Lithium

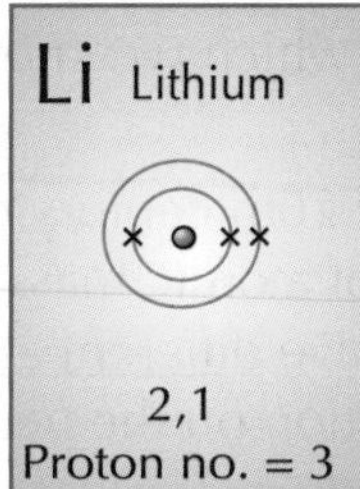

2,1
Proton no. = 3

C Carbon

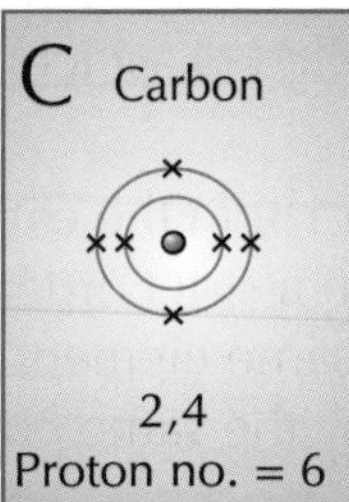

2,4
Proton no. = 6

Ne Neon

2,8
Proton no. = 10

Ca Calcium

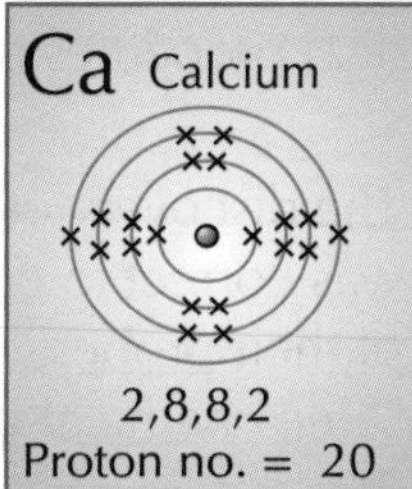

2,8,8,2
Proton no. = 20

Electron shells — one of the most important ideas in chemistry

It's really important to learn the rules for filling electron shells — make sure you practise.

Development of the Periodic Table

We haven't always known as much about chemistry as we do now. Early chemists looked to try and understand patterns in the elements' properties to get a bit of understanding.

In the Early 1800s Elements Were Arranged By Atomic Mass

Until quite recently, there were two obvious ways to categorise elements:

1) Their physical and chemical properties   2) Their relative atomic mass

1) Remember, scientists had no idea of atomic structure or of protons, neutrons or electrons, so there was no such thing as atomic number to them. (It was only in the 20th century after protons and electrons were discovered that it was realised the elements were best arranged in order of atomic number.)

Remember — the relative atomic mass is the average mass of one atom of an element.

2) Back then, the only thing they could measure was relative atomic mass, and so the known elements were arranged in order of atomic mass. When this was done, a periodic pattern was noticed in the properties of the elements. This is where the name 'periodic table' comes from...

3) Early periodic tables were not complete and some elements were placed in the wrong group. This is because elements were placed in the order of relative atomic mass and did not take into account their properties.

Dmitri Mendeleev Left Gaps and Predicted New Elements

1) In 1869, Dmitri Mendeleev overcame some of the problems of early periodic tables by taking 50 known elements and arranging them into his Table of Elements — with various gaps as shown.

Mendeleev's Table of the Elements

H																
Li	Be											B	C	N	O	F
Na	Mg											Al	Si	P	S	Cl
K	Ca	*	Ti	V	Cr	Mn	Fe	Co	Ni	Cu	Zn	*	*	As	Se	Br
Rb	Sr	Y	Zr	Nb	Mo	*	Ru	Rh	Pd	Ag	Cd	In	Sn	Sb	Te	I
Cs	Ba	*	*	Ta	W	*	Os	Ir	Pt	Au	Hg	Tl	Pb	Bi		

2) Mendeleev put the elements mainly in order of atomic mass but did switch that order if the properties meant it should be changed. An example of this can been seen with Te and I — iodine actually has a smaller relative atomic mass but is placed after tellurium as it has similar properties to the elements in that group.

3) Gaps were left in the table to make sure that elements with similar properties stayed in the same groups. Some of these gaps indicated the existence of undiscovered elements and allowed Mendeleev to predict what their properties might be. When they were found and they fitted the pattern it helped confirm Mendeleev's ideas. For example, Mendeleev made really good predictions about the chemical and physical properties of an element he called ekasilicon, which we know today as germanium.

The discovery of isotopes (see page 18) in the early 20th century confirmed that Mendeleev was correct to not place elements in a strict order of atomic mass but to also take account of their properties. Isotopes of the same element have different atomic masses but have the same chemical properties so occupy the same position on the periodic table.

By leaving gaps in the table Mendeleev had the right idea

Make sure you can describe what Mendeleev did to overcome problems with early periodic tables.

The Modern Periodic Table

Chemists were getting pretty close to producing something that you might recognise as a periodic table. The big breakthrough came when the structure of the atom was understood a bit better.

The Periodic Table Helps you to See Patterns in Properties

1) There are 100ish elements, which all materials are made of.
2) In the periodic table the elements are laid out in order of increasing atomic (proton) number. Arranging the elements like this means there are repeating patterns in the properties of the elements (the properties are said to occur periodically, hence the name periodic table).
3) If it wasn't for the periodic table organising everything, you'd have a hard job remembering all those properties.
4) It's a handy tool for working out which elements are metals and which are non-metals. Metals are found to the left and non-metals to the right.

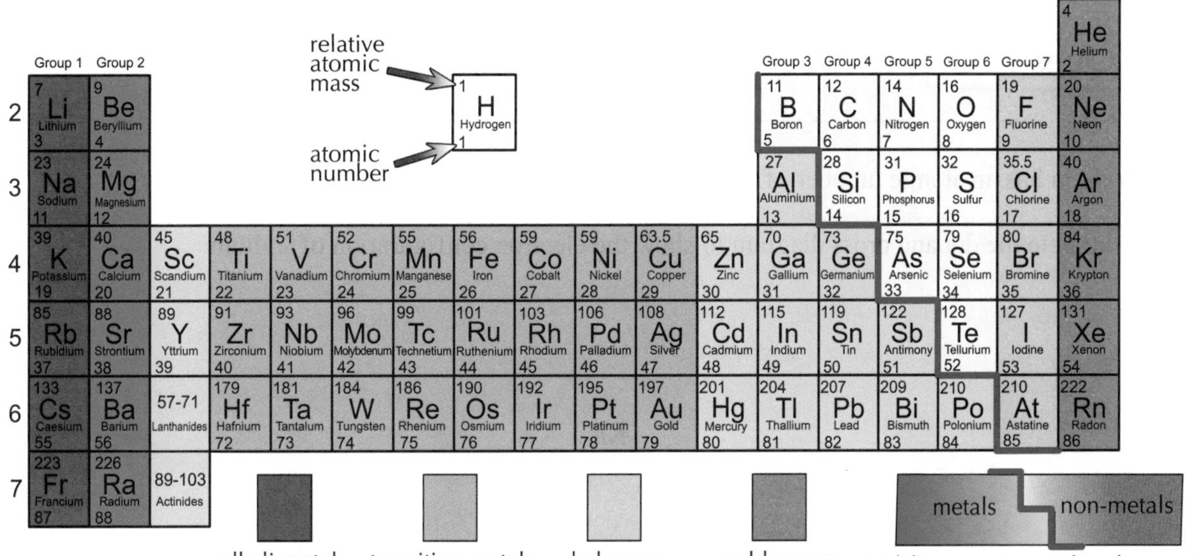

5) Elements with similar properties form columns.
6) These vertical columns are called groups.
7) The group number tells you how many electrons there are in the outer shell. For example, Group 1 elements all have one electron in their outer shell and Group 7 all have seven electrons in their outer shell. The exception to the rule is Group 0, for example helium has two electrons in its outer shell. This is useful as the way atoms react depends upon the number of electrons in their outer shell. So all elements in the same group are likely to react in a similar way.
8) If you know the properties of one element, you can predict properties of other elements in that group — and in the exam, you might be asked to do this. For example the Group 1 elements are Li, Na, K, Rb, Cs and Fr. They're all metals and they react in a similar way (see pages 34-40).
9) You can also make predictions about trends in reactivity. E.g. in Group 1, the elements react more vigorously as you go down the group. And in Group 7, reactivity decreases as you go down the group.
10) The rows are called periods. Each new period represents another full shell of electrons.

The modern periodic table is vital for understanding chemistry

This is a good example of how science progresses. A scientist has a basically good (though incomplete) hypothesis (see page 1), other scientists question it and bring more evidence to the table. The hypothesis may be modified or even scrapped to take account of available evidence. Only when all of the available evidence supports a hypothesis will it be accepted.

Warm-Up & Exam Questions

The last few pages have been tough with lots of information to learn. Luckily, here are some questions to get your head around to help you test your understanding.

Warm-Up Questions

1) Describe J J Thomson's model of the atom.
2) How many electrons can be held in the following:
   a) the first shell of an atom
   b) the second shell of an atom
3) What discovery supported Mendeleev's decision not to place elements in order of relative atomic mass?

Exam Questions

1 Sodium has an atomic number of 11. 

1.1 Complete the dot and cross diagram to show the electron configuration of sodium.

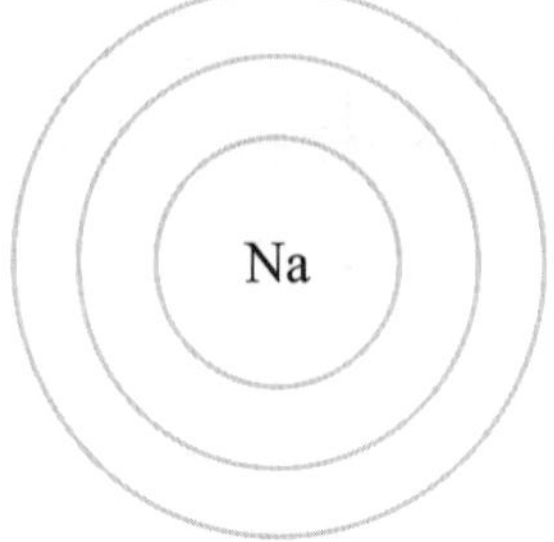

[1 mark]

1.2 Sodium is in Group 1.
Name another element that would have the same number of outer shell electrons.

[1 mark]

1.3 How many electrons does sodium need to lose so that it has a full outer shell?

[1 mark]

2 The periodic table contains all the known elements arranged in order. 

2.1 How were elements generally ordered in early periodic tables?

[1 mark]

2.2 How are the elements arranged in the modern periodic table?

[1 mark]

2.3 If two elements are in the same group of the periodic table what will they have in common?
Explain your answer.

[1 mark]

3* Describe how the theory of atomic structure has changed throughout history. 

[6 marks]

Metals and Non-Metals

Metals are used for all sorts of things so they're really important in modern life.

Most Elements are Metals

1) Metals are elements which can form positive ions when they react.
2) They're towards the bottom and to the left of the periodic table.
3) Most elements in the periodic table are metals.
4) Non-metals are at the far right and top of the periodic table.
5) Non-metals don't generally form positive ions when they react.

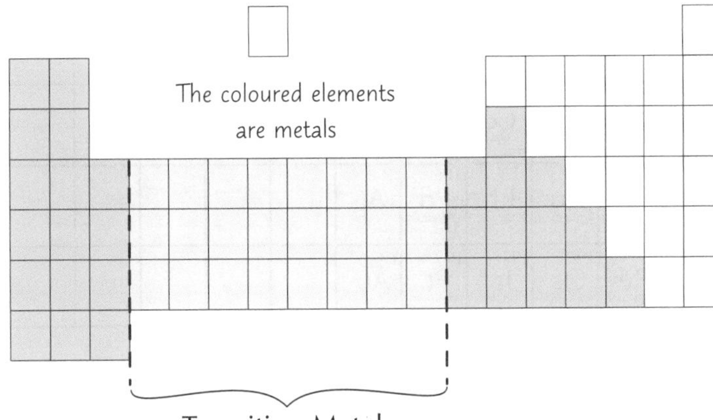
The coloured elements are metals
Transition Metals

Only the white elements are non-metals.

The Electronic Structure of Atoms Affects How They Will React

1) Atoms generally react to form a full outer shell. They do this via losing, gaining or sharing electrons.
2) Metals to the left of the periodic table don't have many electrons to remove and metals towards the bottom of the periodic table have outer electrons which are a long way from the nucleus so feel a weaker attraction. Both these effects mean that not much energy is needed to remove the electrons so it's feasible for the elements to react to form positive ions with a full outer shell.
3) For non-metals, forming positive ions is much more difficult. This is because they are either to the right of the periodic table — where they have lots of electrons to remove to get a full outer shell, or towards the top — where the outer electrons are close to the nucleus so feel a strong attraction. It's far more feasible for them to either share or gain electrons to get a full outer shell.

Metals and Non-Metals Have Different Physical Properties

1) All metals have metallic bonding which causes them to have similar basic physical properties.
    - They're strong (hard to break), but can be bent or hammered into different shapes (malleable).
    - They're great at conducting heat and electricity.
    - They have high boiling and melting points.
2) As non-metals don't have metallic bonding, they don't tend to exhibit the same properties as metals. They tend to be dull looking, more brittle, aren't always solids at room temperature, don't generally conduct electricity and often have a lower density.

Non-metals form a variety of different structures so have a wide range of chemical properties.

Metals have quite different properties from non-metals

So, all metals can conduct electricity and heat and be bent into shape, whereas mainly non-metals can't.

Transition Metals

Transition metals make up the big clump of elements in the middle of the periodic table.

Transition Metals can be Found Between **Group 2** and **Group 3**

1) Transition metals are in the centre of the periodic table.

Group 1 Group 2 — Group 3 Group 4 Group 5 Group 6 Group 7 Group 0

Here they are, right in the middle of Group 2 and Group 3.

48 Ti Titanium 22	51 V Vanadium 23	52 Cr Chromium 24	55 Mn Manganese 25	56 Fe Iron 26	59 Co Cobalt 27	59 Ni Nickel 28	63.5 Cu Copper 29
91 Zr Zirconium 40	93 Nb Niobium 41	96 Mo Molybdenum 42	98 Tc Technetium 43	101 Ru Ruthenium 44	103 Rh Rhodium 45	106 Pd Palladium 46	108 Ag Silver 47
178 Hf Hafnium 72	181 Ta Tantalum 73	184 W Tungsten 74	186 Re Rhenium 75	190 Os Osmium 76	192 Ir Iridium 77	195 Pt Platinum 78	197 Au Gold 79

2) Transition metals are typical metals, and have the properties you would expect of a 'proper' metal — they're good conductors of heat and electricity, and they're very dense, strong and shiny.
3) Transition metals also have some pretty special properties...

Transition metals can have more than one ion.

For example, copper forms Cu^{+} and Cu^{2+} ions. Cobalt forms Co^{2+} and Co^{3+} ions.

Transition metal ions are often coloured, and so compounds that contain them are colourful.

For example, potassium chromate(VI), which is yellow, potassium(VII) manganate is purple.

Transition metal compounds often make good catalysts (things that speed up the rate of a reaction — see p.119).

For example, a nickel based catalyst is used in the hydrogenation of alkenes (p.139), and an iron catalyst is used in the Haber process for making ammonia (p.188).

Transition metals are pretty useful elements...

Transition metals have similar properties to 'typical metals' but also have a few more unusual characteristics. These include having multiple ions, coloured compounds, and the ability to act as catalysts.

Group 1 Elements

Group 1 elements are known as the alkali metals — these are silvery solids that have to be stored in oil (and handled with forceps) as they're very reactive.

The **Group 1** Elements are **Reactive, Soft** Metals

1) The alkali metals are lithium, sodium, potassium, rubidium, caesium and francium.
2) The alkali metals are all soft and have low density. The first three in the group are less dense than water.
3) They all have one electron in their outer shell. This makes them very reactive and gives them similar properties.

Trends of Alkali Metals:

The trends for the alkali metals as you go down Group 1 include:

1) Increasing reactivity — the outer electron is more easily lost as the attraction between the nucleus and electron decreases, because the electron is further away from the nucleus the further down the group you go.
2) Lower melting and boiling points.
3) Higher relative atomic mass.

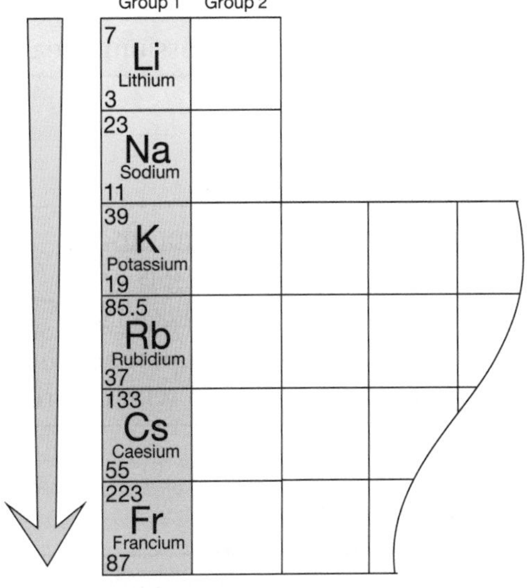

Alkali Metals Form **Ionic Compounds** with **Non-Metals**

1) The Group 1 elements don't need much energy to lose their one outer electron to form a full outer shell, so they readily form 1+ ions.
2) It's so easy for them to lose their outer electron that they only ever react to form ionic compounds. These compounds are usually white solids that dissolve in water to form colourless solutions.

There's more on ionic compounds on page 50.

Details of the reactions of Group 1 metals with non-metals are on the next page.

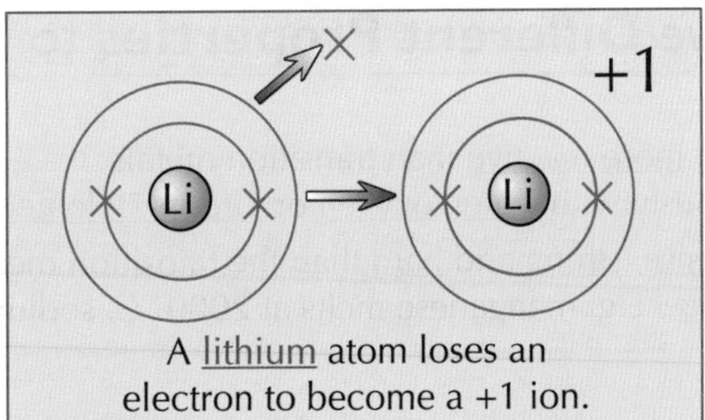

A lithium atom loses an electron to become a +1 ion.

The properties of Group 1 metals change as you go down the group

In the exam you might be given a trend and then asked to predict properties of other Group 1 metals. For example, the reactivity of Group 1 metals increases as you go down the group, so you know that potassium will react more vigorously than sodium.

Group 1 Elements

You met Group 1 metals on the previous page, so now it's time to learn about some of their reactions...

Reactions of Alkali Metals with Non-Metals:

Reaction with water

1) When Group 1 metals are put in water, they react vigorously to produce hydrogen gas and metal hydroxides — salts that dissolve in water to produce alkaline solutions.

$$2Na_{(s)} + 2H_2O_{(l)} \rightarrow 2NaOH_{(aq)} + H_{2(g)}$$

sodium + water → sodium hydroxide + hydrogen

All the Group 1 metals react with water in a similar way.

2) The more reactive (lower down in the group) an alkali metal is, the more violent the reaction.
3) The amount of energy given out by the reaction increases down the group — the reaction with potassium releases enough energy to ignite hydrogen.

Reaction with chlorine

1) Group 1 metals react vigorously when heated in chlorine gas to form white metal chloride salts.
2) As you go down the group, reactivity increases so the reaction with chlorine gets more vigorous.

$$2Na_{(s)} + Cl_{2(g)} \rightarrow 2NaCl_{(s)}$$

sodium + chlorine → sodium chloride

Reaction with oxygen

The Group 1 metals can react with oxygen to form a metal oxide. Different types of oxide will form depending on the Group 1 metal:

- Lithium reacts to form lithium oxide (Li_2O).
- Sodium reacts to form a mixture of sodium oxide (Na_2O) and sodium peroxide (Na_2O_2).
- Potassium reacts to form a mixture of potassium peroxide (K_2O_2) and potassium superoxide (KO_2).

The reactions with oxygen are why Group 1 metals tarnish in the air — the metal reacts with oxygen in the air to form a dull metal oxide layer.

Group 1 Metals Have Different Properties to Transition Metals

1) Group 1 metals are much more reactive than transition metals — they react more vigorously with water, oxygen or Group 7 elements for example.
2) They're also much less dense, strong and hard than the transition metals, and have much lower melting points. E.g. manganese melts at 2000 °C, sodium melts at 98 °C.

So now you know why they are called alkali metals...

...because they react with water to give alkaline solutions. Make sure you also know how they react with chlorine and oxygen. Memorise a couple of example equations for these reactions.

Group 7 Elements

The Group 7 elements are known as the halogens. Similarly to the alkali metals, halogens also show trends down the group. However, these trends are a bit different...

The Halogens are All Non-Metals with Coloured Vapours

- Fluorine is a very reactive, poisonous yellow gas.
- Chlorine is a fairly reactive, poisonous dense green gas.

- Bromine is a dense, poisonous, red-brown volatile liquid.
- Iodine is a dark grey crystalline solid or a purple vapour.

- They all exist as molecules which are pairs of atoms.

F_2 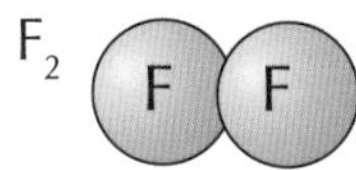

Cl_2 

Br_2 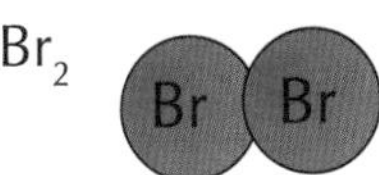

I_2 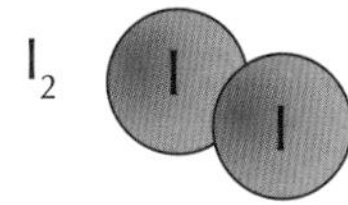

Learn These Trends:

As you go down Group 7, the halogens:

1) become less reactive — it's harder to gain an extra electron, because the outer shell's further from the nucleus.
2) have higher melting and boiling points.
3) have higher relative atomic masses.

All the Group 7 elements react in similar ways. This is because they all have seven electrons in their outer shell.

You can use these trends to predict properties of halogens. For example, you know that iodine will have a higher boiling point than chlorine as it's further down the group in the periodic table.

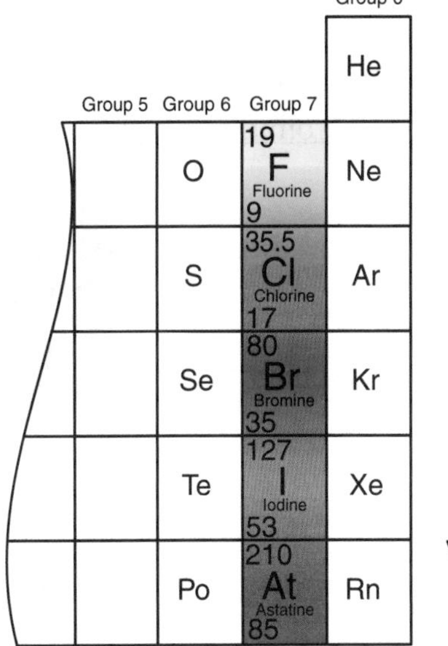

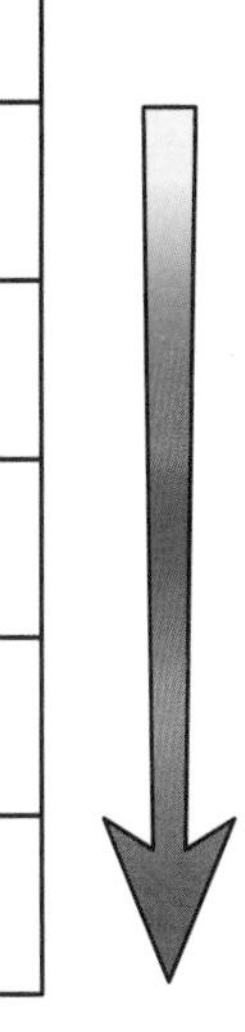

Halogens can Form Molecular Compounds

Halogen atoms can share electrons via covalent bonding (see page 53) with other non-metals so as to achieve a full outer shell. For example HCl, PCl_5, HF and CCl_4 contain covalent bonds. The compounds that form when halogens react with non-metals all have simple molecular structures.

Halogens all exist as molecules with two atoms

Just like alkali metals, you may be asked to predict the properties of a halogen from a given trend down the group. Make sure you understand why their electronic structure means halogens react in similar ways.

Group 7 Elements

Halogens Form **Ionic Bonds** with **Metals**

1) The halogens form <u>1– ions</u> called <u>halides</u> (F^-, Cl^-, Br^- and I^-) when they bond with <u>metals</u>, for example Na^+Cl^- or $Fe^{3+}Br^-_3$.
2) The compounds that form have <u>ionic structures</u>.
3) The diagram shows the bonding in sodium chloride, NaCl.

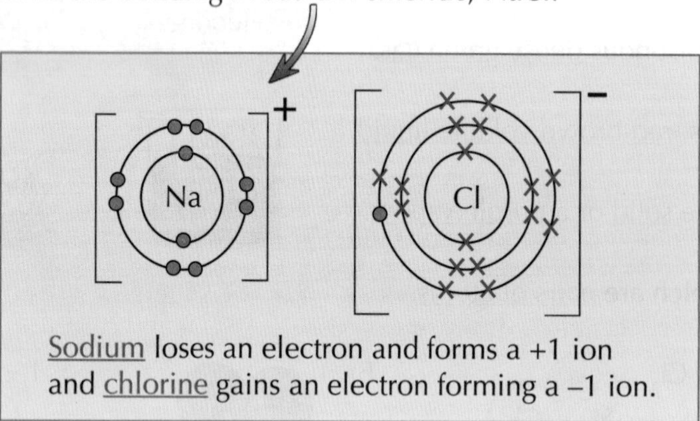

<u>Sodium</u> loses an electron and forms a +1 ion and <u>chlorine</u> gains an electron forming a –1 ion.

More Reactive Halogens Will **Displace** Less Reactive Ones

A <u>displacement reaction</u> can occur between a more reactive halogen and the salt of a less reactive one.

For example, <u>chlorine</u> can displace <u>bromine</u> and <u>iodine</u> from an aqueous <u>solution</u> of its salt (a <u>bromide</u> or <u>iodide</u>). <u>Bromine</u> will also displace <u>iodine</u> because of the <u>trend</u> in <u>reactivity</u>.

$$\underset{\text{Pale Green}}{Cl_{2\,(g)}} + 2KI_{(aq)} \rightarrow \underset{\text{Brown}}{I_{2\,(aq)}} + 2KCl_{(aq)}$$

$$\underset{\text{Pale Green}}{Cl_{2\,(g)}} + 2KBr_{(aq)} \rightarrow \underset{\text{Orange}}{Br_{2\,(aq)}} + 2KCl_{(aq)}$$

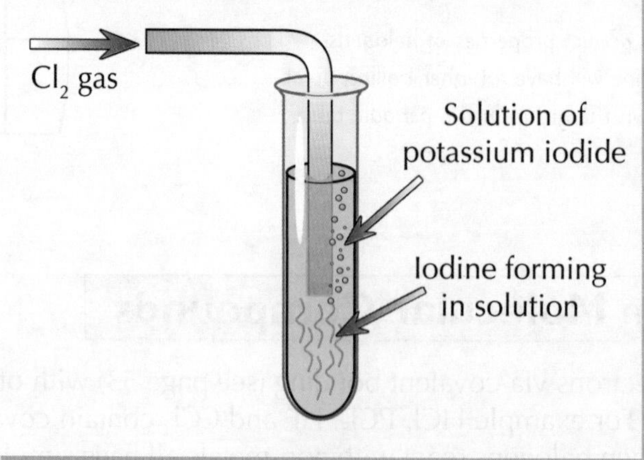

Halogens all have different reactivities

Once more, you don't have to be a mind-reader to be able to guess the kind of thing they could ask you about halogens in the exam. Something to do with displacement reactions seems very possible — will iodine displace bromine from some compound or other, for instance. Make sure you learn the trends.

Group 0 Elements

The noble gases don't react with very much and you can't even see them — makes them, a bit dull really.

Group 0 Elements are All Inert, Colourless Gases

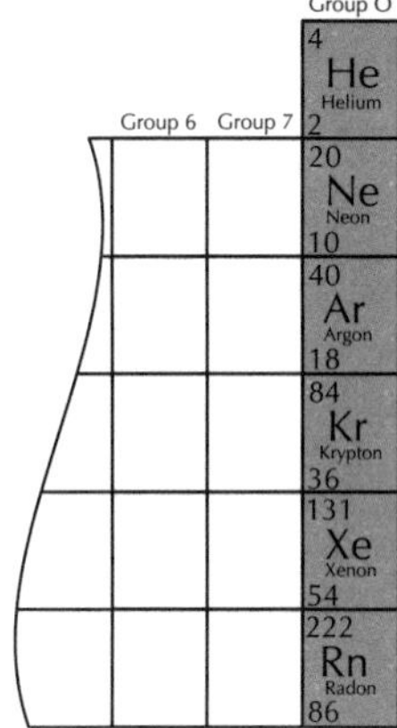

1) Group 0 elements are called the noble gases and include the elements helium, neon and argon (plus a few others).
2) They all have eight electrons in their outer energy level, apart from helium which has two, giving them a full outer-shell. As their outer shell is energetically stable they don't need to give up or gain electrons to become more stable. This means they are more or less inert — they don't react with much at all.
3) They exist as monatomic gases — single atoms not bonded to each other.
4) All elements in Group 0 are colourless gases at room temperature.
5) As the noble gases are inert they're non-flammable — they won't set on fire.

Helium only has electrons in the first shell, which only needs 2 to be filled.

There are Patterns in the Properties of the Noble Gases

1) The boiling points of the noble gases increase as you move down the group along with increasing relative atomic mass.
2) The increase in boiling point is due to an increase in the number of electrons in each atom leading to greater intermolecular forces between them which need to be overcome. There's more on intermolecular forces for small molecules on page 55.
3) In the exam you may be given the boiling point of one noble gas and asked to estimate the value for another one. So make sure you know the pattern.

Noble Gas
helium
neon
argon
krypton
xenon
radon

Increasing boiling point

EXAMPLE:

Neon is a gas at 25°C. Predict what state helium is at this temperature.

Helium has a lower boiling than neon as it is further up the group.
So, helium must also be a gas at 25°C.

EXAMPLE:

Radon and krypton have boiling points of –62 °C and –153 °C respectively. Predict the boiling point of xenon.

Xenon comes in between radon and krypton in the group so you can predict that its boiling point would be halfway between their boiling points: $(-153) + (-62) = -215$

$-215 \div 2 = -107.5 \approx -108°C$

So, xenon should have a boiling point of about –108 °C

The actual boiling point of xenon is –108 °C — just as predicted.

Just like other groups of elements, the noble gases follow patterns...

Although they're unreactive and hard to see, they're actually pretty useful. It took a while to discover them, but we now know all about them, including the trend in their boiling points.

Warm-Up & Exam Questions

These questions are all about the groups of the periodic table that you need to know about. Treat the exam questions like the real thing — don't look back through the book until you've finished.

Warm-Up Questions

1) What kind of ions do metals form?
2) Where are the transition metals found in the periodic table?
3) Give three physical properties of most metals.
4) Give three typical properties of transition metals.
5) Which product is formed when lithium reacts with oxygen?
6) Do halide ions have a positive or a negative charge?
7) In which group of the periodic table are the noble gases?

Exam Questions

1 **Table 1** shows some of the physical properties of four of the halogens. 

Table 1

Halogen	Properties			
	Atomic number	**Colour**	**Physical state at room temperature**	**Reactivity**
Fluorine	9	yellow	gas	....
Chlorine	17	green	gas	....
Bromine	35	red-brown	liquid	....
Iodine	53	dark grey	solid	....

1.1 **Table 1** has a column for reactivity. Write an **X** in the row of the halogen with the **highest** reactivity and a **Y** in the row of the halogen with the **lowest** reactivity.

[2 marks]

1.2 Which halogen in **Table 1** has the highest melting point?

[1 mark]

2 **Figure 1** shows the periodic table. 

2.1 Element **X** is found in the centre of the periodic table. What name is given to the elements found in this part of the table?

[1 mark]

2.2 Element **Y** does not conduct electricity. Predict whether element **Y** will be found to the left or the right of line **A** in **Figure 1**. Explain your answer.

[2 marks]

Figure 1

line **A**

X

Exam Questions

3 Chlorine is a Group 7 element.
Its electron arrangement is shown in the diagram below. 

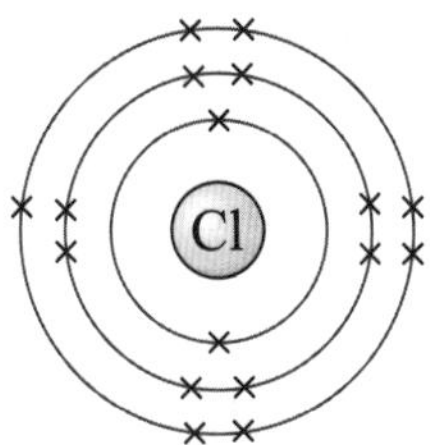

3.1 Chlorine is very reactive and forms compounds with metals.
What type of bonds form between chlorine and metals?

[1 mark]

When chlorine is bubbled through potassium iodide solution a reaction occurs.
The equation below shows the reaction.

$$Cl_{2(g)} + 2KI_{(aq)} \rightarrow I_{2(aq)} + 2KCl_{(aq)}$$

3.2 State what type of reaction this is.

[1 mark]

3.3 Which is the less reactive halogen in this reaction? Explain your answer.

[2 marks]

3.4 Chlorine gas can also react with potassium bromide.
Using your knowledge of Group 7 elements predict the products of this reaction and their states.
Give an explanation for you answer.

[4 marks]

3.5 None of the elements in Group 0 will react with potassium iodide or potassium bromide.
Using your knowledge of the electronic structure of the Group 0 elements,
explain why no reaction occurs.

[2 marks]

4 Group 1 elements are metals.
They include lithium, sodium and potassium. 

4.1 Explain why the Group 1 elements react vigorously with water.

[1 mark]

4.2 The Group 1 elements all react with water at a different rate. Explain why this is.

[3 marks]

4.3 Give the two products formed when potassium reacts with water.

[2 marks]

4.4 During these reactions a solution is formed.
Is this solution acidic, neutral or alkaline?

[1 marks]

Revision Summary for Topic 1

Well, that wraps up Topic 1 — time to put yourself to the test and find out how much you really know.

- Try these questions and tick off each one when you get it right.
- When you've done all the questions under a heading and are completely happy with it, tick it off.

Atomic Structure, Compounds and Formulas (p.16-21) ☐

1) Sketch an atom. Label the nucleus and the electrons. ☐
2) What is the relative mass of a neutron? ☐
3) True or false? Elements only contain atoms with the same number of protons. ☐
4) Define relative atomic mass. ☐
5) How similar are the properties of a compound to the elements that it's made from? ☐
6) Give the formula for:
   a) carbon dioxide b) sodium carbonate c) hydrochloric acid ☐
7) Balance these equations:
   a) $Mg + O_2 \rightarrow MgO$ b) $H_2SO_4 + NaOH \rightarrow Na_2SO_4 + H_2O$ ☐

Mixtures and Separation (p.24-29) ☐

8) What is the difference between a compound and a mixture? ☐
9) What is the name of the pattern of spots generated by paper chromatography? ☐
10) Give the name of a method used to separate an insoluble solid from a liquid. ☐
11) Describe how different substances are separated by fractional distillation. ☐

Atomic Structure and the History of the Periodic Table (p.31-35) ☐

12) Who demonstrated that the plum pudding model was wrong? ☐
13) What did Bohr's nuclear model of the atom suggest? ☐
14) Give the electronic structure of the following elements as numbers:
    a) helium b) carbon c) sodium ☐
15) Why did Mendeleev leave gaps in his Table of Elements? ☐
16) What name is given to the vertical columns in the periodic table? ☐
17) How are the group number and the number of electrons in the outer shell of an element related? ☐

Groups of the Periodic Table (p.37-43) ☐

18) Where are non-metals on the periodic table? ☐
19) Give three properties which are specific to transition metals. ☐
20) State three trends in Group 1 elements as you go down the group. ☐
21) State the products of the reaction of sodium and water. ☐
22) State the differences between Group 1 elements and transition metals for the following properties:
    a) hardness b) reactivity c) melting points ☐
23) How do the boiling points of halogens change as you go down the group? ☐
24) How many atoms are in each molecule of a halogen? ☐
25) What is the charge of the ions that halogens form when they react with metals? ☐
26) What is the trend in boiling point as you go down Group 0? ☐

Ions

Ions crop up all over the place in chemistry. You're gonna have to be able to explain how they form and predict the charges of simple ions formed by elements in Groups 1, 2, 6 and 7.

Ions are Made When Electrons are Transferred

1) Ions are charged particles — they can be single atoms (e.g. Cl^-) or groups of atoms (e.g. NO_3^-).

2) When atoms lose or gain electrons to form ions, all they're trying to do is get a full outer shell like a noble gas (also called a "stable electronic structure"). Atoms with full outer shells are very stable.

Remember that the noble gases are in Group 0 of the periodic table.

3) When metals form ions, they lose electrons from their outer shell to form positive ions.

4) When non-metals form ions, they gain electrons into their outer shell to form negative ions.

5) The number of electrons lost or gained is the same as the charge on the ion. E.g. If 2 electrons are lost the charge is 2+. If 3 electrons are gained the charge is 3–.

Ionic Bonding — Transfer of Electrons

1) When a metal and a non-metal react together, the metal atom loses electrons to form a positively charged ion and the non-metal gains these electrons to form a negatively charged ion.

2) These oppositely charged ions are strongly attracted to one another by electrostatic forces. This attraction is called an ionic bond.

3) The diagram below shows the formation of an ionic bond between sodium and chlorine to form the ionic compound sodium chloride.

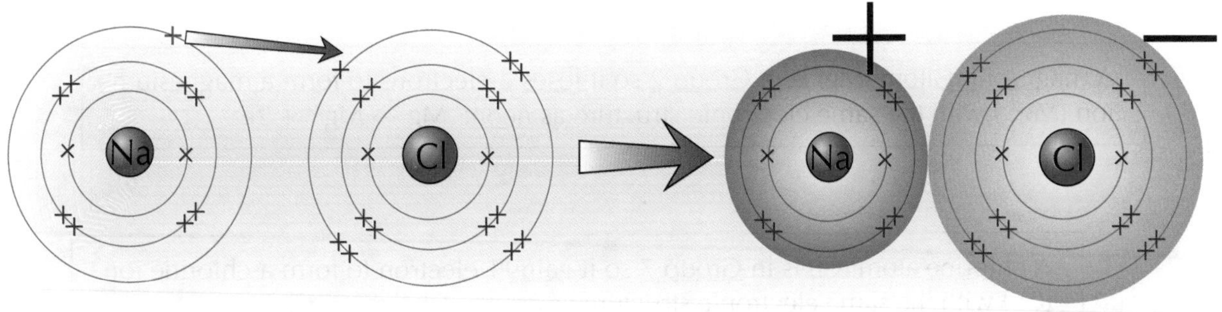

Metals and non-metals form ionic bonds

Some atoms gain electrons and some atoms lose electrons, but they all do it to gain a full outer shell. An ionic bond is just the attraction between the electrostatic charges of the newly formed ions.

Ions

You don't need to remember what elements form which ions — the periodic table is here to help you...

Groups 1 & 2 and 6 & 7 are the Most Likely to Form Ions

1) The elements that most readily form ions are those in Groups 1, 2, 6 and 7.
2) Group 1 and 2 elements are metals and they lose electrons to form positive ions (cations).
3) Group 6 and 7 elements are non-metals. They gain electrons to form negative ions (anions).
4) Elements in the same group all have the same number of outer electrons. So they have to lose or gain the same number to get a full outer shell. And this means that they form ions with the same charges.

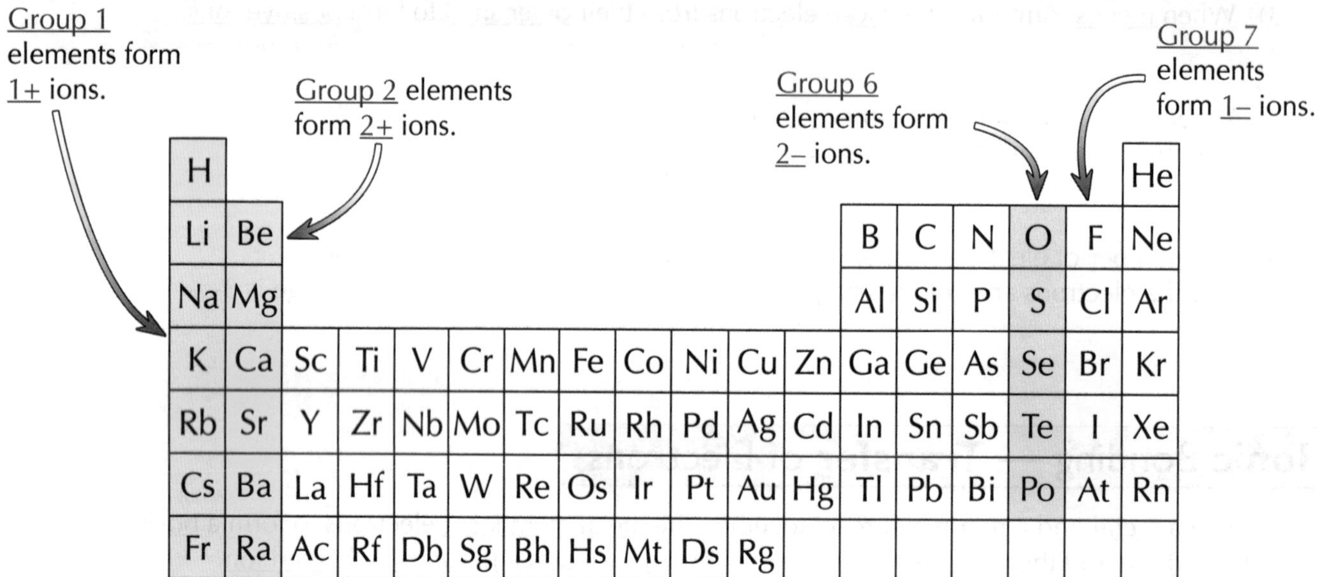

Have a look back at page 33 for how to work out electronic structures.

A sodium atom (Na) is in Group 1 so it loses 1 electron to form a sodium ion (Na^+) with the same electronic structure as neon: $Na \rightarrow Na^+ + e^-$.

A magnesium atom (Mg) is in Group 2 so it loses 2 electrons to form a magnesium ion (Mg^{2+}) with the same electronic structure as neon: $Mg \rightarrow Mg^{2+} + 2e^-$.

A chlorine atom (Cl) is in Group 7 so it gains 1 electron to form a chloride ion (Cl^-) with the same electronic structure as argon: $Cl + e^- \rightarrow Cl^-$.

An oxygen atom (O) is in Group 6 so it gains 2 electrons to form an oxide ion (O^{2-}) with the same electronic structure as neon: $O + 2e^- \rightarrow O^{2-}$.

Ions

Dot and Cross Diagrams Show How Ionic Bonds are Formed

Dot and cross diagrams show the arrangement of electrons in an atom or ion. Each electron is represented by a dot or a cross. So these diagrams can show which atom the electrons in an ion originally came from.

Sodium Chloride (NaCl)

The sodium atom gives up its outer electron, becoming an Na^+ ion. The chlorine atom picks up the electron, becoming a Cl^- (chloride) ion.

Here, the dots represent the Na electrons and the crosses represent the Cl electrons (all electrons are really identical, but this is a good way of following their movement).

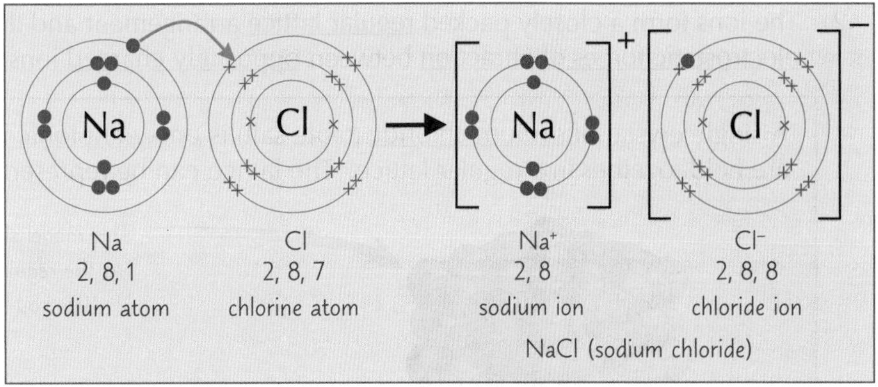

Magnesium Oxide (MgO)

The magnesium atom gives up its two outer electrons, becoming an Mg^{2+} ion. The oxygen atom picks up the electrons, becoming an O^{2-} (oxide) ion.

Here we've only shown the outer shells of electrons on the dot and cross diagram — it makes it much simpler to see what's going on.

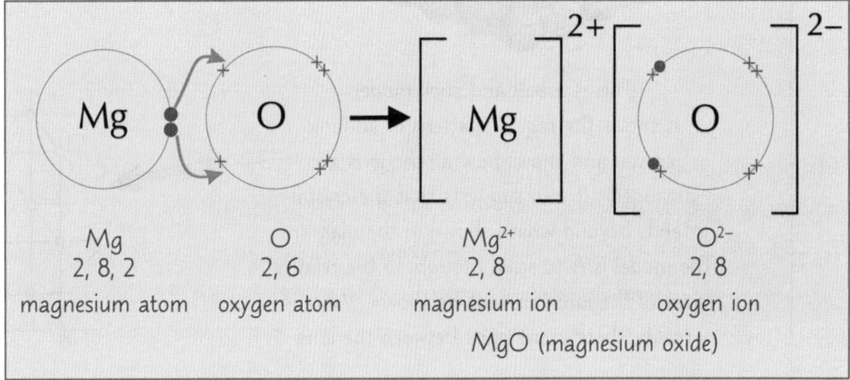

Magnesium Chloride ($MgCl_2$)

The magnesium atom gives up its two outer electrons, becoming an Mg^{2+} ion.
The two chlorine atoms pick up one electron each, becoming two Cl^- (chloride) ions.

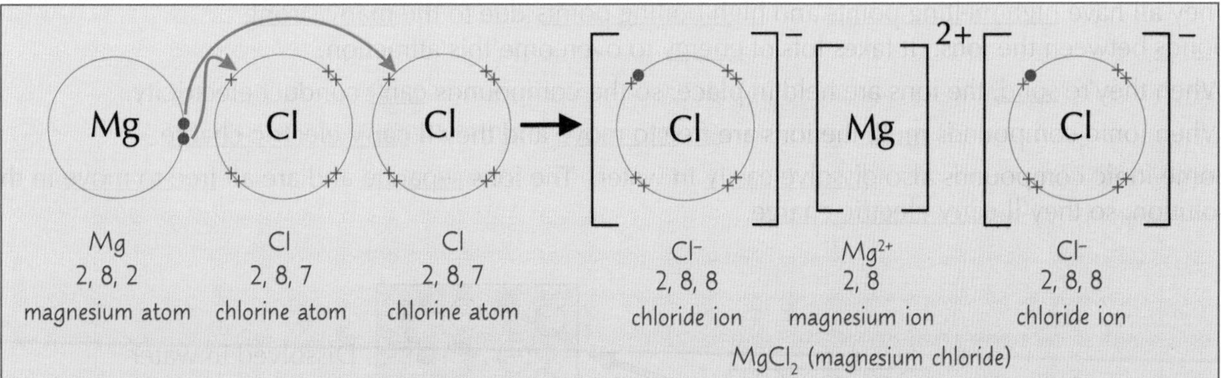

Dot and cross diagrams are useful for showing how ionic compounds are formed, but they don't show the structure of the compound, the size of the ions or how they're arranged.

Show the electronic structure of ions with square brackets

Whether or not you're able to reproduce the drawings on this page all comes down to how well you've understood ionic bonding. (So if you're struggling, try reading the previous few pages again.)

Ionic Compounds

An ionic compound is any compound that only contains ionic bonds...

Ionic Compounds Have A Regular Lattice Structure

The electrostatic attraction between the oppositely charged ions is ionic bonding.

1) Ionic compounds have a structure called a giant ionic lattice.
2) The ions form a closely packed regular lattice arrangement and there are very strong electrostatic forces of attraction between oppositely charged ions, in all directions in the lattice.

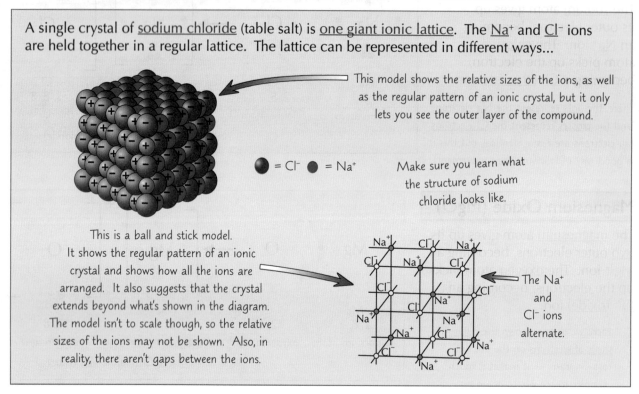

A single crystal of sodium chloride (table salt) is one giant ionic lattice. The Na^+ and Cl^- ions are held together in a regular lattice. The lattice can be represented in different ways...

This model shows the relative sizes of the ions, as well as the regular pattern of an ionic crystal, but it only lets you see the outer layer of the compound.

● = Cl^- ● = Na^+

Make sure you learn what the structure of sodium chloride looks like.

This is a ball and stick model. It shows the regular pattern of an ionic crystal and shows how all the ions are arranged. It also suggests that the crystal extends beyond what's shown in the diagram. The model isn't to scale though, so the relative sizes of the ions may not be shown. Also, in reality, there aren't gaps between the ions.

The Na^+ and Cl^- ions alternate.

Ionic Compounds All Have Similar Properties

1) They all have high melting points and high boiling points due to the many strong bonds between the ions. It takes lots of energy to overcome this attraction.
2) When they're solid, the ions are held in place, so the compounds can't conduct electricity.
3) When ionic compounds melt, the ions are free to move and they'll carry electric charge.
4) Some ionic compounds also dissolve easily in water. The ions separate and are all free to move in the solution, so they'll carry electric charge.

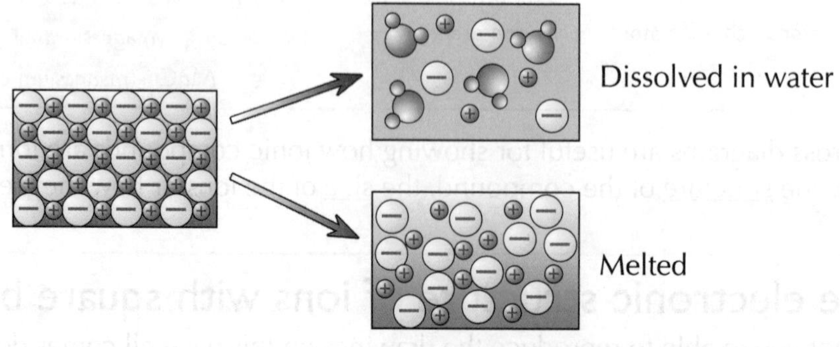

Ionic Compounds

Look at Charges to Find the Formula of an Ionic Compound

1) You might have to work out the empirical formula of an ionic compound from a diagram of the compound.
2) If it's a dot and cross diagram, count up how many atoms there are of each element. Write this down to give you the empirical formula.
3) If you're given a 3D diagram of the ionic lattice, use it to work out what ions are in the ionic compound.
4) You'll then have to balance the charges of the ions so that the overall charge on the compound is zero.

What's the empirical formula of the ionic compound shown below?

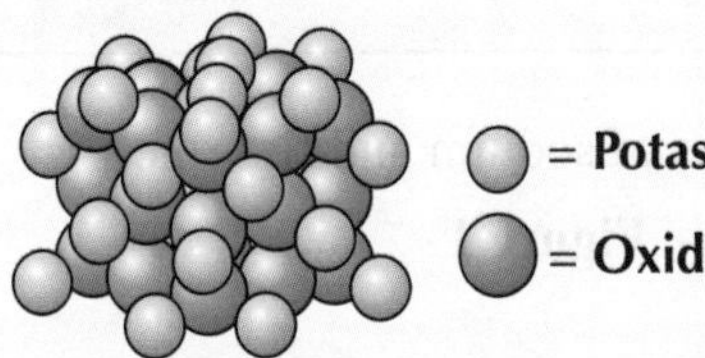

1) Look at the diagram to work out what ions are in the compound. — The compound contains potassium and oxide ions.
2) Work out what charges the ions will form. — Potassium is in Group 1 so forms 1+ ions. Oxygen is in Group 6 so forms 2– ions.
3) Balance the charges so the charge of the empirical formula is zero. — A potassium ion only has a 1+ charge, so you'll need two of them to balance out the 2– charge of an oxide ion. The empirical formula is K_2O.

What's the empirical formula of the ionic compound shown below?

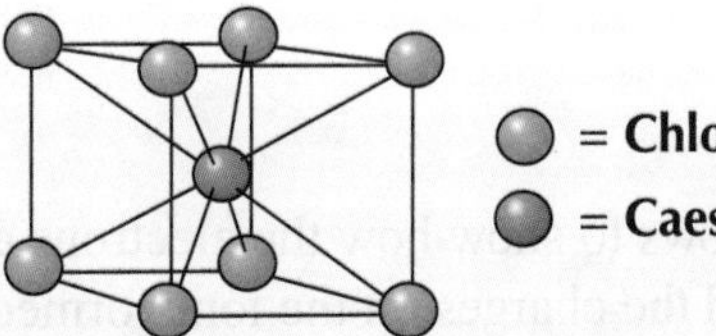

1) Look at the diagram to work out what ions are in the compound. — The compound contains caesium ions and chloride ions.
2) Work out what charges the ions will form. — Caesium is in Group 1 so forms 1+ ions. Chlorine is in Group 7 so forms 1– ions.
3) Balance the charges so the charge of the empirical formula is zero. — A caesium ion has a 1+ charge, so you only need one to balance out the 1– charge of the chloride ion. The empirical formula is CsCl.

Ionic compounds have regular lattice structures

As long as you can find the charge of the ions in an ionic compound, you can work out the empirical formula. Try practising with different ionic compounds.

Warm-Up & Exam Questions

Congratulations you got to the end of the pages on ions — time to make sure you know them back to front...

Warm-Up Questions

1) What is an ion?
2) What is the charge on a ion formed from a Group 2 element?
3) What is the charge on a ion formed from a Group 7 element?
4) Sodium chloride has a giant ionic structure. Does it have a high or a low boiling point?
5) Why do some ionic compounds conduct electricity when dissolved in water?
6) What is the formula of the compound containing Al^{3+} and OH^- ions only?

Exam Questions

1 **Figure 1** shows the electronic structures of sodium and fluorine.

Figure 1

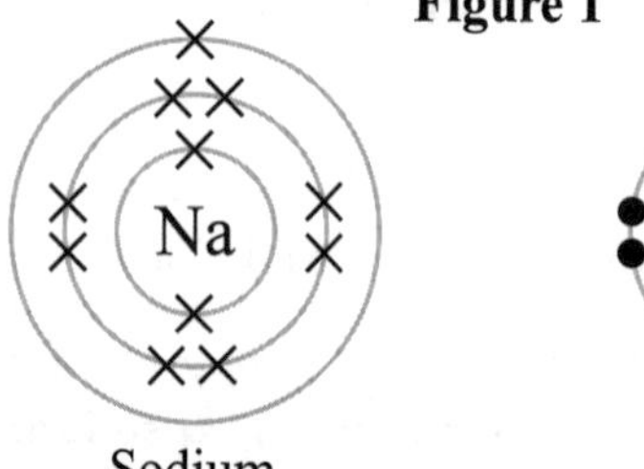

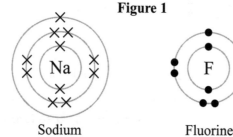

1.1 Describe what will happen when sodium and fluorine react, in terms of electrons.

[2 marks]

1.2 When sodium and fluorine react they form an ionic compound.
Describe the structure of an ionic compound

[3 marks]

2 When lithium reacts with oxygen it forms the ionic compound Li_2O.

2.1 Name the compound formed.

[1 mark]

2.2 Complete **Figure 2** below using arrows to show how the electrons are transferred when Li_2O is formed.
Show the electron arrangements and the charges on the ions formed.

Figure 2

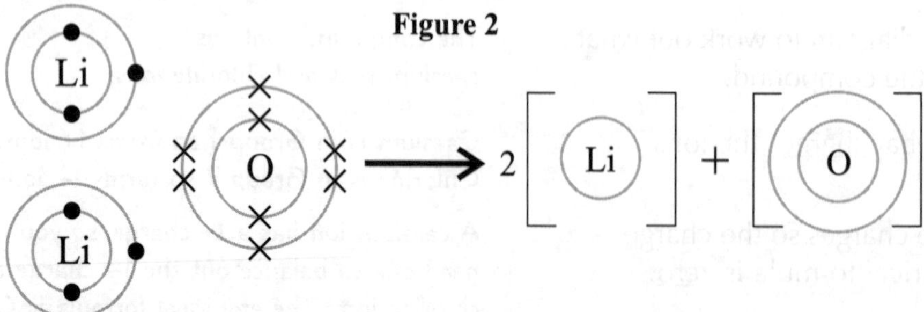

[3 marks]

2.3 Explain why Li_2O conducts electricity when molten.

[2 marks]

2.4 Lithium forms an ionic compound with chlorine.
What is the formula of this compound? Explain why this is.

[2 marks]

Covalent Bonding

Some elements bond ionically (see page 47) but others form strong covalent bonds.
This is where atoms share electrons with each other so that they've got full outer shells.

Covalent Bonds — Sharing Electrons

1) When non-metal atoms bond together, they share pairs of electrons to make covalent bonds.
2) The positively charged nuclei of the bonded atoms are attracted to the shared pair of electrons by electrostatic forces, making covalent bonds very strong.
3) Atoms only share electrons in their outer shells (highest energy levels).
4) Each single covalent bond provides one extra shared electron for each atom.
5) Each atom involved generally makes enough covalent bonds to fill up its outer shell. Having a full outer shell gives them the electronic structure of a noble gas, which is very stable.
6) Covalent bonding happens in compounds of non-metals (e.g. H_2O) and in non-metal elements (e.g. Cl_2).

There are Different Ways of Drawing Covalent Bonds

1) You can use dot and cross diagrams to show the bonding in covalent compounds.
2) Electrons drawn in the overlap between the outer orbitals of two atoms are shared between those atoms.
3) Dot and cross diagrams are useful for showing which atoms the electrons in a covalent bond come from, but they don't show the relative sizes of the atoms, or how the atoms are arranged in space.

You don't have to draw the orbitals in dot and cross diagrams. The important thing is that you get all the dots and crosses in the right places.

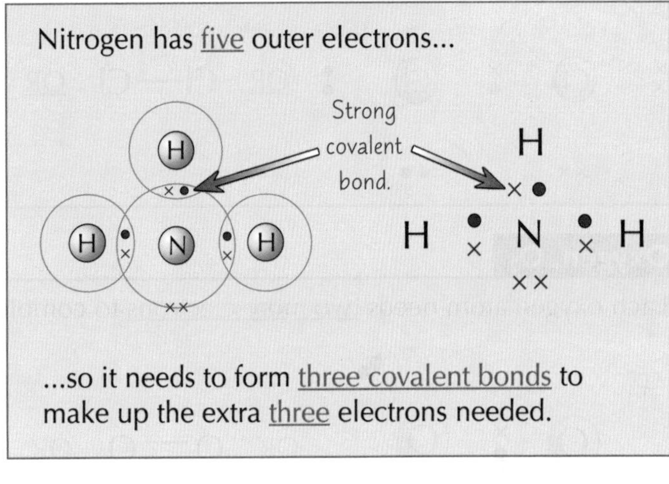

Nitrogen has five outer electrons...
H
N
Strong covalent bond.
...so it needs to form three covalent bonds to make up the extra three electrons needed.

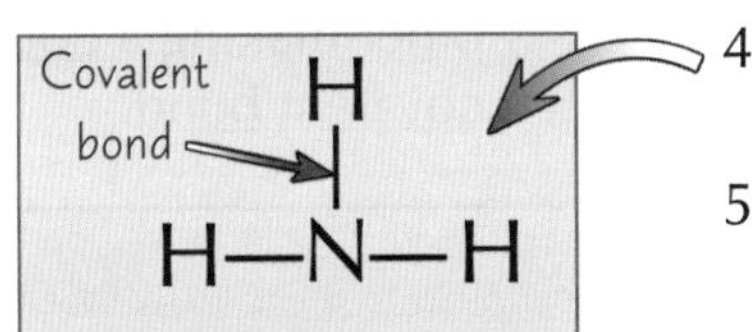

4) The displayed formula of ammonia (NH_3) shows the covalent bonds as single lines between atoms.
5) This is a great way of showing how atoms are connected in large molecules. However, they don't show the 3D structure of the molecule, or which atoms the electrons in the covalent bond have come from.
6) The 3D model of ammonia shows the atoms, the covalent bonds and their arrangement in space next to each other. But 3D models can quickly get confusing for large molecules where there are lots of atoms to include. They don't show where the electrons in the bonds have come from, either.

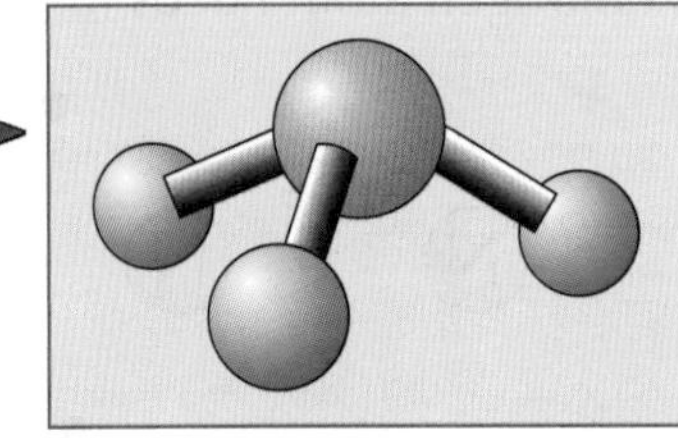

7) You can find the molecular formula of a simple molecular compound from any of these diagrams by counting up how many atoms of each element there are.

A molecular formula shows you how many atoms of each element are in a molecule.

A diagram of the molecule ethane is shown on the right. Use the diagram to find the molecular formula of ethane.

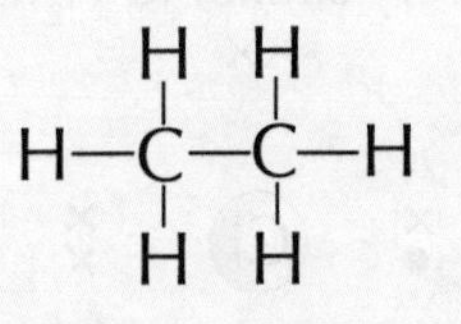

In the diagram, there are two carbon atoms and six hydrogen atoms. So the molecular formula is C_2H_6.

Covalent Bonding

Learn Some Examples of **Simple Molecular** Substances

Simple molecular substances are made up of molecules containing a few atoms joined together by covalent bonds. Here are some common examples that you should know...

Hydrogen, H_2

Hydrogen atoms have just one electron. They only need one more to complete the first shell...

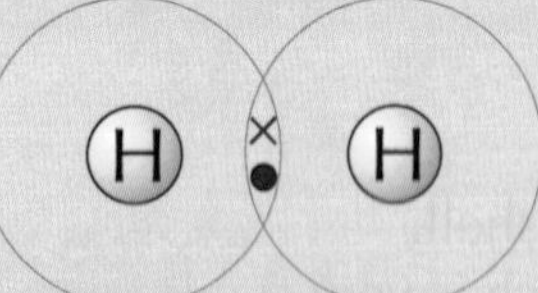

OR H—H OR H ×• H

...so they often form single covalent bonds, either with other hydrogen atoms or with other elements, to achieve this.

Chlorine, Cl_2

Each chlorine atom needs just one more electron to complete the outer shell...

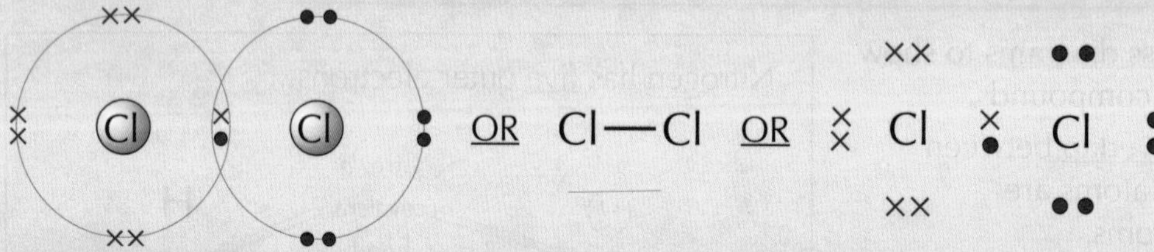

...so two chlorine atoms can share one pair of electrons and form a single covalent bond.

Oxygen, O_2

Each oxygen atom needs two more electrons to complete its outer shell...

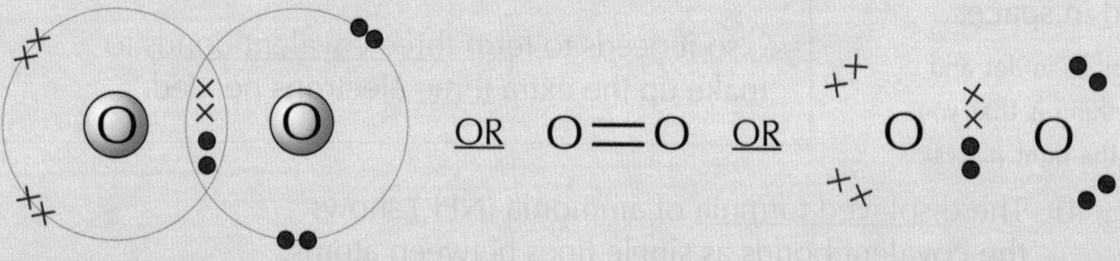

...so in oxygen gas two oxygen atoms share two pairs of electrons with each other making a double covalent bond.

Nitrogen, N_2

A nitrogen atom needs three more electrons to complete its outer shell...

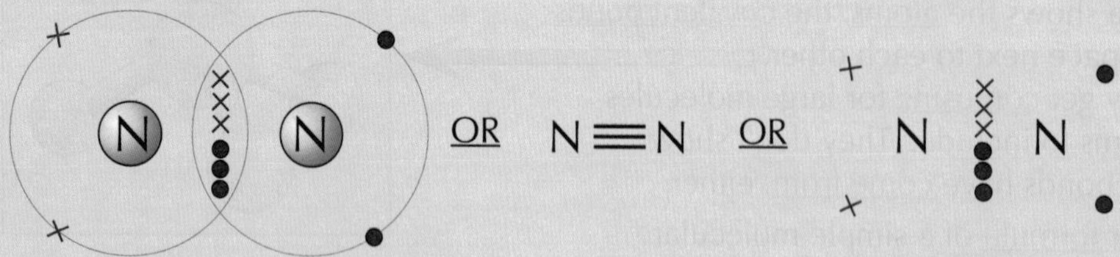

...so two nitrogen atoms share three pairs of electrons to fill their outer shells. This creates a triple bond.

Hydrogen chloride, HCl

This is very similar to H_2 and Cl_2...

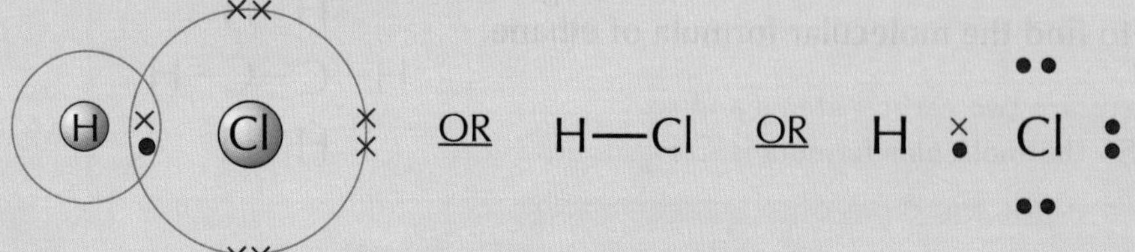

...again, both atoms only need one more electron to complete their outer shells.

Covalent Bonding

Methane, CH_4

Carbon has <u>four outer electrons</u>, which is <u>half</u> a full shell...

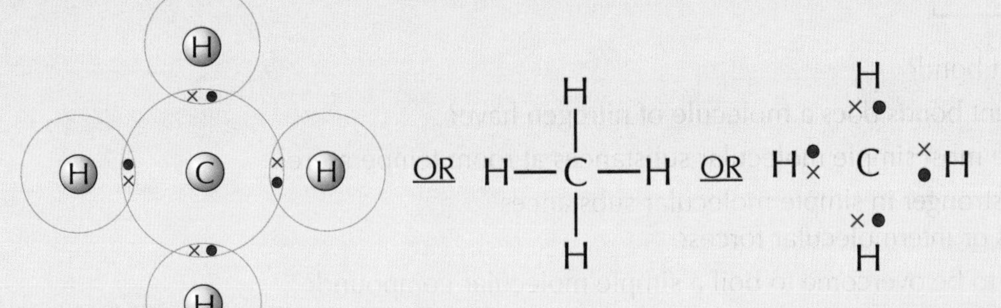

It can form <u>four covalent bonds</u> with <u>hydrogen</u> atoms to fill up its outer shell.

Water, H_2O

Make sure you can also draw the dot and cross diagram of ammonia, NH_3, which is on page 53.

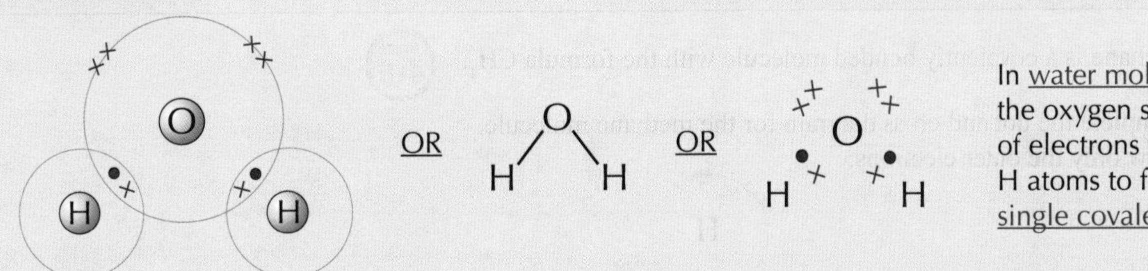

In <u>water molecules</u>, the oxygen shares a pair of electrons with two H atoms to form two <u>single covalent bonds</u>.

Properties of **Simple Molecular** Substances

1) Substances containing <u>covalent bonds</u> usually have <u>simple molecular structures</u>, like the examples shown above and on the previous page.
2) The atoms within the molecules are held together by <u>very strong covalent bonds</u>. By contrast, the forces of attraction <u>between</u> these molecules are <u>very weak</u>.
3) To melt or boil a simple molecular compound, you only need to break these <u>feeble intermolecular forces</u> and <u>not</u> the covalent bonds. So the melting and boiling points are <u>very low</u>, because the molecules are <u>easily parted</u> from each other.
4) Most molecular substances are <u>gases or liquids</u> at room temperature.
5) As molecules get <u>bigger</u>, the strength of the intermolecular forces <u>increases</u>, so <u>more energy</u> is needed to break them, and the melting and boiling points <u>increase</u>.
6) Molecular compounds <u>don't conduct electricity</u>, simply because they <u>aren't charged</u>, so there are <u>no free electrons</u> or ions.

Weak intermolecular forces

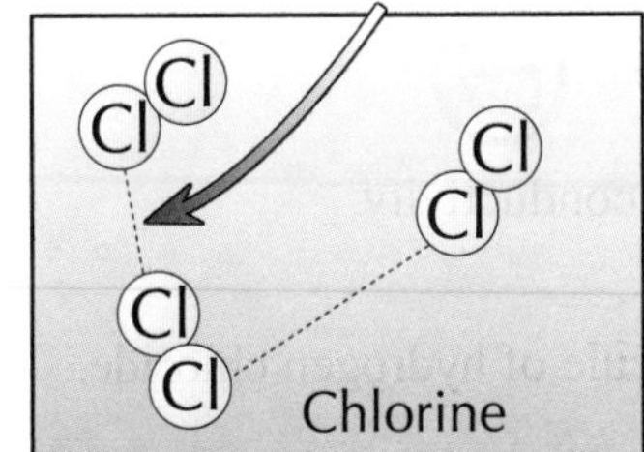

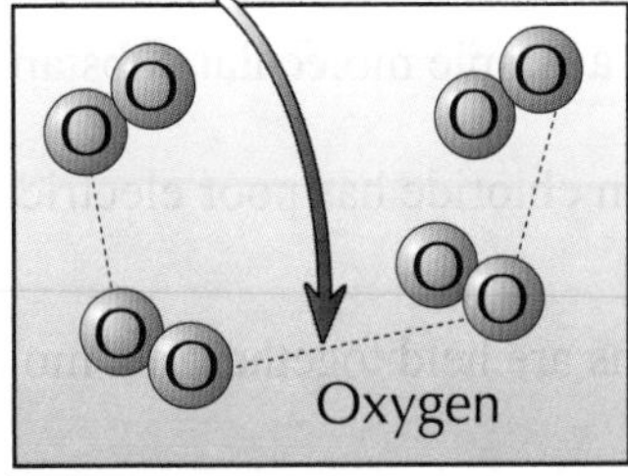

EXAM TIP

Covalent bonding involves sharing electrons

You might be asked to draw a <u>dot and cross diagram</u> for a simple molecule in the exam. The ones shown on the previous couple of pages are good ones to learn (oh, and ammonia is handy too).

Warm-Up & Exam Questions

The questions on this page are all about covalent bonding. Go through them and if you have any problems, make sure you look back at the relevant pages again until you've got to grips with it.

Warm-Up Questions

1) What is a covalent bond?
2) How many covalent bonds does a molecule of nitrogen have?
3) In which states are most simple molecular substances at room temperature?
4) Which forces are stronger in simple molecular substances — covalent bonds or intermolecular forces?
5) What forces need to be overcome to boil a simple molecular compound?

Exam Questions

1 Methane is a covalently bonded molecule with the formula CH_4. 

Complete the dot and cross diagram for the methane molecule.
Show only the outer electrons.

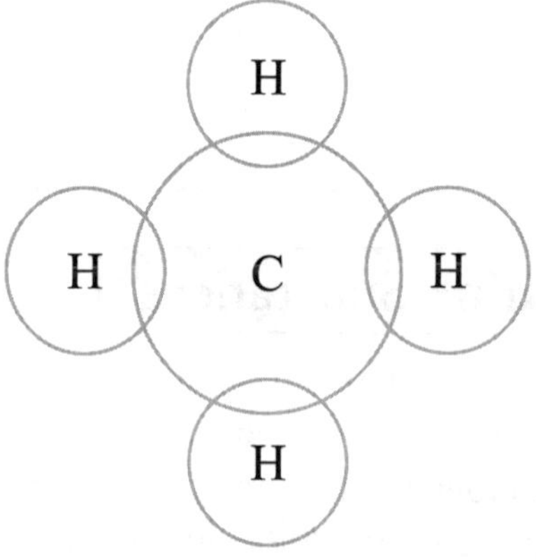

[2 marks]

2 Dot and cross diagrams can be used to show the position of electrons in covalent molecules. 

2.1 Draw a dot and cross diagram for oxygen (O_2). Only show the outer electrons.

[2 marks]

2.2 Nitrogen is in Group 5 of the periodic table.
How many bonds does it need to make to gain a full outer shell?

[1 mark]

3 Hydrogen chloride is a simple molecular substance. 

3.1 Explain why hydrogen chloride has poor electrical conductivity.

[1 mark]

3.2 Explain how the atoms are held together in a molecule of hydrogen chloride.

[2 marks]

3.3 A molecule of hydrogen chloride has a stronger bond than a molecule of chlorine (Cl_2).
However, hydrogen chloride boils at –85 °C, whereas chlorine boils at –34 °C.
Suggest and explain why chlorine has a higher boiling point than hydrogen chloride.

[3 marks]

Polymers

Most polymers are held together with covalent bonds. However, these molecules are a bit different to the simple covalent compounds you have already met...

Polymers Are Long Chains of Repeating Units

1) In a polymer, lots of small units are linked together to form a long molecule that has repeating sections. All the atoms in a polymer are joined by strong covalent bonds.

Instead of drawing out a whole long polymer molecule (which can contain thousands or even millions of atoms), you can draw the shortest repeating section, called the repeating unit, like this:

This polymer is called 'poly(ethene)'.

$$\left(\begin{array}{cc} H & H \\ | & | \\ -C- & -C- \\ | & | \\ H & H \end{array} \right)_n$$

The bonds through the brackets join up to the next repeating unit.

'*n*' is a large number. It tells you that the unit's repeated lots of times.

The bit in brackets is the repeating unit.

To find the molecular formula of a polymer, write down the molecular formula of the repeating unit in brackets, and put an 'n' outside.
So for poly(ethene), the molecular formula of the polymer is $(C_2H_4)_n$.

2) The intermolecular forces between polymer molecules are larger than between simple covalent molecules, so more energy is needed to break them. This means most polymers are solid at room temperature.

3) The intermolecular forces are still weaker than ionic or covalent bonds, so they generally have lower boiling points than ionic or giant molecular compounds.

There's more about how polymers are made on p.141 and p.148 and more about their properties and uses on p.173.

Polymers are very large molecules

Make sure you can work out the molecular formula of a polymer. To do this you need to be comfortable with what the repeating unit of a polymer is. There's more on the repeating units of polymers on page 142.

Giant Covalent Structures

Polymers and simple molecular substances aren't the only compounds held together by covalent bonds. Giant covalent structures are too — some are found in nature, while others can be made in a lab.

Giant Covalent Structures Are Macromolecules

1) In giant covalent structures, all the atoms are bonded to each other by strong covalent bonds.
2) They have very high melting and boiling points as lots of energy is needed to break the covalent bonds between the atoms.
3) They don't contain charged particles, so they don't conduct electricity — not even when molten (except for a few weird exceptions such as graphite, see next page).
4) The main examples are diamond and graphite, which are both made from carbon atoms only, and silicon dioxide (silica).

Diamond

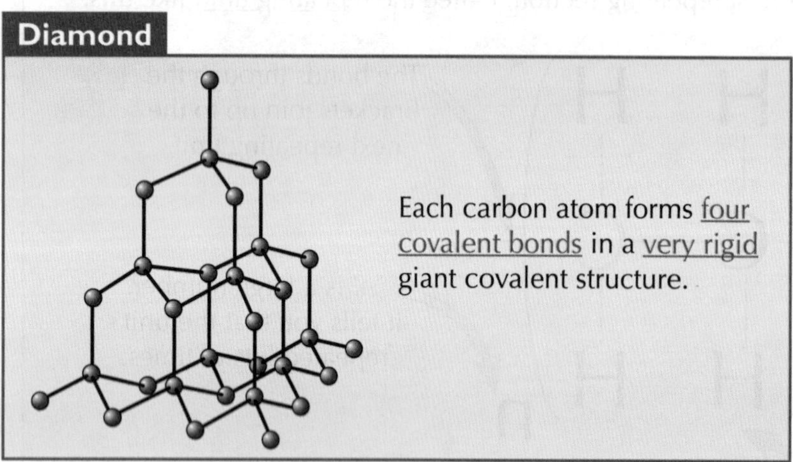

Each carbon atom forms four covalent bonds in a very rigid giant covalent structure.

There's more about diamond and graphite, as well as other types of carbon structure on the next page.

Graphite

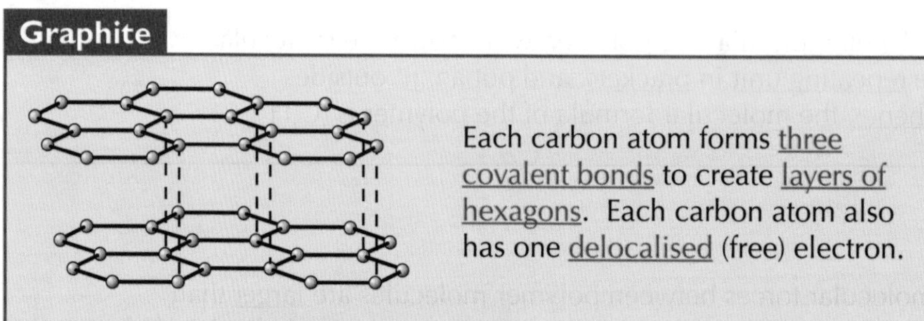

Each carbon atom forms three covalent bonds to create layers of hexagons. Each carbon atom also has one delocalised (free) electron.

Silicon dioxide

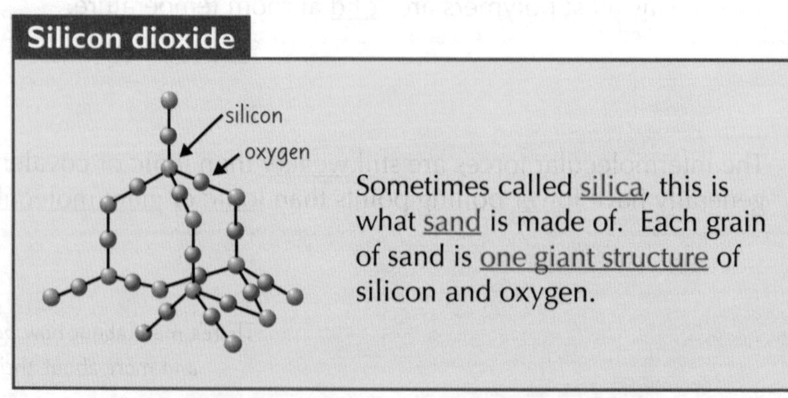

Sometimes called silica, this is what sand is made of. Each grain of sand is one giant structure of silicon and oxygen.

Giant covalent structures have high melting and boiling points

To melt or boil a giant covalent structure, you have to break very strong covalent bonds. The high melting and boiling points of diamond, graphite and silicon are caused by the strength of their covalent bonds.

Allotropes of Carbon

Allotropes are just different structural forms of the same element in the same physical state, e.g. they're all solids. Carbon has quite a few allotropes with lots of different properties.

Diamond is Very Hard

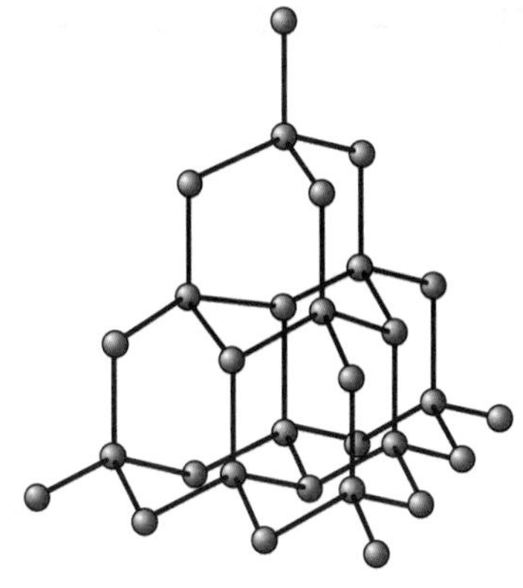

1) Diamond has a giant covalent structure, made up of carbon atoms that each form four covalent bonds. This makes diamond really hard.
2) Those strong covalent bonds take a lot of energy to break and give diamond a very high melting point.
3) It doesn't conduct electricity because it has no free electrons or ions.

Graphite Contains Sheets of Hexagons

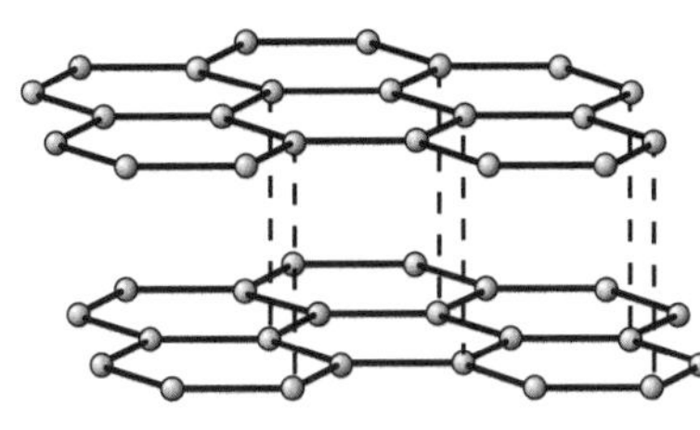

1) In graphite, each carbon atom only forms three covalent bonds, creating sheets of carbon atoms arranged in hexagons.
2) There aren't any covalent bonds between the layers — they're only held together weakly, so they're free to move over each other. This makes graphite soft and slippery, so it's ideal as a lubricating material.
3) Graphite's got a high melting point — the covalent bonds in the layers need loads of energy to break.
4) Only three out of each carbon's four outer electrons are used in bonds, so each carbon atom has one electron that's delocalised (free) and can move. So graphite conducts electricity and thermal energy.

Graphene is One Layer of Graphite

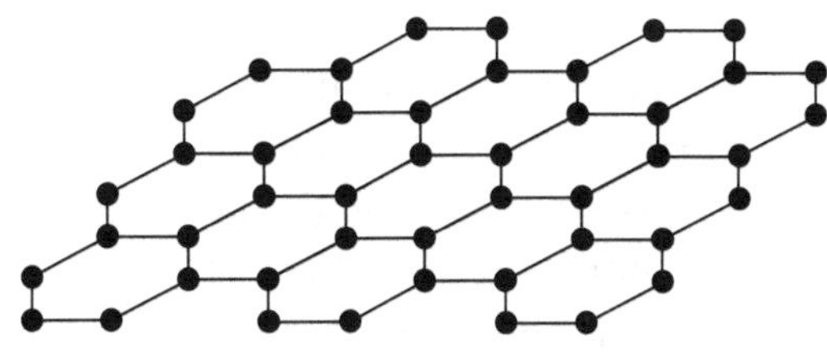

1) Graphene is a sheet of carbon atoms joined together in hexagons.
2) The sheet is just one atom thick, making it a two-dimensional substance.

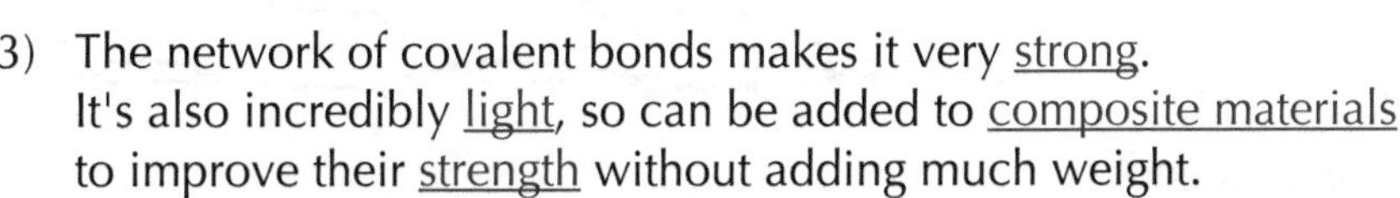

3) The network of covalent bonds makes it very strong. It's also incredibly light, so can be added to composite materials to improve their strength without adding much weight.
4) Like graphite, it contains delocalised electrons so can conduct electricity through the whole structure. This means it has the potential to be used in electronics.

Diamond, graphite and graphene contain exactly the same atoms

The different substances on this page are made purely from carbon — there's no difference at all in their atoms. The difference in properties is all down to the way the atoms are held together.

Allotropes of Carbon

Diamond, graphite, and graphene aren't the only allotropes of carbon.
In fact, we are discovering and making new allotropes of carbon all the time.

Fullerenes Form Spheres and Tubes

1) Fullerenes are molecules of carbon, shaped like closed tubes or hollow balls.
2) They're mainly made up of carbon atoms arranged in hexagons, but can also contain pentagons (rings of five carbon atoms) or heptagons (rings of seven carbon atoms).

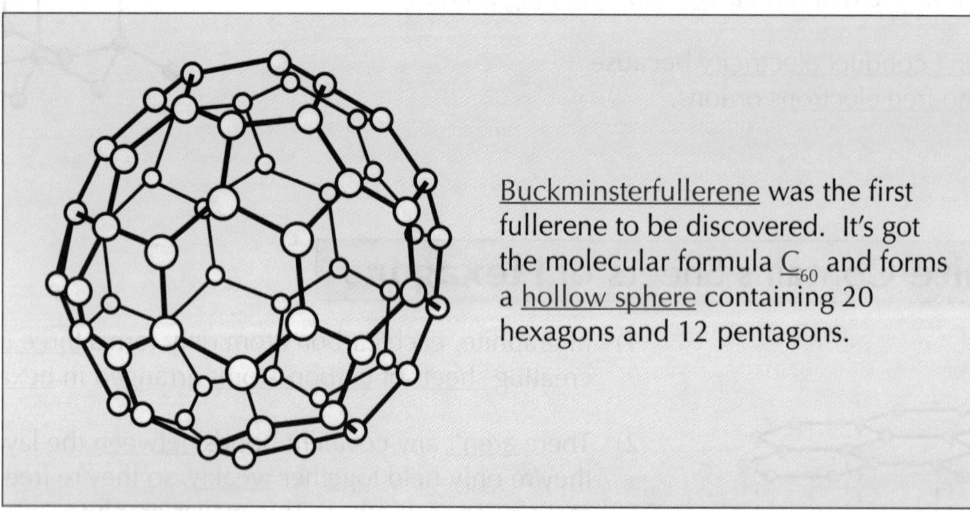

Buckminsterfullerene was the first fullerene to be discovered. It's got the molecular formula C_{60} and forms a hollow sphere containing 20 hexagons and 12 pentagons.

3) Fullerenes can be used to 'cage' other molecules. The fullerene structure forms around another atom or molecule, which is then trapped inside. This could be used to deliver a drug into the body.
4) Fullerenes have a huge surface area, so they could help make great industrial catalysts — individual catalyst molecules could be attached to the fullerenes (the bigger the surface area the better).

Catalysts speed up the rates of chemical reactions without being used up themselves (see p.119).

5) Fullerenes also make great lubricants.

1) Fullerenes can form nanotubes — tiny carbon cylinders.
2) The ratio between the length and the diameter of nanotubes is very high.
3) Nanotubes can conduct both electricity and thermal energy (heat).
4) They also have a high tensile strength (they don't break when they're stretched).
5) Technology that uses very small particles such as nanotubes is called nanotechnology. Nanotubes can be used in electronics or to strengthen materials without adding much weight, such as in tennis racket frames.

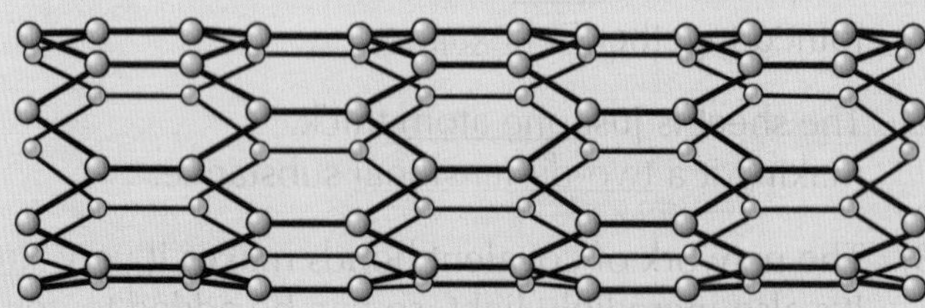

Fullerenes are really useful materials

Before you move on, make sure you can explain the properties of the allotropes of carbon.
While you are at it, don't forget to learn the potential uses of fullerenes that are mentioned on this page.

Metallic Bonding

Ever wondered what gives a metal its properties? Most of it comes down to bonding...

Metallic Bonding Involves **Delocalised Electrons**

1) Metals also consist of a giant structure.
2) The electrons in the outer shell of the metal atoms are delocalised (free to move around). There are strong forces of electrostatic attraction between the positive metal ions and the shared negative electrons.
3) These forces of attraction hold the atoms together in a regular structure and are known as metallic bonding. Metallic bonding is very strong.

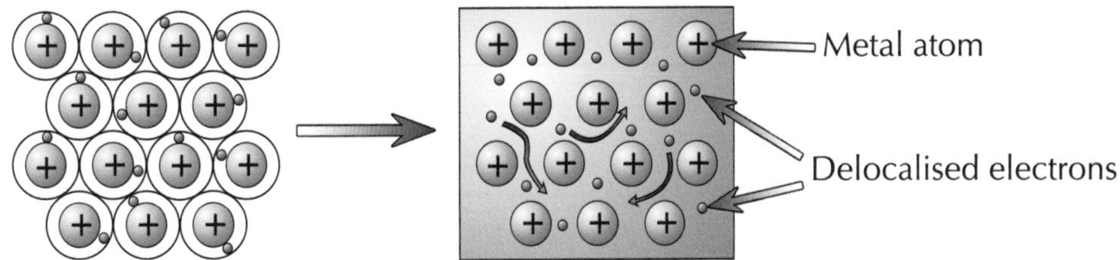

4) Substances that are held together by metallic bonding include metallic elements and alloys (see below).
5) It's the delocalised electrons in the metallic bonds which produce all the properties of metals...

Most Metals are **Solid** at **Room Temperature**

The electrostatic forces between the metal atoms and the delocalised sea of electrons are very strong, so need lots of energy to be broken. This means that most compounds with metallic bonds have very high melting and boiling points, so they're generally solid at room temperature.

Metals are **Good Conductors** of **Electricity** and **Heat**

The delocalised electrons carry electrical charge and thermal (heat) energy through the whole structure, so metals are good conductors of electricity and heat.

Most Metals are **Malleable**

The layers of atoms in a metal can slide over each other, making metals malleable — this means that they can be bent or hammered or rolled into flat sheets.

Alloys are **Harder** Than **Pure Metals**

1) Pure metals often aren't quite right for certain jobs — they're often too soft when they're pure so are mixed with other elements to make them harder. Most of the metals we use everyday are alloys — a mixture of two or more metals or a metal and another element. Alloys are harder and so more useful than pure metals.
2) Different elements have different sized atoms. So when another element is mixed with a pure metal, the new metal atoms will distort the layers of metal atoms, making it more difficult for them to slide over each other. This makes alloys harder than pure metals.

Metallic bonding is what makes metals, well... metals

You should understand what metallic bonding is, and be able to relate how it causes metals to have the properties they do. Also, don't forget to learn the differences between alloys and pure metals.

Warm-Up & Exam Questions

Lots of information to learn on the previous few pages — here are some questions to test yourself on.

Warm-Up Questions

1) How are the repeating units in a polymer bonded together?
2) At room temperature, what state are most polymers in?
3) Describe the differences in the hardness and electrical conductivity of diamond and graphite.
4) Which fullerene was the first to be discovered?
5) Explain why most metals are malleable.

Exam Questions

1 Silicon carbide has a giant covalent structure and is a solid at room temperature. 

1.1 Explain, in terms of its bonding and structure, why silicon carbide has a high melting point.

[2 marks]

1.2 Give **one** other example of a substance with a giant covalent structure.

[1 mark]

2 Graphite, diamond and fullerenes are entirely made from carbon but have different properties. 

2.1 Why does the structure of graphite make it a useful lubricant?

[2 marks]

2.2 Using your knowledge of the structure of diamond, suggest why it is useful as a cutting tool.

[2 marks]

2.3 Suggest **one** possible use for fullerenes.

[1 mark]

2.4 Explain why graphite is able to conduct electricity.

[1 mark]

3 **Figure 1** shows the arrangement of atoms in pure iron. 

Figure 1

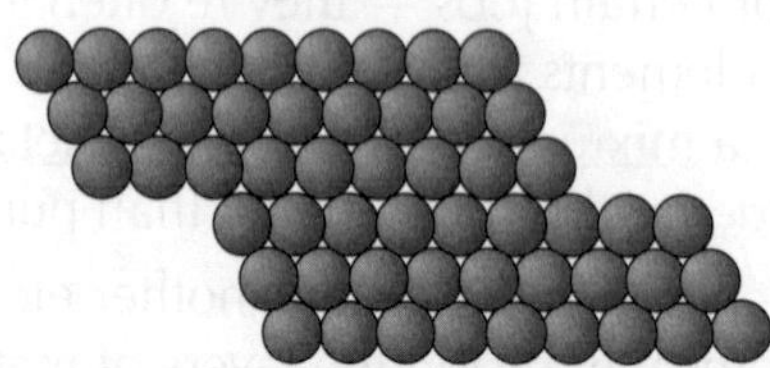

3.1 Steel is an alloy of iron and carbon.
Sketch a similar diagram to show the arrangement of atoms in steel.

[2 marks]

3.2 Steel is harder than iron. Explain why.

[3 marks]

States of Matter

You can explain quite a bit of stuff in chemistry if you can get your head round this lot.

The Three States of Matter — Solid, Liquid and Gas

Materials come in three different forms — solid, liquid and gas. These are the three states of matter. Which state something is at a certain temperature (solid, liquid or gas) depends on how strong the forces of attraction are between the particles of the material. How strong the forces are depends on THREE THINGS:

a) the material (the structure of the substance and the type of bonds holding the particles together),

b) the temperature,

c) the pressure.

The particles could be atoms, ions or molecules.

You can use a model called particle theory to explain how the particles in a material behave in each of the three states of matter by considering each particle as a small, solid, inelastic sphere.

Solids

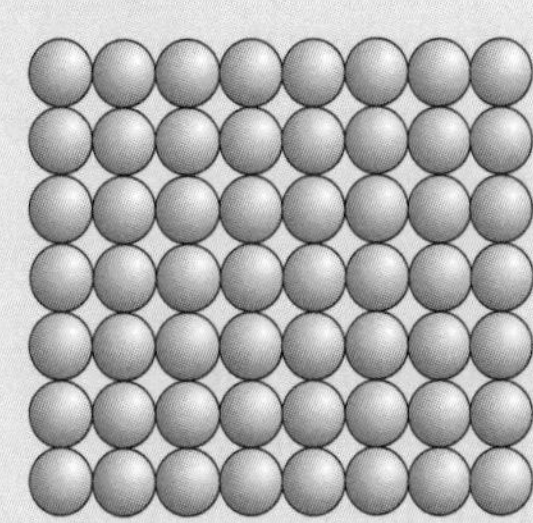

1) In solids, there are strong forces of attraction between particles, which holds them close together in fixed positions to form a very regular lattice arrangement.
2) The particles don't move from their positions, so all solids keep a definite shape and volume, and don't flow like liquids.
3) The particles vibrate about their positions — the hotter the solid becomes, the more they vibrate (causing solids to expand slightly when heated).

Liquids

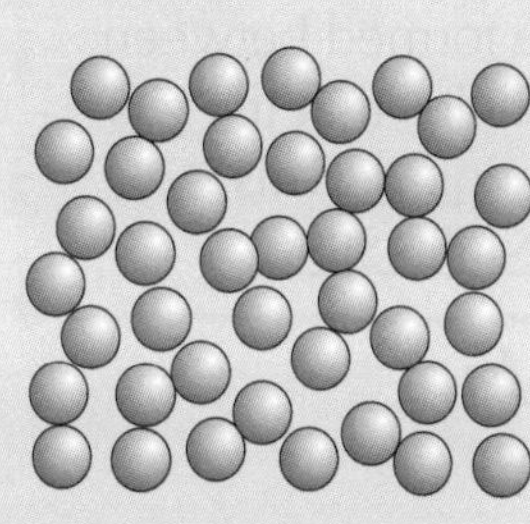

1) In liquids, there's a weak force of attraction between the particles. They're randomly arranged and free to move past each other, but they tend to stick closely together.
2) Liquids have a definite volume but don't keep a definite shape, and will flow to fill the bottom of a container.
3) The particles are constantly moving with random motion. The hotter the liquid gets, the faster they move. This causes liquids to expand slightly when heated.

Gases

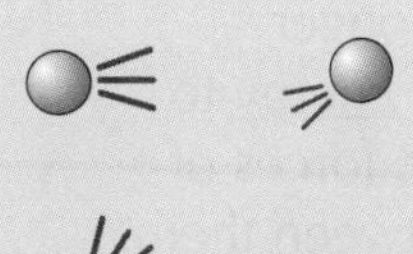

1) In gases, the force of attraction between the particles is very weak — they're free to move and are far apart. The particles in gases travel in straight lines.
2) Gases don't keep a definite shape or volume and will always fill any container.
3) The particles move constantly with random motion. The hotter the gas gets, the faster they move. Gases either expand when heated, or their pressure increases.

Particle theory is a great model for explaining the three states of matter, but it isn't perfect. In reality, the particles aren't solid or inelastic and they aren't spheres — they're atoms, ions or molecules. Also, the model doesn't show the forces between the particles, so there's no way of knowing how strong they are.

States of Matter

Substances Can Change from One State to Another

Physical changes don't change the particles — just their arrangement or their energy.

1) When a solid is heated, its particles gain more energy.

2) This makes the particles vibrate more, which weakens the forces that hold the solid together.

3) At a certain temperature, called the melting point the particles have enough energy to break free from their positions. This is called MELTING and the solid turns into a liquid.

4) When a liquid is heated, again the particles get even more energy.

5) This energy makes the particles move faster, which weakens and breaks the bonds holding the liquid together.

6) At a certain temperature, called the boiling point, the particles have enough energy to break their bonds. This is BOILING (or evaporating). The liquid becomes a gas.

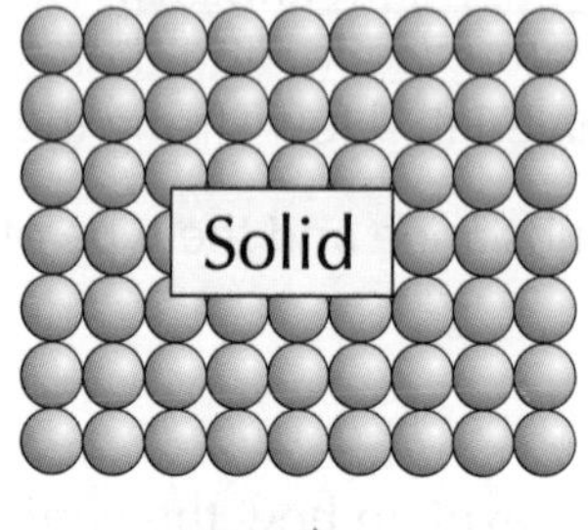

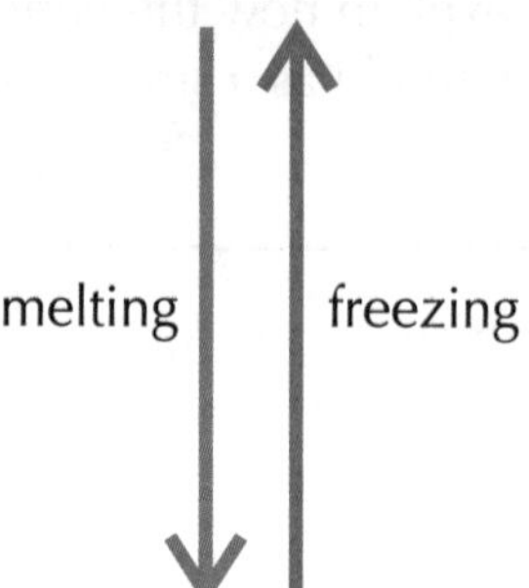

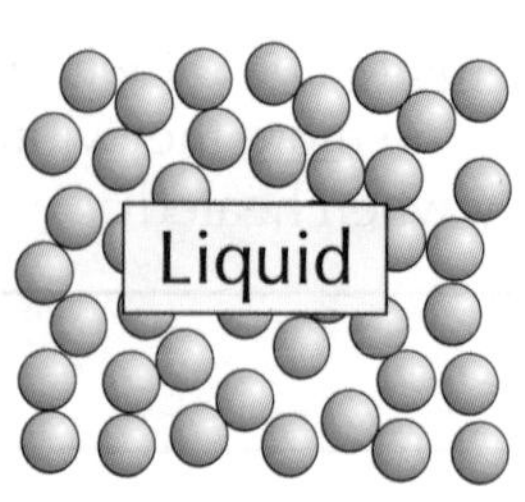

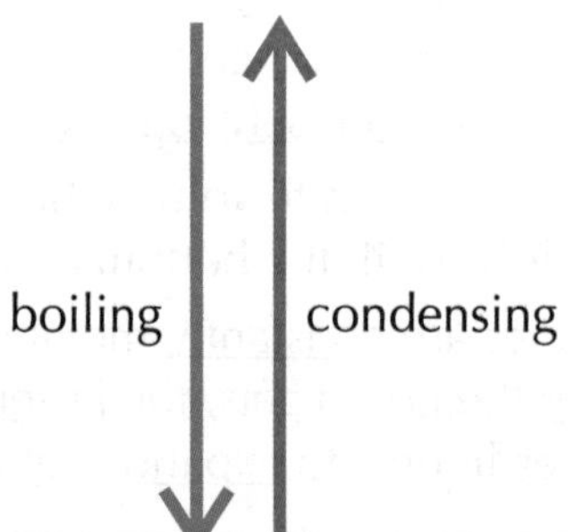

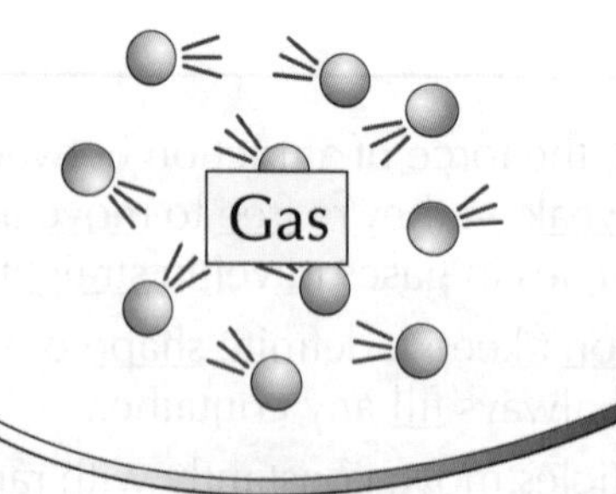

7) As a gas cools, the particles no longer have enough energy to overcome the forces of attraction between them.

8) Bonds form between the particles.

9) At the boiling point, so many bonds have formed between the gas particles that the gas becomes a liquid. This is called CONDENSING.

10) When a liquid cools, the particles have less energy, so move around less.

11) There's not enough energy to overcome the attraction between the particles, so more bonds form between them.

12) At the melting point, so many bonds have formed between the particles that they're held in place. The liquid becomes a solid. This is FREEZING.

So, the amount of energy needed for a substance to change state depends on how strong the forces between particles are. The stronger the forces, the more energy is needed to break them, and so the higher the melting and boiling points of the substance.

States of Matter

The state a substance is in can be really important in working out how it will react. This page is all about how chemists show what the state of a substance is and how they predict what state a substance will be in.

State Symbols Tell You the State of a Substance in an Equation

You saw on page 20 how a chemical reaction can be shown using a word equation or symbol equation. Symbol equations can also include state symbols next to each substance — they tell you what physical state the reactants and products are in:

(s) — solid (l) — liquid (g) — gas (aq) — aqueous

'Aqueous' means 'dissolved in water'.

Here are a couple of examples:

Aqueous hydrogen chloride reacts with solid calcium carbonate to form aqueous calcium chloride, liquid water and carbon dioxide gas:

$$2HCl_{(aq)} + CaCO_{3(s)} \rightarrow CaCl_{2(aq)} + H_2O_{(l)} + CO_{2(g)}$$

Chlorine gas reacts with aqueous potassium iodide to form aqueous potassium chloride and solid iodine:

$$Cl_{2(g)} + 2KI_{(aq)} \rightarrow 2KCl_{(aq)} + I_{2(s)}$$

You Might Need to Predict the State of a Substance

1) You might be asked to predict what state a substance is in at a certain temperature.
2) If the temperature's below the melting point of substance, it'll be a solid. If it's above the boiling point, it'll be a gas. If it's in between the two points, then it's a liquid.

Which of the molecular substances in the table is a liquid at room temperature (25 °C)?

	Melting point	Boiling point
oxygen	–219 °C	–183 °C
nitrogen	–210 °C	–196 °C
bromine	–7 °C	59 °C

Oxygen and nitrogen have boiling points below 25 °C, so will both be gases at room temperature. So the answer's bromine. It melts at –7 °C and boils at 59 °C. So, it'll be a liquid at room temperature.

The bulk properties such as the melting point of a material depend on how lots of atoms interact together. An atom on its own doesn't have these properties.

Physical changes are reversible

Make sure you can describe what happens to particles, and the forces between them, as a substance is heated and cooled. Don't forget to learn the technical terms for each state change.

Nanoparticles

Using nanoparticles is known as nanoscience. Many new uses of nanoparticles are being developed.

Nanoparticles Are Really Really Really Really Tiny

Particles are put into categories depending on their diameter. For example:

Coarse particles (PM_{10}) have a diameter between 2500 nm (2.5×10^{-6} m) and 10 000 nm (1×10^{-5} m). They're also called dust.

1 nm = 0.000 000 001 m (Like I said — really tiny.)

Fine particles ($PM_{2.5}$) have a diameter between 100 nm (1×10^{-7} m) and 2500 nm (2.5×10^{-6} m).

Nanoparticles have a diameter between 1 nm (1×10^{-9} m) and 100 nm (1×10^{-7} m). These are particles that contain only a few hundred atoms.

A typical atom has a diameter of about 0.1 nm (1×10^{-10} m).

Nanoparticles Have a Large Surface Area to Volume Ratio

1) The surface area to volume ratio is an important factor as it can affect the way that a particle behaves.

$$\text{surface area to volume ratio} = \text{surface area} \div \text{volume}$$

2) As particles decrease in size, the size of their surface area increases in relation to their volume. This causes the surface area to volume ratio to increase.
3) You can see this happening by using two cubes as an example:

EXAMPLE: Find the surface area to volume ratio for each of the cubes below.

The green cube on the right has sides of length 100 nm.
Each face has a surface area of 100 nm × 100 nm = 10 000 nm^2
The cube has six faces, so the total surface area is 6 × 10 000 nm^2 = 60 000 nm^2
The volume of the cube is 100 nm × 100 nm × 100 nm = 1 000 000 nm^3
The surface area to volume ratio = surface area ÷ volume = 60 000 ÷ 1 000 000 = 0.06 nm^{-1}

100 nm

The pink cube on the right has sides of length 10 nm.
Each face has a surface area of 10 nm × 10 nm = 100 nm^2
The cube has six faces, so the total surface area is 6 × 100 nm^2 = 600 nm^2
The volume of the cube is 10 nm × 10 nm × 10 nm = 1000 nm^3
The surface area to volume ratio = surface area ÷ volume = 600 ÷ 1000 = 0.6 nm^{-1}

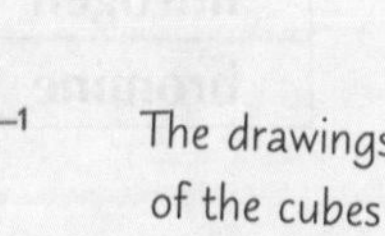

10 nm

The drawings of the cubes aren't to scale.

As you decrease the side of any cube by a factor of ten, the surface area to volume ratio will always increase by a factor of ten.

4) Nanoparticles have a very high surface area to volume ratio — this means the surface area is very large compared to the volume.
5) This can cause the properties of a material to be different depending on whether it's a nanoparticle or whether it's in bulk. For example, you'll often need less of a material that's made up of nanoparticles to work as an effective catalyst compared to a material made up of 'normal' sized particles (containing billions of atoms rather than a few hundred).

Nanoparticles

Nanoparticles Can Be **Used** in **Lots** of Things

Finding new ways to use nanoparticles is a really important area of scientific research.
Here are some of the uses that have already been developed...

1) They have a huge surface area to volume ratio, so they could help make new catalysts.
2) Nanomedicine is a hot topic. The idea is that tiny particles (such as fullerenes from page 60) are absorbed more easily by the body than most particles. This means they could deliver drugs right into the cells where they're needed.
3) Some nanoparticles conduct electricity, so they can be used in tiny electric circuits for computer chips.
4) Silver nanoparticles have antibacterial properties. They can be added to polymer fibres that are then used to make surgical masks and wound dressings and they can also be added to deodorants.
5) Nanoparticles are also being used in cosmetics. For example, they're used to improve moisturisers without making them really oily.

The **Effects** of Nanoparticles on **Health** Aren't Fully **Understood**

1) Although nanoparticles are useful, the way they affect the body isn't fully understood, so it's important that any new products are tested thoroughly to minimise the risks.
2) Some people are worried that products containing nanoparticles have been made available before the effects on human health have been investigated properly, and that we don't know what the long-term impacts on health will be.
3) As the long-term impacts aren't known, many people believe that products containing nanoscale particles should be clearly labelled, so that consumers can choose whether or not to use them.

Nanoparticles are being used in sun creams as they have been shown to be better than the materials in traditional sun creams at protecting skin from harmful UV rays.
They also give better skin coverage than traditional sun creams.
But it's not yet clear whether the nanoparticles can get into your body, and, if they do, whether they might damage cells.
It's also possible that when they are washed away they might damage the environment.

There are so many possible uses for nanotechnology

You may be asked why nanoparticles are used for certain applications, given their properties.
They are pretty amazing, but don't forget — they may have an as yet unknown dark side too.

Warm-Up & Exam Questions

Reckon you know all there is to know about this section? Have a go at these questions and see how you get on. If you get stuck on something — just flick back and give it another read through.

Warm-Up Questions

1) What three factors are the strength of the forces between particles dependent on?
2) What state of matter has a fixed lattice arrangement?
3) What happens to the particles in a gas when it is heated?
4) Write down how you would show that sodium ions (Na^+) are in solution in a symbol equation.
5) What range of diameters do coarse particles have?
6) How does the surface area to volume ratio change as particles get smaller?

Exam Questions

1.1 **Figure 1** shows a vessel in a distillery. The walls of the vessel are solid copper.

Figure 1

Complete the sentences about solids using words from the box. Each word may be used once, more than once or not at all.

weak	move	colder	hotter	random
strong	expand	heavier	dissolve	regular

In solids, there are .... forces of attraction between particles, which hold them in fixed positions in a .... arrangement.

The particles don't .... from their positions, so solids keep their shape.

The .... the solid becomes, the more the particles in the solid vibrate.

[4 marks]

1.2 Inside the vessel, liquid ethanol is turned into ethanol gas. Describe the changes in arrangement, movement and energy of the particles when the liquid ethanol is heated to become a gas.

[2 marks]

2 Some people are concerned about the use of nanoparticles in medicines and cosmetics. Suggest one reason why some people are concerned about the use of nanoparticles and suggest what could be done to minimise the risks.

[2 marks]

3 This question is on states of matter.

3.1 Using particle theory, explain why gases fill their containers.

[2 marks]

3.2 Chlorine has a melting point of –101.5 °C and a boiling point of –34 °C. Predict what state chlorine will be in at –29 °C.

[1 mark]

Revision Summary for Topic 2

That wraps up Topic 2 — time to put yourself to the test and find out how much you really know.

- Try these questions and tick off each one when you get it right.
- When you've done all the questions under a heading and are completely happy with it, tick it off.

Ions and Ionic Compounds (p.47-51)

1) Describe how an ionic bond forms.
2) Which group of the periodic table contains elements which form ions with the following charges?
   a) +1
   b) –1
3) Sketch dot and cross diagrams to show the formation of:
   a) sodium chloride b) magnesium oxide c) magnesium chloride
4) Describe the structure of a crystal of sodium chloride.
5) List the main properties of ionic compounds.

Covalent Bonding (p.53-55)

6) Describe how covalent bonds form.
7) Sketch dot and cross diagrams showing the bonding in a molecule of:
   a) hydrogen b) water c) ammonia
8) Explain why simple molecular compounds typically have low melting and boiling points.

Covalent Structures and Metallic Bonding (p.57-61)

9) Describe the structure of a polymer.
10) What type of bond occurs between carbon atoms in diamond?
11) Describe the structure of graphene.
12) Explain how fullerenes could be used to deliver drugs into the body.
13) What is metallic bonding?
14) Explain why metals have the following properties:
    a) good conduction of heat and electricity
    b) solid at room temperature

States of Matter and Nanoparticles (p.63-67)

15) Name the three states of matter.
16) What is the name of the temperature at which a liquid becomes a gas?
17) How does the strength of the forces between particles influence the temperature at which a substance changes state?
18) What is the state symbol of a solid substance?
19) What is nanoscience?
20) Give three uses of nanoparticles.

Relative Formula Mass

Calculating relative formula mass is straight forward enough, but things can get a bit more confusing when you start working out the percentage compositions of compounds.

Compounds Have a Relative Formula Mass, M_r

If you have a compound like $MgCl_2$ then it has a relative formula mass, M_r, which is just the relative atomic masses of all the atoms in the molecular formula added together.

You can find the relative atomic mass (A_r) of an element from the periodic table — it's the same as its mass number. See page 18 for more.

EXAMPLE:

Find the relative formula mass of $MgCl_2$.

1) Look up the relative atomic masses of all the elements in the compound on the periodic table. (In the exams, you might be given the A_r you need in the question.)
   A_r of Mg = 24 and the A_r of Cl = 35.5.
2) Add up all the relative atomic masses of the atoms in the compound.
   Mg + (2 × Cl) = 24 + (2 × 35.5) = 95 So M_r of $MgCl_2$ = 95

There are two chlorine atoms in $MgCl_2$, so the relative atomic mass of chlorine needs to be multiplied by 2.

You Can Calculate the % Mass of an Element in a Compound

This is actually dead easy — so long as you've learnt this formula:

$$\text{Percentage mass of an element in a compound} = \frac{A_r \times \text{number of atoms of that element}}{M_r \text{ of the compound}} \times 100$$

EXAMPLE:

Find the percentage mass of sodium in sodium carbonate, Na_2CO_3.

A_r of sodium = 23, A_r of carbon = 12, A_r of oxygen = 16

M_r of Na_2CO_3 = (2 × 23) + 12 + (3 × 16) = 106

$$\text{Percentage mass of sodium} = \frac{A_r \times \text{number of atoms of that element}}{M_r \text{ of the compound}} \times 100 = \frac{23 \times 2}{106} \times 100 = 43\%$$

You might also come across more complicated questions where you need to work out the percentage mass.

EXAMPLE:

A mixture contains 20% iron ions by mass. What mass of iron chloride ($FeCl_2$) would you need to provide the iron ions in 50 g of the mixture? A_r of Fe = 56, A_r of Cl = 35.5.

1) Find the mass of iron in the mixture.

The mixture contains 20% iron by mass, so in 50 g there will be $50 \times \frac{20}{100} = 10$ g of iron.

2) Calculate the percentage mass of iron in iron chloride.

$$\text{Percentage mass of iron} = \frac{A_r \times \text{number of atoms of that element}}{M_r \text{ of the compound}} \times 100 = \frac{56}{56 + (2 \times 35.5)} \times 100 = 44.09...\%$$

3) Calculate the mass of iron chloride that contains 10 g of iron.

Iron chloride contains 44.09% iron by mass, so there will be 10 g of iron in $10 \div \frac{44.09...}{100} = 23$ g

So you need 23 g of iron chloride to provide the iron in 50 g of the mixture.

Relative formula mass — add up all the relative atomic masses

You'll get a periodic table in the exams, which could be handy for these sorts of calculations — the relative atomic mass of an element is the bigger number next to the element's symbol.

The Mole and Mass

The mole can be pretty confusing. I think it's the word that puts people off. It's difficult to see the relevance of the word "mole" to anything but a small burrowing animal.

"The Mole" is Simply the Name Given to an **Amount** of a Substance

1) Just like "a million" is this many: 1 000 000; or "a billion" is this many: 1 000 000 000, so "the Avogadro constant" is this many: 602 000 000 000 000 000 000 000 or 6.02×10^{23}. And that's all it is. Just a number.

2) One mole of any substance is just an amount of that substance that contains an Avogadro number of particles — so 6.02×10^{23} particles. The particles could be atoms, molecules, ions or electrons.

The symbol for the unit 'moles' is 'mol'.

3) So why is such a long number like the Avogadro constant used? The answer is that the mass of that number of atoms or molecules of any substance is exactly the same number of grams as the relative atomic mass (A_r) or relative formula mass (M_r) of the element or compound.

4) In other words, one mole of atoms or molecules of any substance will have a mass in grams equal to the relative formula mass (A_r or M_r) for that substance. Here are some examples:

Carbon has an A_r of 12.
So one mole of carbon weighs exactly 12 g.

Nitrogen gas, N_2, has an M_r of 28 (2 × 14).
So one mole of N_2 weighs exactly 28 g.

Carbon dioxide, CO_2, has an M_r of 44 (12 + [2 × 16]).
So one mole of CO_2 weighs exactly 44 g.

5) This means that 12 g of carbon, or 28 g of N_2, or 44 g of CO_2, all contain the same number of particles, namely one mole or 6.023×10^{23} atoms or molecules.

The Mole and Mass

So you now you know what a mole is, you probably want to know how to work out the number of moles in a given mass of a substance, right? Well you're in luck...

Nice Formula to Find the Number of Moles in a Given Mass:

$$\text{Number of moles} = \frac{\text{mass in g (of an element or compound)}}{M_r \text{ (of the element or compound)}}$$

EXAMPLE: **How many moles are there in 66 g of carbon dioxide (CO_2)?**

1) Calculate the M_r of carbon dioxide. M_r of $CO_2 = 12 + (16 \times 2) = 44$
2) Use the formula above to find out how many moles there are. No. of moles = Mass (g) ÷ M_r = 66 ÷ 44 = 1.5 mol

You can rearrange the equation above using this handy formula triangle.
You could use it to find the mass of a known number of moles of a substance, or to find the M_r of a substance from a known mass and number of moles. Just cover up the thing you want to find with your finger and write down what's left showing.

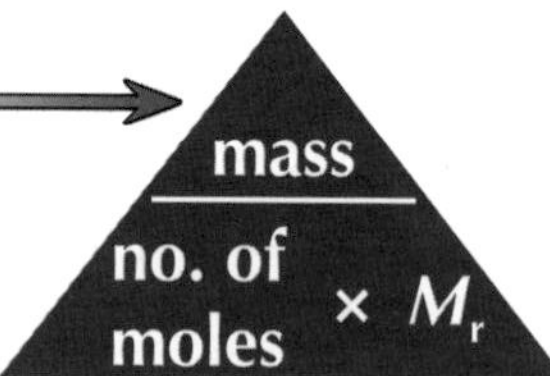

EXAMPLE: **What mass of carbon is there in 4 moles of carbon dioxide?**

There are 4 moles of carbon in 4 moles of CO_2.
Cover up 'mass' in the formula triangle. That leaves you with 'no. of moles × M_r'.
So the mass of 4 moles of carbon = 4 × 12 = 48 g

In a Chemical Reaction, Mass is Always Conserved

1) During a chemical reaction no atoms are destroyed and no atoms are created.
2) This means there are the same number and types of atoms on each side of a reaction equation.
3) Because of this, no mass is lost or gained — we say that mass is conserved during a reaction. For example:

$$2Li + F_2 \rightarrow 2LiF$$

In this reaction, there are 2 lithium atoms and 2 fluorine atoms on each side of the equation.

4) By adding up the relative formula masses of the substances on each side of a balanced symbol equation, you can see that mass is conserved. The total M_r of all the reactants equals the total M_r of the products.

EXAMPLE:

Show that mass is conserved in this reaction: $2Li + F_2 \rightarrow 2LiF$.

1) Add up the relative formula masses on the left-hand side of the equation.
$2 \times M_r(Li) + 2 \times M_r(F) = (2 \times 7) + (2 \times 19) = 14 + 38 = 52$
2) Add up the relative formula masses on the right-hand side of the equation.
$2 \times M_r(LiF) = 2 \times (7 + 19) = 2 \times 26 = 52$

The total M_r on the left hand side of the equation is equal to the total M_r on the right hand side, so mass is conserved.

There's more about balanced symbol equations on p.21.

The Mole and Mass

If the Mass **Seems to Change**, There's Usually a **Gas** Involved

In some experiments, you might observe a change of mass of an unsealed reaction vessel during a reaction. There are usually two explanations for this:

Explanation 1:

1) If the mass increases, it's probably because one of the reactants is a gas that's found in air (e.g. oxygen) and all the products are solids, liquids or aqueous.

2) Before the reaction, the gas is floating around in the air. It's there, but it's not contained in the reaction vessel, so you can't account for its mass.

3) When the gas reacts to form part of the product, it becomes contained inside the reaction vessel — so the total mass of the stuff inside the reaction vessel increases.

For example, when a metal reacts with oxygen in an unsealed container, the mass of the container increases. The mass of the metal oxide produced equals the total mass of the metal and the oxygen that reacted from the air.

$$\text{metal}_{(s)} + \text{oxygen}_{(g)} \rightarrow \text{metal oxide}_{(s)}$$

Explanation 2:

1) If the mass decreases, it's probably because one of the products is a gas and all the reactants are solids, liquids or aqueous.

2) Before the reaction, all the reactants are contained in the reaction vessel.

3) If the vessel isn't enclosed, then the gas can escape from the reaction vessel as it's formed. It's no longer contained in the reaction vessel, so you can't account for its mass — the total mass of the stuff inside the reaction vessel decreases.

For example, when a metal carbonate thermally decomposes to form a metal oxide and carbon dioxide gas, the mass of the reaction vessel will decrease if it isn't sealed. But in reality, the mass of the metal oxide and the carbon dioxide produced will equal the mass of the metal carbonate that decomposed.

$$\text{metal carbonate}_{(s)} \rightarrow \text{metal oxide}_{(s)} + \text{carbon dioxide}_{(g)}$$

Remember from the particle model on page 63 that a gas will expand to fill any container it's in. So if the reaction vessel isn't sealed, the gas expands out from the vessel, and escapes into the air around.

Mass is ALWAYS conserved

There is quite a bit to take in on these pages, but it's all really important in everything you do in chemistry. Make sure you learn the equation on the previous page — it'll crop up again in the rest of the Topic.

Warm-Up & Exam Questions

If you feel like you need some practice with relative masses and mole calculations, then you're in luck. Here's a nice bunch of questions for you to have a go at.

Warm-Up Questions

1) What name is given to the sum of the relative atomic masses of the atoms in a molecule?
2) What is the name given to the particular number of particles equal to one mole of a substance?
3) Write down the definition of a mole.
4) What is the mass of one mole of iron?
5) What is the mass of one mole of oxygen gas?

Exam Questions

1 Boron can form a number of covalent compounds.
Use the A_r values B = 11, O = 16, F = 19 and H = 1 to calculate the relative formula masses of these boron compounds:

1.1 BF_3

[1 mark]

1.2 $B(OH)_3$

[1 mark]

2 A student was asked to calculate the number of moles and the masses of different compounds she would be using in her lab practical.
State the formula used to work out the number of moles from the mass of a compound.

Grade 4-6

[1 mark]

3 A teacher has a 140 g sample of potassium hydroxide (KOH).
Calculate, in grams, how much more KOH the teacher needs to have a 4 mole sample.

[3 marks]

4 A scientist burnt 300 g of metal **X** in an unsealed beaker. She then weighed the contents of the beaker and found that it now weighed 500 g.
All of the metal **X** reacted to form a single product, a metal oxide.

4.1 Explain why the mass appeared to increase following the reaction.

[1 mark]

4.2 Analysis of the metal oxide shows that it contains 40% oxygen by mass.
What is the percentage mass of metal **X** in the oxide?

[1 mark]

4.3 The scientist plans to use the metal oxide as part of a mixture.
The final mixture contains 24% metal **X** by mass, which all comes from the metal oxide.
What mass of the metal oxide was used to make 8.0 g of the mixture?

[2 marks]

The Mole and Equations

Remember the 'number of moles = mass ÷ M_r' equation from page 72? This is where it comes into its own.

The Numbers in Chemical Equations Mean Different Things

Remember those balanced equations back on page 21? Well, the big numbers in front of the chemical formulas of the reactants and products tell you how many moles of each substance takes part or is formed during the reaction. For example:

$$Mg_{(s)} + 2HCl_{(aq)} \rightarrow MgCl_{2(aq)} + H_{2(g)}$$

In this reaction, 1 mole of magnesium and 2 moles of hydrochloric acid react together to form 1 mole of magnesium chloride and 1 mole of hydrogen gas.

The little numbers tell you how many atoms of each element there are in each of the substances.

You Can Balance Equations Using Reacting Masses

If you know the masses of the reactants and products that took part in a reaction, you can work out the balanced symbol equation for the reaction. Here are the steps you should take:

1) Divide the mass of each substance by its relative formula mass to find the number of moles.
2) Divide the number of moles of each substance by the smallest number of moles in the reaction.
3) If any of the numbers aren't whole numbers, multiply all the numbers by the same amount so that they all become whole numbers.
4) Write the balanced symbol equation for the reaction by putting these numbers in front of the chemical formulas.

EXAMPLE:

8.1 g of zinc oxide (ZnO) reacts completely with 0.60 g of carbon to form 2.2 g of carbon dioxide and 6.5 g of zinc. Write a balanced symbol equation for this reaction. $A_r(C) = 12$, $A_r(O) = 16$, $A_r(Zn) = 65$.

1) Work out M_r for each of the substances in the reaction:
   ZnO: 65 + 16 = 81   C: 12 CO_2: 12 + (2 × 16) = 44   Zn: 65
2) Divide the mass of each substance by its M_r to calculate how many moles of each substance reacted or were produced:
   ZnO: $\frac{8.1}{81} = 0.10$ mol   C: $\frac{0.60}{12} = 0.050$ mol
   CO_2: $\frac{2.2}{44} = 0.050$ mol   Zn: $\frac{6.5}{65} = 0.10$
3) Divide by the smallest number of moles, which is 0.050:
   ZnO: $\frac{0.10}{0.050} = 2.0$ C: $\frac{0.050}{0.050} = 1.0$
   CO_2: $\frac{0.050}{0.050} = 1.0$ Zn: $\frac{0.10}{0.050} = 2.0$

These numbers give the ratio of the amounts of each substance in the reaction equation.

4) The numbers are all whole numbers, so you can write out the balanced symbol equation straight away:

$$2ZnO + C \rightarrow CO_2 + 2Zn$$

Practise, Practise, Practise

You can check your final answer to questions like the one in the example above by making sure there's the same number of atoms of each element on both sides of the equation.

The Mole and Equations

Reactions don't go on forever — you need stuff in the reaction flask that can react. If one reactant gets completely used up in a reaction before the rest, then the reaction will stop. That reactant's called limiting.

Reactions Stop When One Reactant is Used Up

When some magnesium carbonate ($MgCO_3$) is placed into a beaker of hydrochloric acid, you can tell a reaction is taking place because you see lots of bubbles of gas being given off.

After a while, the amount of fizzing slows down and the reaction eventually stops...

1) The reaction stops when all of one of the reactants is used up. Any other reactants are in excess. They're usually added in excess to make sure that the other reactant is used up.

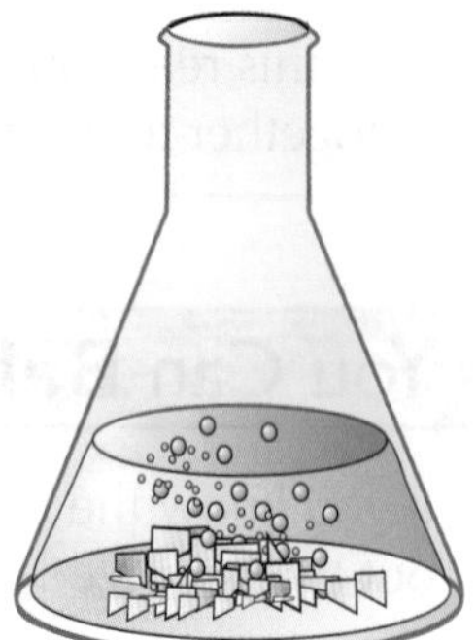

2) The reactant that's used up in a reaction is called the limiting reactant (because it limits the amount of product that's formed).

3) The amount of product formed is directly proportional to the amount of limiting reactant. For example, if you halve the amount of limiting reactant, the amount of product formed will also halve. If you double the amount of limiting reactant, the amount of product will double (as long as it is still the limiting reactant).

4) This is because if you add more reactant there will be more reactant particles to take part in the reaction, which means more product particles.

The Amount of Product Depends on the Limiting Reactant

You can calculate the mass of a product formed in a reaction by using the mass of the limiting reactant and the balanced reaction equation.

1) Write out the balanced equation.
2) Work out relative formula masses (M_r) of the reactant and product you want.
3) Find out how many moles there are of the substance you know the mass of.
4) Use the balanced equation to work out how many moles there'll be of the other substance. In this case, that's how many moles of product will be made of this many moles of reactant.
5) Use the number of moles to calculate the mass.

The Mole and Equations

Here's an **Example** Mass Calculation:

EXAMPLE: **Calculate the mass of aluminium oxide formed when 135 g of aluminium is burned in air.**

1) Write out the balanced equation: $4Al + 3O_2 \rightarrow 2Al_2O_3$
2) Calculate the relative formula masses: Al: 27 Al_2O_3: (2 × 27) + (3 × 16) = 102 — You don't have to find M_r of oxygen because it's in excess.
3) Calculate the number of moles of aluminium in 135 g: Moles = $\frac{\text{mass}}{M_r} = \frac{135}{27} = 5$
4) Look at the ratio of moles in the equation: 4 moles of Al react to produce 2 moles of Al_2O_3 — half the number of moles are produced. So 5 moles of Al will react to produce 2.5 moles of Al_2O_3

   If the question asked for the number of moles of aluminium oxide formed, you'd stop here.
5) Calculate the mass of 2.5 moles of aluminium oxide: mass = moles × M_r = 2.5 × 102 = 255 g

The mass of product (in this case aluminium oxide) is called the yield of a reaction. Masses you calculate in this way are called theoretical yields. In practice you never get 100% of the yield, so the amount of product you get will be less than you calculated.

One Mole of **Any Gas** Occupies **24 dm³** at 20 °C

At the same temperature and pressure, equal numbers of moles of any gas will occupy the same volume.

At room temperature and pressure (r.t.p. = 20 °C and 1 atm) one mole of any gas occupies 24 dm³.
You can use this formula to find the volume of a known mass of any gas at r.t.p.:

$$\text{Volume of gas} = \frac{\text{Mass of gas}}{M_r \text{ of gas}} \times 24$$

Volume of gas in dm³; mass of gas in g.

EXAMPLE:

What's the volume of 319.5 g of chlorine at room temperature and pressure? $M_r(Cl_2) = 71$.

Volume = $\frac{\text{Mass}}{M_r} \times 24 = \frac{319.5}{71} \times 24 = 108$ dm³

You Can Calculate **Volumes** of Gases in **Reactions**

For reactions between gases, you can use the volume of one gas to find the volume of another.

EXAMPLE: **How much carbon dioxide is formed when 30 dm³ of oxygen reacts with carbon monoxide? $2CO_{(g)} + O_{2(g)} \rightarrow 2CO_{2(g)}$**

1 mole of O_2 → 2 moles of CO_2 so 1 volume of O_2 → 2 volumes of CO_2

So 30 dm³ of O_2 → (2 × 30 dm³) = 60 dm³ of CO_2

Remember 24 dm³ at 20 °C

The important thing here is that reactions are limited by the reactant which gets used up first.
Once you've got that in the bag, keep practising until the calculations are an absolute breeze...

Warm-Up & Exam Questions

Think you know everything there is to about moles, limiting reactants and the volumes of gases? Time to put that to the test with some questions...

Warm-Up Questions

1) What do the numbers in front of the reactants and products in a balanced chemical equation stand for?
2) State why you might want to make sure one reactant is in excess when carrying out a chemical reaction.
3) In a reaction, what happens to the amount of product formed if the limiting reactant is halved?
4) How many moles of fluorine gas are in 2.4 dm³ at r.t.p?

Exam Questions

1 3.5 g of Li reacts completely with 4 g of O_2 to produce 7.5 g of Li_2O.
$A_r(Li) = 7$, $M_r(O_2) = 32$, $M_r(Li_2O) = 30$.

1.1 Calculate how many moles of each substance reacted or was produced.

[2 marks]

1.2 Use your answer to part **1.1** to write a balanced symbol equation for this reaction.

[2 marks]

2 Sulfuric acid reacts with sodium hydrogen carbonate to produce aqueous sodium sulfate, water and carbon dioxide. The balanced equation for this reaction is:

$$H_2SO_{4(aq)} + 2NaHCO_{3(s)} \rightarrow Na_2SO_{4(aq)} + 2H_2O_{(l)} + 2CO_{2(g)}$$

2.1 A student reacted 6.0 g of solid $NaHCO_3$ with an excess of sulfuric acid. Calculate the theoretical yield of Na_2SO_4 for this reaction.

[5 marks]

2.2 Sodium hydrogen carbonate is the limiting reactant in the example in **2.1**. Describe what is meant by the limiting reactant.

[1 mark]

3 Calcium carbonate decomposes under heating to form calcium oxide and carbon dioxide. The balanced equation for this reaction is shown below:

$$CaCO_{3(s)} \rightarrow CaO_{(s)} + CO_{2(g)}$$

25 g of $CaCO_3$ were heated by a student.
Assuming that all the calcium carbonate decomposed, calculate the volume occupied at r.t.p. by the quantity of CO_2 produced by this reaction.
1 mol of gas occupies 24 dm³ at r.t.p.

[4 marks]

Solutions

This page deals with calculating the concentrations of solutions.

Concentration is a Measure of How Crowded Things Are

1) Lots of reactions in chemistry take place between substances that are dissolved in a solution. The amount of a substance (e.g. the mass or the number of moles) in a certain volume of a solution is called its concentration.

2) The more solute (the substance that's dissolved) there is in a given volume, the more concentrated the solution.

3) One way to measure the concentration of a solution is by calculating the mass of a substance in a given volume of solution. The units will be units of mass/units of volume. Here's how to calculate the concentration of a solution in g/dm^3:

$$\underset{\text{in g/dm}^3}{\text{concentration}} = \frac{\text{mass of solute (in g)}}{\text{volume of solvent (in dm}^3\text{)}}$$

4) You can calculate the mass of a solute in solution by rearranging the formula above to: mass = conc. × volume.

5) Concentration can also be given in mol/dm^3:

$$\underset{\text{in mol/dm}^3}{\text{concentration}} = \frac{\text{number of moles of solute (in mol)}}{\text{volume of solvent (in dm}^3\text{)}}$$

EXAMPLES:

1) **What's the concentration, in g/dm^3, of a solution of sodium chloride where 30 g of sodium chloride is dissolved in 0.2 dm^3 of water?**

concentration $= \frac{30}{0.2} = 150\ g/dm^3$

$1\ dm^3 = 1000\ cm^3$

2) **What's the concentration, in mol/dm^3, of a solution with 2 moles of salt in 500 cm^3?**
   - Convert the volume to dm^3 by dividing by 1000: $500\ cm^3 \div 1000 = 0.5\ dm^3$
   - Now you've got the number of moles and the volume in the right units, just stick them in the formula: Concentration $= \frac{2}{0.5} = 4\ mol/dm^3$

Concentration can be expressed in two ways

You'll be using the calculations above on the next page to work out concentrations of solutions from reacting volumes. So it's worth that extra bit of practice to make sure you're on top of them.

Concentration Calculations

If you've ever found yourself wondering how to find the concentration of a solution from the concentration of a different solution and a couple of volume values, then here's how to do it.

You Might Be Asked to **Calculate** the **Concentration**

Titrations are experiments that let you find the volumes of two solutions that are needed to react together completely. If you know the concentration of one of the solutions, you can use the volumes from the titration experiment, along with the reaction equation, to find the concentration of the other solution.

Find the Concentration in **mol/dm³**

You might remember the formula for working out the concentration of a substance in mol/dm³ from page 79. Well, here it is in a handy formula triangle. It's dead useful in titration calculations.

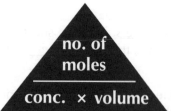

no. of moles
conc. × volume

A student started with 30.0 cm³ of sulfuric acid (H_2SO_4) of unknown concentration in a flask. She found by titration that it took an average of 25.0 cm³ of 0.100 mol/dm³ sodium hydroxide (NaOH) to neutralise the sulfuric acid. Find the concentration of the acid in mol/dm³. The balanced symbol equation for the reaction is:

$$2NaOH + H_2SO_4 \rightarrow Na_2SO_4 + 2H_2O$$

1) Work out how many moles of the "known" substance you have using the formula: no. of moles = conc. × volume. — Make sure the volume is in dm³.
   0.100 mol/dm³ × (25.0 ÷ 1000) dm³ = 0.00250 moles of NaOH
2) Use the reaction equation to work out how many moles of the "unknown" stuff you must have had.
   Using the equation, you can see that two moles of sodium hydroxide reacts with one mole of sulfuric acid. So 0.00250 moles of NaOH must have reacted with 0.00250 ÷ 2 = 0.00125 moles of H_2SO_4.
3) Work out the concentration of the "unknown" stuff.
   Concentration = number of moles ÷ volume
   = 0.00125 mol ÷ (30.0 ÷ 1000) dm³ = 0.041666... mol/dm³ = 0.0417 mol/dm³
   Again, make sure the volume is in dm³. Don't round your answer until right at the end.

All measurements have some uncertainty to them. Titration experiments will often be repeated, and then the average (mean) of these repeated measurements will be calculated. The range of the results can also be found, and can be used to give you an idea of how uncertain the mean value is (see p.14).

Converting **mol/dm³ to g/dm³**

1) To find the concentration in g/dm³, start by finding the concentration in mol/dm³ using the steps above.
2) Then, convert the concentration in mol/dm³ to g/dm³ using the equation mass = moles × M_r.

You can also convert from g/dm³ to mol/dm³ by rearranging this equation to: moles = mass ÷ M_r

EXAMPLE:

What's the concentration, in g/dm³, of the sulfuric acid solution in the example above?

1) Work out the relative formula mass for the acid.
   $M_r(H_2SO_4)$ = (2 × 1) + 32 + (4 × 16) = 98
2) Convert the concentration in moles (that you've already worked out) into concentration in grams. So, in 1 dm³:
   Mass in grams = moles × relative formula mass = 0.041666... × 98 = 4.08333... g
   So the concentration in g/dm³ = 4.08 g/dm³

So many calculations...

These calculations seem pretty scary but if you practise them you'll realise there is nothing to be afraid of.

Warm-Up & Exam Questions

These pesky calculations for titration experiments can be quite tricky. Here are some questions to test yourself out on. If you have any problems, have a look back through the section.

Warm-Up Questions

1) What is the concentration, in g/dm^3, of a solution that contains 20.0 g of solute in 160 cm^3 of solvent?
2) How many moles of hydrochloric acid are there in 25 cm^3 of a 0.1 mol/dm^3 solution?
3) A solution of sodium carbonate, Na_2CO_3, has a concentration of 0.025 mol/dm^3. What is the concentration of this solution in g/dm^3?

Exam Questions

1 In a titration, 30.4 cm^3 of a solution of 1.00 mol/dm^3 sodium hydroxide was required to neutralise 25.0 cm^3 of a solution of sulfuric acid.

1.1 Calculate the number of moles of sodium hydroxide used in the titration.

[1 mark]

1.2 Work out the number of moles of sulfuric acid used in the titration.
The equation is: $2NaOH_{(aq)} + H_2SO_{4(aq)} \rightarrow Na_2SO_{4(aq)} + 2H_2O_{(l)}$

[1 mark]

1.3 Calculate the concentration of the sulfuric acid solution in mol/dm^3.

[2 marks]

1.4 Give your answer for **1.3** in g/dm^3.

[2 marks]

2 Jonah is concerned about the amount of acid in soft drinks. He decides to use a titration method to find the acid content of his favourite lemonade. He uses a solution of 0.10 mol/dm^3 sodium hydroxide in titrations with 25 cm^3 samples of the lemonade. His results are shown in the table.

Repeat	Initial burette reading (cm^3)	Final burette reading (cm^3)	Vol. of NaOH needed (cm^3)
1	0.0	9.4	9.4
2	9.4	18.4	9.0
3	18.4	27.4	9.0

Jonah calculates that the average volume of 0.10 mol/dm^3 NaOH needed is 9.0 cm^3.

The equation for the reaction in the titration can be written:

$HA + NaOH \rightarrow NaA + H_2O$, where HA is the acid present in the lemonade.
Calculate the concentration of acid HA present in the lemonade.

[4 marks]

Atom Economy and Percentage Yield

It's important in industrial reactions that as much of the reactants as possible get turned into useful products. This depends on the atom economy and the percentage yield of the reaction.

"Atom Economy" — % of Reactants Forming **Useful Products**

1) A lot of reactions make more than one product. Some of the products will be useful, but others will be waste.

2) The atom economy (or atom utilisation) of a reaction tells you how much of the mass of the reactants is wasted when manufacturing a chemical and how much ends up as the desired product. Learn the equation:

$$\text{atom economy} = \frac{\text{relative formula mass of desired product}}{\text{relative formula mass of all reactants}} \times 100$$

3) 100% atom economy means that all the atoms in the reactants have been turned into the desired product. The higher the atom economy the 'greener' the process.

EXAMPLE:

Calculate the atom economy of the following reaction to produce hydrogen gas:

$$CH_{4(g)} + H_2O_{(g)} \rightarrow CO_{(g)} + 3H_{2(g)}$$

1) Identify the desired product — that's the hydrogen gas.
2) Work out the M_r of all the reactants: $M_r(CH_4) = 12 + (4 \times 1) = 16$; $M_r(H_2O) = (2 \times 1) + 16 = 18$; $16 + 18 = 34$
3) Work out the total M_r of the desired product: $3 \times M_r(H_2) = 3 \times (2 \times 1) = 6$
4) Use the formula to calculate the atom economy: $\frac{6}{34} \times 100 = 17.6\%$

So in this reaction, over 80% of the starting materials are wasted.

Atom economy is really important in industry

You might get asked about an unfamiliar industrial reaction in the exam. Don't panic — whatever example they give you, the same stuff applies.

Atom Economy and Percentage Yield

High Atom Economy is Better for **Profits** and the **Environment**

1) Pretty obviously, if you're making lots of waste, that's a problem.
2) Reactions with low atom economy use up resources very quickly. At the same time, they make lots of waste materials that have to be disposed of somehow. That tends to make these reactions unsustainable — if the raw materials are non-renewable, they'll run out, and the waste has to go somewhere.
3) For the same reasons, low atom economy reactions may not be profitable. Raw materials are often expensive to buy, and waste products can be expensive to remove and dispose of responsibly.
4) The best way around the problem is to find a use for the waste products rather than just throwing them away. There's often more than one way to make the product you want, so the trick is to come up with a reaction that gives useful "by-products" rather than useless ones.
5) The reactions with the highest atom economy are the ones that only have one product. Those reactions have an atom economy of 100%. The more products there are, the lower the atom economy is likely to be.

Atom economy isn't the only factor to be considered when choosing which reaction to use to make a certain product. Things such as the yield (see below), the rate of the reaction (see page 117) and the position of equilibrium for reversible reactions (see page 128) also need to be thought about. A reaction with a low atom economy, that produces useful by-products, might also be used.

Percentage Yield Compares **Actual** and **Theoretical** Yield

The amount of product you get is known as the yield. The more reactants you start with, the higher the actual yield will be — that's pretty obvious. But the percentage yield doesn't depend on the amount of reactants you started with — it's a percentage.

1) Percentage yield is given by the formula:

$$\text{percentage yield} = \frac{\text{mass of product actually made}}{\text{maximum theoretical mass of product}} \times 100$$

This is just the maximum theoretical yield. It can be calculated from the balanced reaction equation (see p.77).

2) Percentage yield is always somewhere between 0 and 100%. 100% yield means that you got all the product you expected to get. 0% yield means that no reactants were converted into product, i.e. no product at all was made.
3) Industrial processes should have as high a percentage yield as possible to reduce waste and reduce costs.
4) If you know the percentage yield and the maximum theoretical yield of the reaction, you can rearrange the equation above to work out what the mass of product actually made would be.

Atom Economy and Percentage Yield

It's nearly impossible to get a 100% yield, this page looks at a few of the reasons why.

Yields are Always **Less Than 100%**

In real life, you never get a 100% yield. Some product or reactant always gets lost along the way — and that goes for big industrial processes as well as school lab experiments. How this happens depends on what sort of reaction it is and what apparatus is being used.

Lots of things can go wrong, but three common problems are:

1. Not **All** Reactants React to Make a **Product**

In reversible reactions (see page 127), the products can turn back into reactants, so the yield will never be 100%.

For example, in the Haber process, at the same time as the reaction $N_2 + 3H_2 \rightarrow 2NH_3$ is taking place, the reverse reaction $2NH_3 \rightarrow N_2 + 3H_2$ is also happening.

This means the reaction $N_2 + 3H_2 \rightarrow 2NH_3$ never goes to completion (the reactants don't all get used up).

2. There Might be **Side Reactions**

The reactants sometimes react differently to how you expect. They might react with gases in the air, or impurities in the reaction mixture, so they end up forming extra products other than the ones you want.

3. You Lose Some Product When You **Separate It** From the Reaction Mixture

When you filter a liquid to remove solid particles, you nearly always lose a bit of liquid or a bit of solid.

- If you want to keep the liquid, you'll lose the bit that remains with the solid and filter paper (as they always stay a bit wet).
- If you want to keep the solid, some of it'll get left behind when you scrape it off the filter paper.

You'll also lose a bit of material when you transfer it from one container to another — even if you manage not to spill it. Some of it always gets left behind on the inside surface of the old container.

A yield of 100% in a reaction is pretty much impossible

No matter what you do you will never get 100% of your theoretical yield, but that doesn't stop chemists trying to design processes to get as close as possible.

Warm-Up & Exam Questions

Atom economy and percentage yield aren't too hard to calculate once you've had some practice. So here are some questions to get you started...

Warm-Up Questions

1) What is the atom economy of the reaction shown? $2SO_2 + O_2 \rightarrow 2SO_3$
2) Why might a reaction with a low atom economy be bad for sustainable development?
3) What is the formula for calculating the percentage yield of a reaction?
4) What is the percentage yield of a reaction which produced 4 g of product if the predicted yield was 5 g?

Exam Questions

1 Which one of the following statements about atom economy is **not** true? 

Tick **one** box.

- ☐ Reactions that only have one product have the highest atom economy.
- ☐ Reactions with a low atom economy tend to be less sustainable than reactions with high atom economies.
- ☐ Reactions with low atom economies will generally produce more waste than reactions with high atom economies.
- ☐ To calculate the atom economy of a reaction, you need to know the mass of products you expected to form, and the mass of products that actually formed.

[1 mark]

2 Ethanol produced by the fermentation of sugar can be converted into ethene, as shown below. The ethene can then be used to make polythene. 

$$C_2H_6O_{(g)} \rightarrow C_2H_{4(g)} + H_2O_{(g)}$$

Calculate the atom economy of this reaction. (A_r values: C = 12, O = 16, H = 1)

[3 marks]

3* Discuss the reasons why yields from chemical reactions are always less than 100%. 

[6 marks]

4 A sample of copper was made by reducing 4.0 g of copper oxide with methane gas. 

When the black copper oxide turned orange-red, the sample was scraped out into a beaker.
Sulfuric acid was added to dissolve any copper oxide that remained.
The sample was then filtered, washed and dried. 2.8 g of copper was obtained.
(A_r values: H = 1, Cu = 63.5, O = 16.) The equation for this reaction is:

$$CH_4 + 4CuO \rightarrow 4Cu + 2H_2O + CO_2$$

4.1 Use the equation to calculate the maximum mass of copper which could be obtained from the reaction (the theoretical yield).

[3 marks]

4.2 Calculate the percentage yield of the reaction.

[2 marks]

Revision Summary for Topic 3

That wraps up Topic 3 — time to put yourself to the test and find out how much you really know.

- Try these questions and tick off each one when you get it right.
- When you've done all the questions under a heading and are completely happy with it, tick it off.

Relative Mass and Moles (p.70-73) ☐

1) How do you calculate the relative formula mass, M_r of a substance? ☐
2) State the value of the Avogadro constant. ☐
3) What is the formula that relates the number of moles of a substance to its mass and M_r? ☐
4) What does conservation of mass mean? ☐
5) Suggest why the mass of a reaction vessel might decrease during a reaction. ☐

Mole Calculations and Volumes of Gases (p.75-77) ☐

6) How can you determine the number of moles of each substance that would react together from the balanced reaction equation? ☐
7) What would increasing the limiting reagent do to the amount of product formed in a reaction? ☐
8) What volume does one mole of any gas occupy at room temperature and pressure? ☐
9) State the values of room temperature and pressure. ☐

Concentrations of Solutions (p.79-80) ☐

10) What is concentration? ☐
11) Give the equation for working out the concentration of a solution in g/dm^3. ☐
12) Give the equation for working out the concentration of a solution in mol/dm^3. ☐

Atom Economy and Percentage Yield (p.82-84) ☐

13) Give the equation for calculating the atom economy of a reaction. ☐
14) Give three reasons why it's better to use reactions that have a high atom economy. ☐
15) What's the atom economy of a reaction that only produces one product? ☐
16) What do you need to find the theoretical yield of a reaction? ☐
17) Give the equation for calculating the percentage yield of a reaction. ☐
18) Give two reasons why it's better to use reactions with high percentage yields. ☐
19) Is a percentage yield of 100% for a reaction possible? ☐

Acids and Bases

Acids and bases crop up everywhere in chemistry — so here's the lowdown on the basics of pH...

The pH Scale Goes From 0 to 14

1) The pH scale is a measure of how acidic or alkaline a solution is.
2) The lower the pH of a solution, the more acidic it is.
   The higher the pH of a solution, the more alkaline it is.
3) A neutral substance (e.g. pure water) has pH 7.

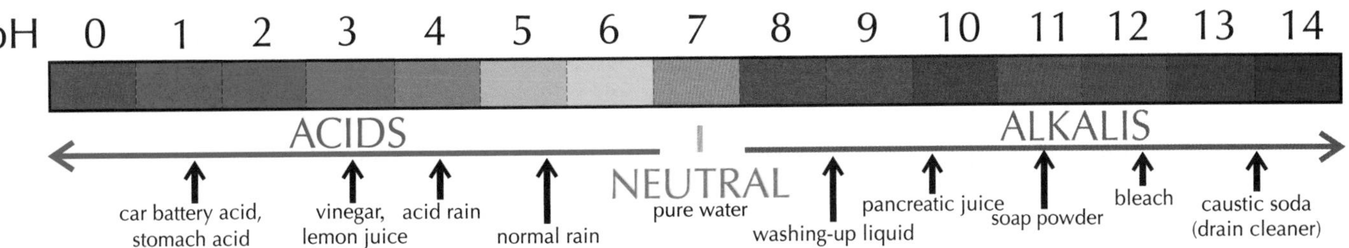

You Can Measure the pH of a Solution

1) An indicator is a dye that changes colour depending on whether it's above or below a certain pH. Some indicators contain a mixture of dyes that means they gradually change colour over a broad range of pH. These are called wide range indicators and they're useful for estimating the pH of a solution. For example, universal indicator gives the colours shown above.
2) A pH probe attached to a pH meter can also be used to measure pH electronically. The probe is placed in the solution you are measuring and the pH is given on a digital display as a numerical value, meaning it's more accurate than an indicator.

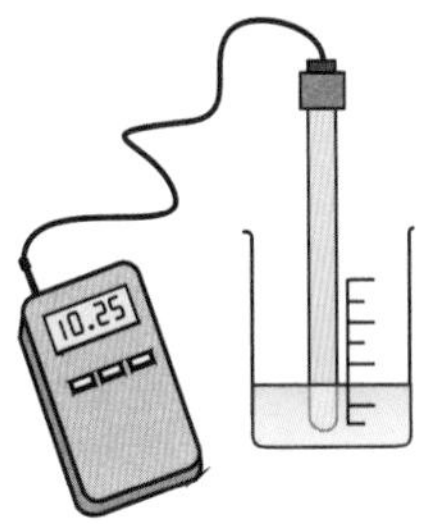

Acids and Bases Neutralise Each Other

1) An acid is a substance that forms an aqueous solution with a pH of less than 7. Acids form H^+ ions in water.
2) A base is any substance that will react with an acid to form a salt.
3) An alkali is a base that dissolves in water to form a solution with a pH greater than 7. Alkalis form OH^- ions in water.

The reaction between acids and bases is called neutralisation:

acid + base → salt + water

Neutralisation between acids and alkalis can be seen in terms of H^+ and OH^- ions like this:

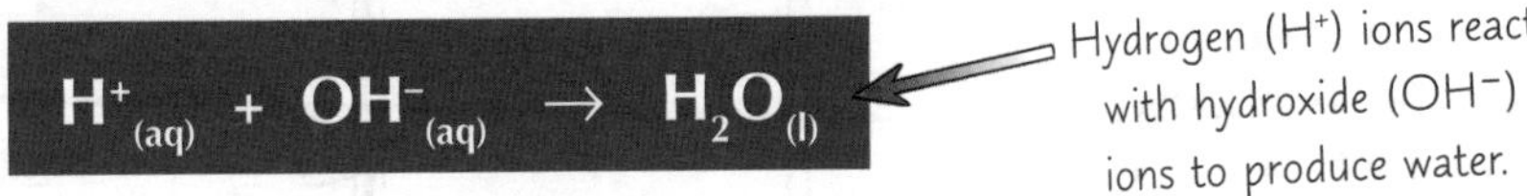

$$H^+_{(aq)} + OH^-_{(aq)} \rightarrow H_2O_{(l)}$$

Hydrogen (H^+) ions react with hydroxide (OH^-) ions to produce water.

When an acid neutralises a base (or vice versa), the products are neutral, i.e. they have a pH of 7. An indicator can be used to show that a neutralisation reaction is over.

Interesting fact — your skin is slightly acidic (pH 5.5)

When you mix an acid with an alkali, hydrogen ions from the acid react with hydroxide ions from the alkali to make water. The leftover bits of the acid and alkali make a salt.

PRACTICAL

Titrations

Titrations are a method of analysing the concentrations of solutions. They're pretty important.

Titrations are Used to Find Out **Concentrations**

Titrations allow you to find out exactly what volume of acid is needed to neutralise a measured volume of alkali — or vice versa. You can then use this data to work out the concentration of the acid or alkali (see page 80). Here's how to do a titration:

1) Say you want to find out the concentration of some alkali. Using a pipette and pipette filler, add a set volume of the alkali to a conical flask. Add two or three drops of indicator too.

2) Use a funnel to fill a burette with some acid of known concentration. Make sure you do this BELOW EYE LEVEL — you don't want to be looking up if some acid spills over. (You should wear safety glasses too.) Record the initial volume of the acid in the burette.

3) Use the burette to add the acid to the alkali a bit at a time, giving the conical flask a regular swirl. Go especially slowly when you think the end-point (colour change) is about to be reached.

4) The indicator changes colour when all the alkali has been neutralised (e.g. phenolphthalein is pink in alkaline conditions, but colourless in acidic conditions).

5) Record the final volume of acid in the burette, and use it, along with the initial reading, to calculate the volume of acid used to neutralise the alkali.

Here's the **Apparatus** You'll Use in a **Titration**:

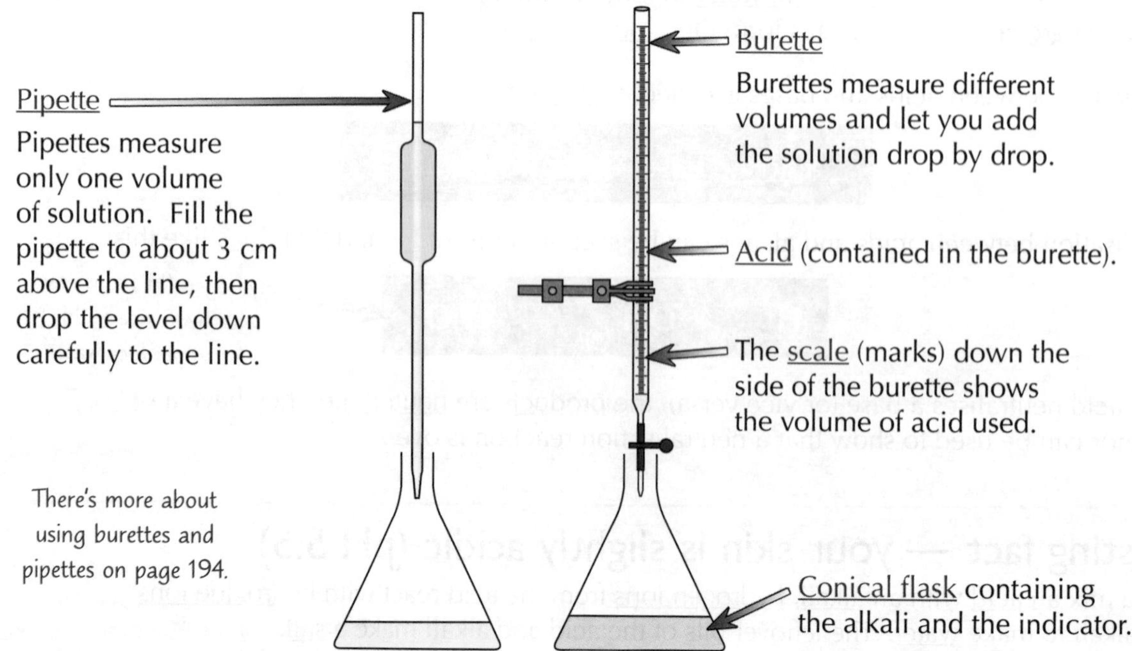
Pipette
Pipettes measure only one volume of solution. Fill the pipette to about 3 cm above the line, then drop the level down carefully to the line.
Burette
Burettes measure different volumes and let you add the solution drop by drop.
Acid (contained in the burette).
The scale (marks) down the side of the burette shows the volume of acid used.
Conical flask containing the alkali and the indicator.

There's more about using burettes and pipettes on page 194.

Titrations

You Should Repeat Titrations and Find a Mean Volume

1) To increase the accuracy of your titration and to spot any anomalous results, you need several consistent readings.
2) The first titration you do should be a rough titration to get an approximate idea of where the solution changes colour (the end-point).
3) You then need to repeat the whole thing a few times, making sure you get (pretty much) the same answer each time (within 0.10 cm³).
4) Finally, calculate a mean of your results, ignoring any anomalous results.

Anomalous results are ones that don't fit in with the rest (see page 7).

Use Single Indicators for Titrations

1) Universal indicator is used to estimate the pH of a solution because it can turn a variety of colours. Each colour indicates a narrow range of pH values.
2) It's made from a mixture of different indicators. The colour gradually changes from red in acidic solutions to violet in alkaline solutions.

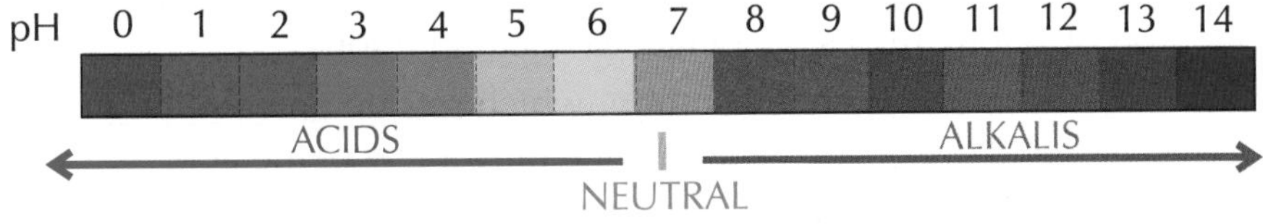

3) But during a titration between an alkali and an acid, you want to see a sudden colour change, at the end-point.
4) So you need to use a single indicator, such as:

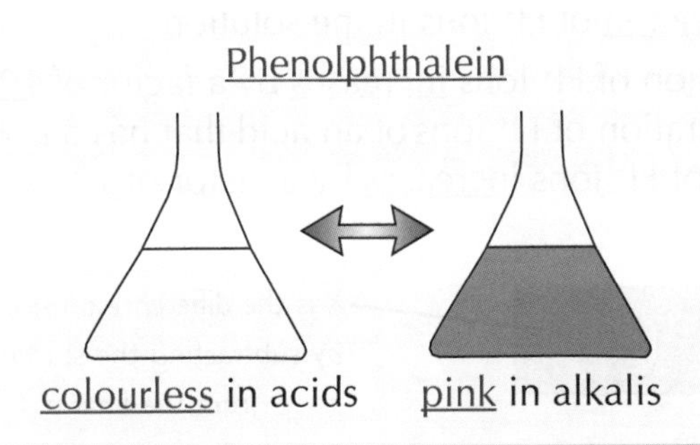

These students are doing a titration using phenolphthalein as the indicator.

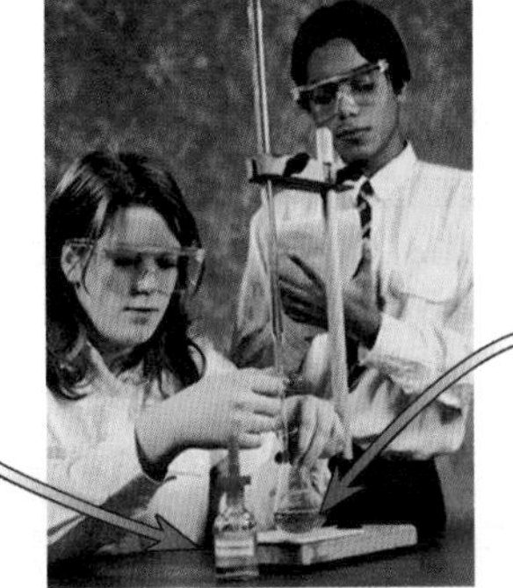

The alkali in the flask is tinted pink. (At the end-point the solution will go colourless.)

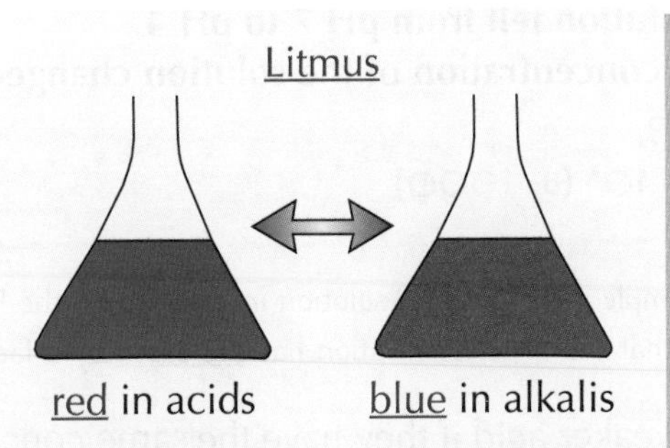

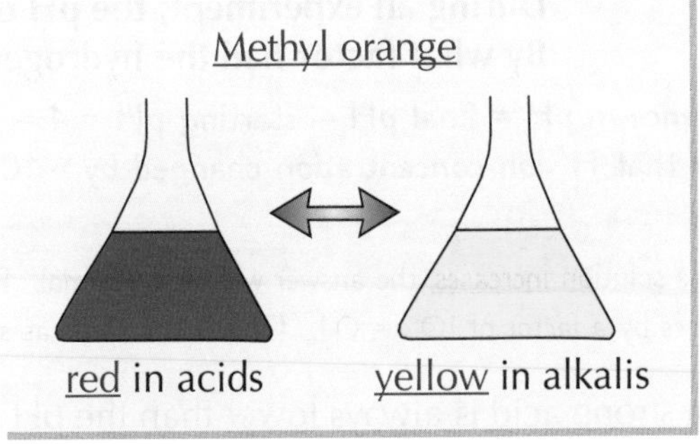

Accuracy is everything...

When you're doing a titration you need to make sure that your results are as accurate as possible. To do this you need to a) go slowly as you reach the end-point, and b) do repeats and find a mean.

Strong Acids, Weak Acids and their Reactions

Right, it's time for strong acids versus weak acids. Brace yourself, it's not the simplest bit of chemistry ever...

Acids **Produce Protons** in **Water**

An H^+ ion is just a proton.

The thing about acids is that they ionise in aqueous solution — they produce hydrogen ions, H^+.

For example:

$HCl \rightarrow H^+ + Cl^-$

$HNO_3 \rightarrow H^+ + NO_3^-$

These acids don't produce hydrogen ions until they meet water. So, for example, hydrogen chloride gas isn't acidic.

Acids Can be **Strong** or **Weak**

1) Strong acids (e.g. sulfuric, hydrochloric and nitric acids) ionise completely in water. All the acid particles dissociate to release H^+ ions.

Strong acid: $HCl \longrightarrow H^+ + Cl^-$

2) Weak acids (e.g. ethanoic, citric and carbonic acids) do not fully ionise in solution. Only a small proportion of acid particles dissociate to release H^+ ions.
3) The ionisation of a weak acid is a reversible reaction, which sets up an equilibrium between the undissociated and dissociated acid. Since only a few of the acid particles release H^+ ions, the position of equilibrium lies well to the left.

Weak acid: $CH_3COOH \rightleftharpoons H^+ + CH_3COO^-$

For more on equilibria turn to p.127.

4) The pH of an acid or alkali is a measure of the concentration of H^+ ions in the solution.
5) For every decrease of 1 on the pH scale, the concentration of H^+ ions increases by a factor of 10. So, an acid that has a pH of 4 has 10 times the concentration of H^+ ions of an acid that has a pH of 5. For a decrease of 2 on the pH scale, the concentration of H^+ ions increases by a factor of 100. The general rule for this is:

$$\text{Factor } H^+ \text{ ion concentration changes by} = 10^{-X}$$

X is the difference in pH, found by subtracting the starting pH from the final pH.

EXAMPLE:

During an experiment, the pH of a solution fell from pH 7 to pH 4. By what factor had the hydrogen ion concentration of the solution changed?

Difference in pH = final pH – starting pH = 4 – 7 = –3.

Factor that H^+ ion concentration changed by = $10^{-(-3)} = 10^3$ (or 1000)

If the pH of the solution increases, the answer will be a decimal. For example if the pH of a solution increases by 1, the H^+ ion concentration changes by a factor of $10^{-1} = 0.1$. (This is the same as saying that the H^+ concentration has decreased by a factor of 10.)

6) The pH of a strong acid is always lower than the pH of a weaker acid if they have the same concentration.

REVISION TIP

Strong acids ionise completely, but weak acids don't

Calculating the factor that hydrogen ion concentration changes by as pH changes can be a bit tricky. The best way to get to grips with it is just to get lots of practice at using the formula.

Strong Acids, Weak Acids and their Reactions

Don't Confuse **Strong** Acids with **Concentrated** Acids

1) Acid strength (i.e. strong or weak) tells you what proportion of the acid molecules ionise in water.
2) The concentration of an acid is different. Concentration measures how much acid there is in a certain volume of water. Concentration is basically how watered down your acid is.
3) The larger the amount of acid there is in a certain volume of liquid, the more concentrated the acid is.
4) So you can have a dilute (not very concentrated) but strong acid, or a concentrated but weak acid.
5) pH will decrease with increasing acid concentration regardless of whether it's a strong or weak acid.

Concentration describes the total number of dissolved acid molecules — not the number of molecules that are ionised to produce hydrogen ions at any given moment.

Metal **Oxides** and Metal **Hydroxides** are **Bases**

1) Some metal oxides and metal hydroxides dissolve in water. These soluble compounds are alkalis. As you saw on page 87, alkalis react with acids in neutralisation reactions.
2) Even bases that won't dissolve in water will still take part in neutralisation reactions with acids.
3) So, all metal oxides and metal hydroxides react with acids to form a salt and water.

Acid + Metal Oxide → Salt + Water

Acid + Metal Hydroxide → Salt + Water

The **Combination** of Metal and Acid Decides the **Salt**

This isn't exactly exciting, but it's pretty important, so try and get the hang of it:

hydrochloric acid	+	copper oxide	→	copper chloride	+	water
$2HCl$	+	CuO	→	$CuCl_2$	+	H_2O
hydrochloric acid	+	sodium hydroxide	→	sodium chloride	+	water
HCl	+	$NaOH$	→	$NaCl$	+	H_2O

sulfuric acid	+	zinc oxide	→	zinc sulfate	+	water
H_2SO_4	+	ZnO	→	$ZnSO_4$	+	H_2O
sulfuric acid	+	calcium hydroxide	→	calcium sulfate	+	water
H_2SO_4	+	$Ca(OH)_2$	→	$CaSO_4$	+	$2H_2O$

nitric acid	+	magnesium oxide	→	magnesium nitrate	+	water
$2HNO_3$	+	MgO	→	$Mg(NO_3)_2$	+	H_2O
nitric acid	+	potassium hydroxide	→	potassium nitrate	+	water
HNO_3	+	KOH	→	KNO_3	+	H_2O

To work out the formula of an ionic compound, you need to balance the charges of the positive and negative ions so the compound is neutral. For more on formulas of ionic compounds, see p.51.

Strong Acids, Weak Acids and their Reactions

Acids and Metal **Carbonates** Produce **Carbon Dioxide**

Metal carbonates are also bases. They will react with acids to produce a salt, water and carbon dioxide.

Acid + Metal Carbonate → Salt + Water + Carbon Dioxide

Here are a few examples of acid and metal carbonate reactions:

hydrochloric acid	+	sodium carbonate	→	sodium chloride	+	water	+	carbon dioxide
$2HCl$	+	Na_2CO_3	→	$2NaCl$	+	H_2O	+	CO_2
sulfuric acid	+	calcium carbonate	→	calcium sulfate	+	water	+	carbon dioxide
H_2SO_4	+	$CaCO_3$	→	$CaSO_4$	+	H_2O	+	CO_2

You can Make **Soluble Salts** Using an **Insoluble Base**

1) You need to pick the right acid, plus an insoluble base such as an insoluble hydroxide, metal oxide or carbonate.

You can also make some salts by reacting a metal with an acid (see page 95 for more).

For example, if you want to make copper chloride, you could mix hydrochloric acid and copper oxide: $CuO_{(s)} + 2HCl_{(aq)} \rightarrow CuCl_{2\,(aq)} + H_2O_{(l)}$

2) Gently warm the dilute acid using a Bunsen burner, then turn off the Bunsen burner.

3) Add the insoluble base to the acid a bit at a time until no more reacts (i.e. the base is in excess). You'll know when all the acid has been neutralised because, even after stirring, the excess solid will just sink to the bottom of the flask.

4) Then filter out the excess solid to get the salt solution (see p.26).

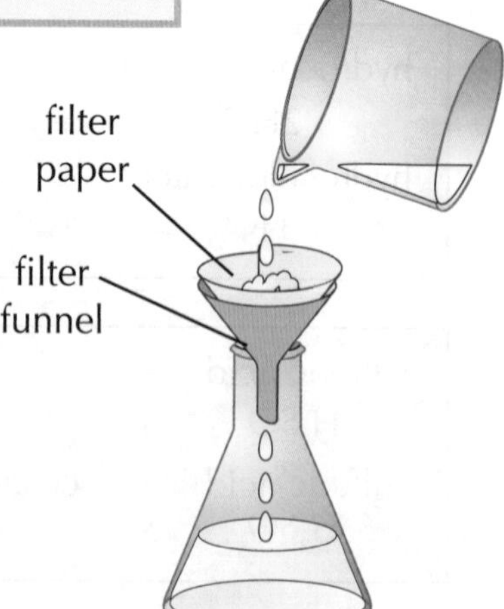

5) To get pure, solid crystals of the salt, gently heat the solution using a water bath or an electric heater to evaporate some of the water (to make it more concentrated) and then stop heating it and leave the solution to cool. Crystals of the salt should form, which can be filtered out of the solution and then dried. This is called crystallisation (see p.26).

Quite a few reactions to learn here...

...but it's not so bad really, because they're all acid + base → salt + water (and sometimes carbon dioxide). Remember, you could be asked to describe how you would make a pure, dry sample of a given soluble salt, so make sure you understand the method and that you could suggest a suitable acid and base to use.

Warm-Up & Exam Questions

So you think you know everything there is to know about acids? Time to put yourself to the test.

Warm-Up Questions

1) What range of values can pH take?
2) What term is used to describe a solution with a pH of 7?
3) What's the difference between a strong acid and a weak acid?
4) If the pH of the solution rises from 2 to 5, what factor has the hydrogen ion concentration of the solution changed by?
5) Name the two substances formed when nitric acid reacts with copper hydroxide.

Exam Questions

1 A student had a sample of acid in a test tube. He gradually added some alkali to the acid. (Grade 4-6)

1.1 Name the type of ion that acids produce in aqueous solutions.

[1 mark]

1.2 What type of reaction took place in the student's experiment? Tick **one** box.

☐ thermal decomposition ☐ neutralisation ☐ redox ☐ combustion

[1 mark]

2 All metal hydroxides are bases. They can react with acids to form a salt and water. (Grade 6-7)

2.1 Sodium hydroxide is a soluble base. What name is given to bases that dissolve in water?

[1 mark]

2.2 Complete and balance the symbol equation given below for the reaction of nitric acid with magnesium hydroxide.

$$2HNO_3 + Mg(OH)_2 \rightarrow \text{.........................} + \text{.........................}$$

[2 marks]

PRACTICAL

3 Silver nitrate is a soluble salt. It can be made by adding an excess of insoluble silver carbonate to nitric acid until no further reaction occurs. (Grade 6-7)

Figure 1

3.1 Give **one** observation that would indicate that the reaction is complete.

[1 mark]

3.2 Once the reaction is complete, the excess silver carbonate can be separated from the silver nitrate solution using the apparatus shown in **Figure 1**. What is this method of separation called?

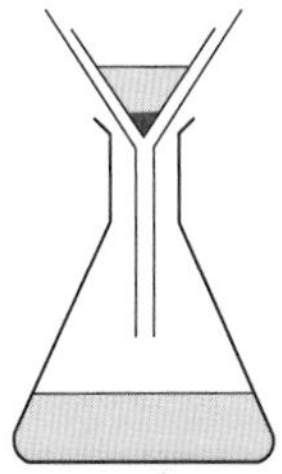

[1 mark]

3.3 Describe how you could produce solid silver nitrate from silver nitrate solution.

[3 marks]

PRACTICAL

4* Describe the method you would use to carry out an accurate titration to find the volume of acid needed to neutralise 25.0 cm^3 of an alkali. Assume you have already done a rough titration. (Grade 6-7)

[6 marks]

Metals and their Reactivity

Metals can be placed in order of reactivity. This can be really useful for predicting their reactions.

The Reactivity Series — How Well a Metal Reacts

1) The reactivity series lists metals in order of their reactivity towards other substances.

2) For metals, their reactivity is determined by how easily they lose electrons — forming positive ions. The higher up the reactivity series a metal is, the more easily it will form positive ions.

3) When metals react with water or acid, they lose electrons and form positive ions. So, the higher a metal is in the reactivity series, the more easily it reacts with water or acid.

4) If you compare the relative reactivity of different metals with either an acid or water and put them in order from the most reactive to the least reactive, the order you get is the reactivity series.

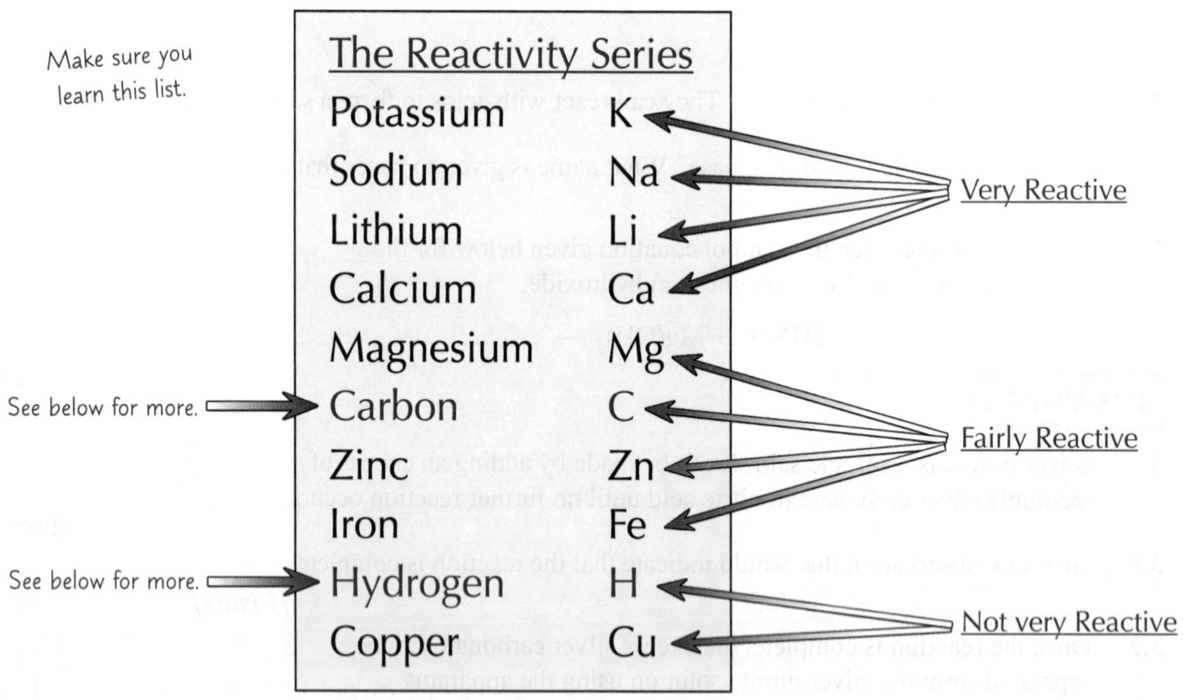

Carbon and Hydrogen

Carbon and hydrogen are non-metals, but they're often included in the reactivity series, because this gives you information about how metals will react with them:

- Metals that are less reactive than carbon can be extracted from their ores by reduction with carbon (see p.96). Metals that are more reactive than carbon cannot be extracted in this way.
- Metals that are more reactive than hydrogen (above it in the reactivity series) will react with acids. Metals less reactive than hydrogen will not react with acids.

Metals and their Reactivity

How **Metals** React With **Acids** Tells You About Their **Reactivity**

Some metals react with acids to produce a salt and hydrogen gas.

Acid + Metal → Salt + Hydrogen

Hydrochloric acid will react to form chloride salts and sulfuric acid will react to form sulfate salts.

1) The speed of reaction is indicated by the rate at which the bubbles of hydrogen are given off.
2) The more reactive the metal, the faster the reaction will go. Very reactive metals like potassium, sodium, lithium and calcium react explosively, but less reactive metals such as magnesium, zinc and iron react less violently. In general, copper won't react with cold, dilute acids.

Magnesium reacts vigorously with cold dilute acids like $HCl_{(aq)}$ and $H_2SO_{4(aq)}$ and produces loads of bubbles.

Both zinc and iron react fairly slowly with dilute acids, but more strongly if you heat them up.

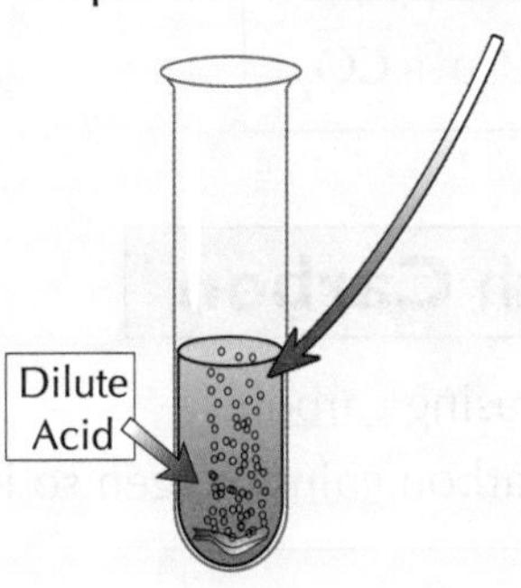

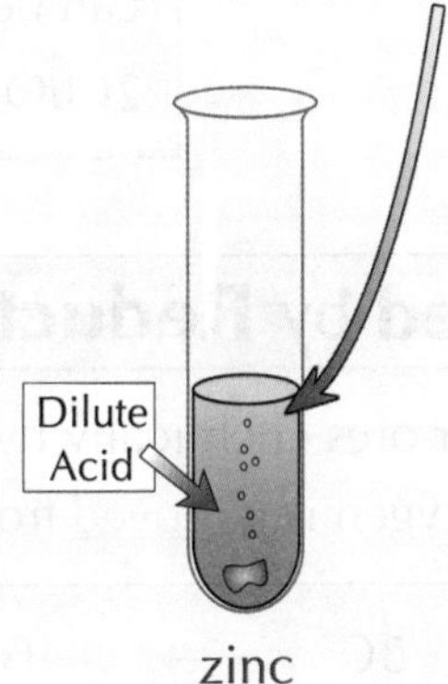

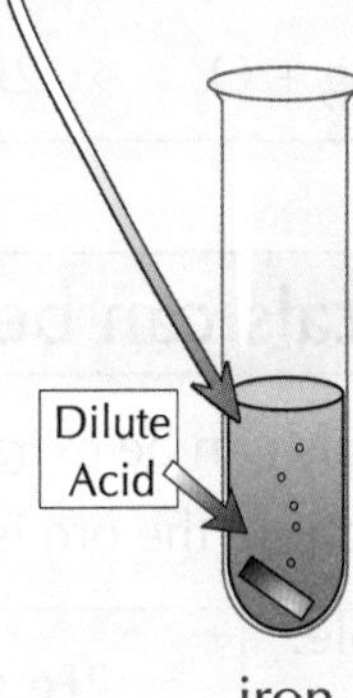

$Mg_{(s)} + 2HCl_{(aq)} \rightarrow MgCl_{2(aq)} + H_{2(g)}$ $Zn_{(s)} + 2HCl_{(aq)} \rightarrow ZnCl_{2(aq)} + H_{2(g)}$ $Fe_{(s)} + 2HCl_{(aq)} \rightarrow FeCl_{2(aq)} + H_{2(g)}$

$Mg_{(s)} + H_2SO_{4(aq)} \rightarrow MgSO_{4(aq)} + H_{2(g)}$ $Zn_{(s)} + H_2SO_{4(aq)} \rightarrow ZnSO_{4(aq)} + H_{2(g)}$ $Fe_{(s)} + H_2SO_{4(aq)} \rightarrow FeSO_{4(aq)} + H_{2(g)}$

You can use the burning splint test (see page 153) to confirm that hydrogen is formed in these reactions.

Metals Also React with **Water**

The reactions of metals with water also show the reactivity of metals.

Metal + Water → Metal Hydroxide + Hydrogen

For example, calcium: $Ca_{(s)} + 2H_2O_{(l)} \rightarrow Ca(OH)_{2(aq)} + H_{2(g)}$

1) The metals potassium, sodium, lithium and calcium will all react with water.
2) Less reactive metals like zinc, iron and copper won't react with water.

You can see more on the reactions of Group 1 metals with water on page 40.

Metals at the top of the reactivity series are highly reactive

You can use the results of experiments like these to work out an order of reactivity for the metals that you used. Just remember, the more vigorous the reaction with acid or water, the more reactive the metal.

Metals and their Reactivity

Metals Often Have to be Separated from their **Oxides**

1) Most metals aren't found in the earth in their pure form. They tend to be fairly reactive, so they're usually found as compounds and have to be extracted before they can be used.
2) Lots of common metals, like iron and aluminium, react with oxygen to form their oxides, which are found in the ground. This process is an example of oxidation. These oxides are often the ores that the metals need to be extracted from.
3) A reaction that separates a metal from its oxide is called a reduction reaction.

An ore is a type of rock that contains metal compounds.

Formation of **Metal Ore**:

Oxidation = Gain of Oxygen

For example, magnesium can be oxidised to make magnesium oxide.

$2Mg + O_2 \rightarrow 2MgO$

Extraction of **Metal**:

Reduction = Loss of Oxygen

For example, copper oxide can be reduced to copper.

$2CuO + C \rightarrow 2Cu + CO_2$

Some Metals can be **Extracted** by **Reduction** with **Carbon**

1) Some metals can be extracted from their ores chemically by reduction using carbon.
2) In this reaction, the ore is reduced as oxygen is removed from it, and carbon gains oxygen so is oxidised.

For example:

$2Fe_2O_3$	+	3C	→	4Fe	+	$3CO_2$
iron(III) oxide	+	carbon	→	iron	+	carbon dioxide

3) As you saw on p.94, the position of the metal in the reactivity series determines whether it can be extracted by reduction with carbon.

- Metals higher than carbon in the reactivity series have to be extracted using electrolysis, which is expensive.
- Metals below carbon in the reactivity series can be extracted by reduction using carbon. For example, iron oxide is reduced in a blast furnace to make iron. This is because carbon can only take the oxygen away from metals which are less reactive than carbon itself is.

The Reactivity Series			
Extracted by using electrolysis.	Potassium	K	more reactive
	Sodium	Na	
	Lithium	Li	
	Calcium	Ca	
	Magnesium	Mg	
	CARBON	C	
Extracted by reduction with carbon.	Zinc	Zn	
	Iron	Fe	
	Hydrogen	H	
	Copper	Cu	less reactive

A few metals are so unreactive that they are found in the earth as the metal itself. For example, gold is mined as its elemental form.

Carbon can't reduce things that are above it in the reactivity series

Make sure you understand the difference between reduction and oxidation and make sure that you can spot which substance has been reduced and which substance has been oxidised in a reaction.

Redox Reactions

Reduction and oxidation doesn't just mean the loss or gain of oxygen — it can also refer to electron transfer.

If Electrons are Transferred, it's a Redox Reaction

Oxidation can mean the addition of oxygen (or a reaction with it), and reduction can be the removal of oxygen. But you can also define oxidation and reduction in terms of electrons:

A loss of electrons is called oxidation.

A gain of electrons is called reduction.

Make sure you learn these definitions.

A handy way to remember this is using the mnemonic OIL RIG:

Oxidation Is Loss, Reduction Is Gain (of electrons)

REDuction and OXidation happen at the same time — hence the name "REDOX".

Some Examples of Redox Reactions:

Metals reacting with acids

1) All reactions of metals with acids (see page 95) are redox reactions.
2) For example, the reaction of iron with dilute sulfuric acid is a redox reaction.
3) The iron atoms lose electrons to become iron(II) ions — they are oxidised by the hydrogen ions. $Fe \rightarrow Fe^{2+} + 2e^-$
4) The hydrogen ions gain electrons to become hydrogen atoms — they are reduced by the iron atoms. $2H^+ + 2e^- \rightarrow H_2$
5) The ionic equation (see page 98) for the redox reaction is: $Fe + 2H^+ \rightarrow Fe^{2+} + H_2$

Ionic equations only show the things that are actually reacting — the sulfate ions from the acid aren't shown here, because nothing happens to them.

Halogen Displacement Reactions

1) A more reactive halogen can displace a less reactive halogen from a salt solution (see page 42).
2) For example, chlorine can displace bromine from potassium bromide solution.
3) The chlorine atoms gain electrons to become chloride ions — they are reduced by the bromide ions. $Cl_2 + 2e^- \rightarrow 2Cl^-$
4) The bromide ions lose electrons to become bromine atoms — they are oxidised by the chlorine atoms. $2Br^- \rightarrow Br_2 + 2e^-$
5) The ionic equation for this redox reaction is: $Cl_2 + 2Br^- \rightarrow 2Cl^- + Br_2$

Same here — the ionic equation doesn't show the potassium ions from the potassium bromide, because nothing happens to them.

Remember OIL RIG — 'Oxidation Is Loss, Reduction Is Gain'

OIL RIG is a pretty handy way of reminding yourself about what goes on during a redox reaction. Make sure that you don't forget that redox reactions are all about the transfer of electrons.

Redox Reactions

Metal Displacement Reactions are Redox Reactions

1) If you put a reactive metal into the solution of a dissolved metal compound, the reactive metal will replace the less reactive metal in the compound (see the reactivity series on page 94).
2) This type of reaction is another example of a displacement reaction.
3) Here's the rule to remember:

A MORE REACTIVE metal will displace a LESS REACTIVE metal from its compound.

For example, if you put iron in a solution of copper sulfate ($CuSO_4$) the more reactive iron will "kick out" the less reactive copper from the solution. You end up with iron sulfate solution ($FeSO_4$) and copper metal:

iron + copper sulfate → iron sulfate + copper

$$Fe_{(s)} + CuSO_{4(aq)} \rightarrow FeSO_{4(aq)} + Cu_{(s)}$$

In this reaction the iron loses 2 electrons to become a 2+ ion — it's oxidised.

$$Fe \rightarrow Fe^{2+} + 2e^-$$

The copper ion gains these 2 electrons to become a copper atom — it's reduced.

$$Cu^{2+} + 2e^- \rightarrow Cu$$

This reaction is used in industry to produce pure copper from copper salt solutions and cheap scrap iron.

3) In metal displacement reactions it's always the metal ion that gains electrons and is reduced. The metal atom always loses electrons and is oxidised.
4) In the exam you could be asked to write word or symbol equations for metal displacement reactions.

Ionic Equations Show Just the **Useful Bits** of Reactions

1) In an ionic equation only the particles that react and the products they form are shown. For example: $Mg_{(s)} + Zn^{2+}_{(aq)} \rightarrow Mg^{2+}_{(aq)} + Zn_{(s)}$
2) This just shows the displacement of zinc ions by magnesium metal. Here's what the full equation of the above reaction would be if you'd started off with magnesium metal and zinc chloride solution:

$$Mg_{(s)} + ZnCl_{2(aq)} \rightarrow MgCl_{2(aq)} + Zn_{(s)}$$

3) If you write out the equation so you can see all the ions, you'll see that the chloride ions don't change in the reaction — they're spectator ions.

$$Mg_{(s)} + Zn^{2+}_{(aq)} + 2Cl^-_{(aq)} \rightarrow Mg^{2+}_{(aq)} + 2Cl^-_{(aq)} + Zn_{(s)}$$

4) You don't include spectator ions in the ionic equation, so to get from this to the ionic equation, you cross out the chloride ions.
5) The ionic equation just concentrates on the substances which are oxidised or reduced.

A displacement reaction is a type of redox reaction

Try writing some ionic equations now — write the ionic equation for the reaction between zinc and iron chloride ($FeCl_2$). What's being oxidised? What's being reduced?

Warm-Up & Exam Questions

Hoping to test your knowledge with some testing chemistry questions? You're in luck...

Warm-Up Questions

1) What determines the reactivity of a metal?
2) Magnesium is reacted with dilute hydrochloric acid. Give the name of the salt formed.
3) Which two substances are formed when a metal reacts with water?
4) This equation shows how zinc reacts with copper chloride: $Zn_{(s)} + CuCl_{2(aq)} \rightarrow ZnCl_{2(aq)} + Cu_{(s)}$
   a) Write an ionic equation for this reaction.
   b) State which species in this reaction is being oxidised.

Exam Questions

1 **Figure 1** shows part of the reactivity series of metals. Hydrogen has also been included in this reactivity series.

Figure 1

Potassium K
Sodium Na
Calcium Ca
Magnesium Mg
Zinc Zn
Iron Fe
HYDROGEN H
Copper Cu

1.1 Name **one** metal from **Figure 1** that is more reactive than magnesium.

[1 mark]

1.2 Name **one** metal from **Figure 1** which would not react with acids at all.

[1 mark]

1.3 A student places a small piece of zinc into dilute acid. The mixture produces bubbles of hydrogen gas fairly slowly. Use this information to predict what would happen if the student repeated the experiment using iron.

[2 marks]

2 Iron can be extracted by the reduction of iron(III) oxide (Fe_2O_3) with carbon (C), to produce iron and carbon dioxide.

Grade 6-7

2.1 Write a balanced symbol equation for this reaction.

[2 marks]

2.2 Explain why the iron(III) oxide is described as being reduced during this reaction.

[1 mark]

3 A student adds a piece of copper to some iron sulfate solution and a piece of iron to some copper sulfate solution. **Figure 2** shows what the test tubes looked like just after he added the metals. Predict what he will observe in both tubes, **A** and **B**, after 2 hours. Explain your answer.

Figure 2

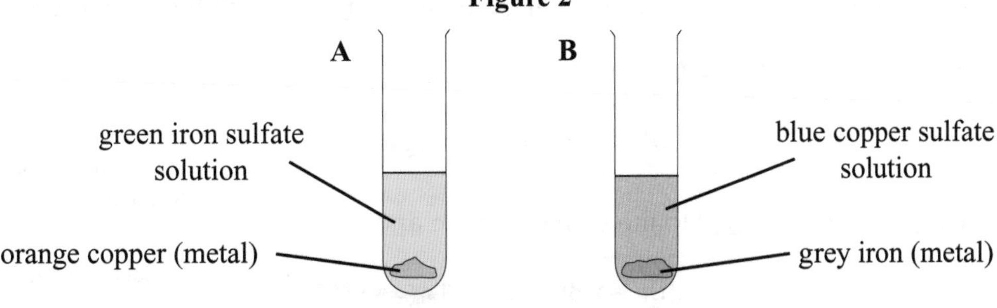

[6 marks]

Electrolysis

You need to know the ins and outs of how electrolysis works. So buckle up, here we go...

Electrolysis Means 'Splitting Up with Electricity'

1) During electrolysis, an electric current is passed through an electrolyte. The electrolyte is a molten or dissolved ionic compound (it must be molten or dissolved so that the ions are free to move).
2) The ions move towards the electrodes, where they react, and the compound decomposes.
3) The positive ions in the electrolyte will move towards the cathode (–ve electrode) and gain electrons (they are reduced).
4) The negative ions in the electrolyte will move towards the anode (+ve electrode) and lose electrons (they are oxidised).
5) This creates a flow of charge through the electrolyte as ions travel to the electrodes.
6) As ions gain or lose electrons, they form the uncharged element and are discharged from the electrolyte.

An electrolyte is just a liquid or solution that can conduct electricity. An electrode is a solid that conducts electricity and is submerged in the electrolyte.

Electrolysis of Molten Ionic Solids Forms Elements

1) An ionic solid can't be electrolysed because the ions are in fixed positions and can't move.
2) Molten ionic compounds can be electrolysed because the ions can move freely and conduct electricity.
3) Molten ionic compounds are always broken up into their elements. A good example of this is the electrolysis of molten lead bromide:

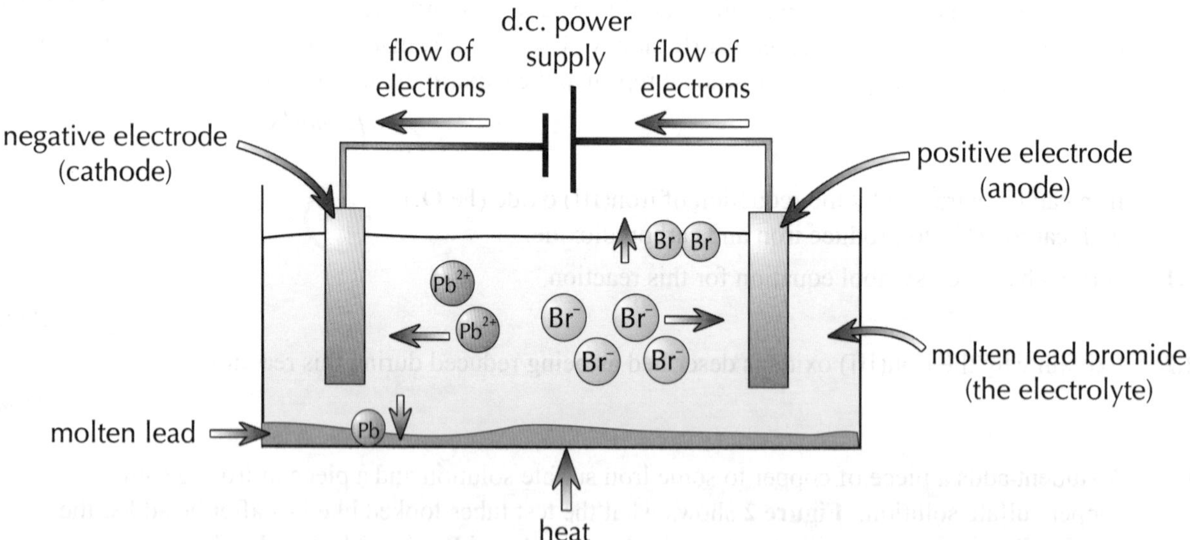

4) The electrodes should be made of an inert material, so they don't react with the electrolyte.
5) Positive metal ions are reduced to the element at the cathode:

$$Pb^{2+} + 2e^- \rightarrow Pb$$

6) Negative non-metal ions are oxidised to the element at the anode:

$$2Br^- \rightarrow Br_2 + 2e^-$$

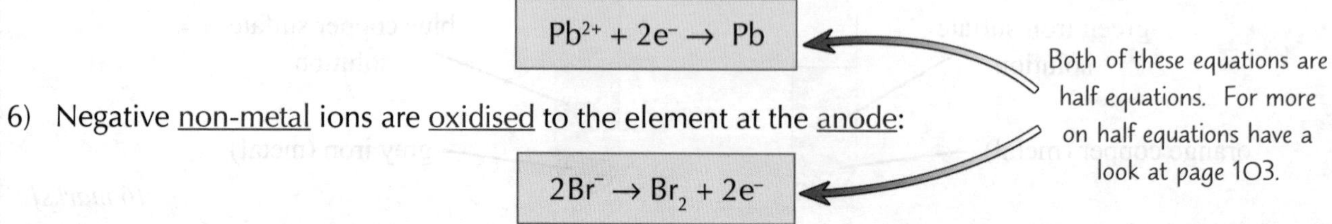

Both of these equations are half equations. For more on half equations have a look at page 103.

Electrolysis

Metals can be Extracted From Their Ores Using Electrolysis

If a metal is too reactive to be reduced with carbon (page 96) or reacts with carbon, then electrolysis can be used to extract it. Extracting metals via this method is very expensive as lots of energy is required to melt the ore and produce the required current. For example:

1) Aluminium is extracted from the ore bauxite by electrolysis. Bauxite contains aluminium oxide, Al_2O_3.
2) Aluminium oxide has a very high melting temperature so it's mixed with cryolite to lower the melting point.
3) The molten mixture contains free ions — so it'll conduct electricity.
4) The positive Al^{3+} ions are attracted to the negative electrode where they each pick up three electrons and turn into neutral aluminium atoms. These then sink to the bottom of the electrolysis tank.
5) The negative O^{2-} ions are attracted to the positive electrode where they each lose two electrons. The neutral oxygen atoms will then combine to form O_2 molecules.

Cryolite is an aluminium based compound with a lower melting point than aluminium oxide.

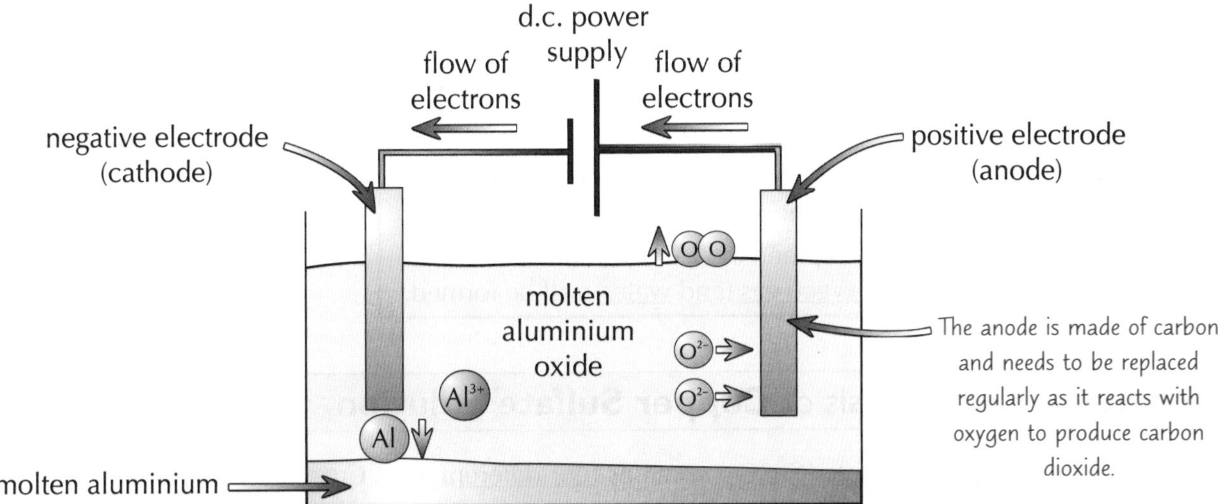

At the Negative Electrode:

1) Metals form positive ions, so they're attracted to the negative electrode.
2) So aluminium is produced at the negative electrode:

$$Al^{3+} + 3e^- \rightarrow Al$$

This is reduction — a gain of electrons.

At the Positive Electrode:

1) Non-metals form negative ions, so they're attracted to the positive electrode.
2) So oxygen is produced at the positive electrode:

$$2O^{2-} \rightarrow O_2 + 4e^-$$

This is oxidation — a loss of electrons.

Overall Equation:

aluminium oxide → aluminium + oxygen

$$2Al_2O_{3(l)} \rightarrow 4Al_{(l)} + 3O_{2(g)}$$

Electrolysis of Aqueous Solutions

When carrying out electrolysis on an aqueous solution you have to factor in the ions in water.

It May be Easier to **Discharge Ions** from **Water** than the **Solute**

1) In aqueous solutions, as well as the ions from the ionic compound, there will be hydrogen ions (H^+) and hydroxide ions (OH^-) from the water: $H_2O_{(l)} \rightleftharpoons H^+_{(aq)} + OH^-_{(aq)}$

2) Which ions are discharged at the electrodes when the solution is electrolysed will depend on the relative reactivity of all the ions in the solution.

Cathode

- At the cathode, if H^+ ions and metal ions are present, hydrogen gas will be produced if the metal ions form an elemental metal that is more reactive than hydrogen (e.g. sodium ions).
- If the metal ions form an elemental metal that is less reactive than hydrogen (e.g. copper ions), a solid layer of the pure metal will be produced instead, which will coat the cathode.

Anode

- At the anode, if OH^- and halide ions (Cl^-, Br^-, I^-) are present, molecules of chlorine, bromine or iodine will be formed.
- If no halide ions are present, then the OH^- ions from the water will be discharged and oxygen gas (and water) will be formed.

Example 1: Electrolysis of **Copper Sulfate Solution**

A solution of copper(II) sulfate ($CuSO_4$) contains four different ions: Cu^{2+}, SO_4^{2-}, H^+ and OH^-.

Copper metal is less reactive than hydrogen, so at the cathode copper metal is produced and coats the electrode:

$$Cu^{2+} + 2e^- \rightarrow Cu$$

There aren't any halide ions present, so at the anode oxygen and water are produced: The oxygen can be seen as bubbles.

$$4OH^- \rightarrow O_2 + 2H_2O + 4e^-$$

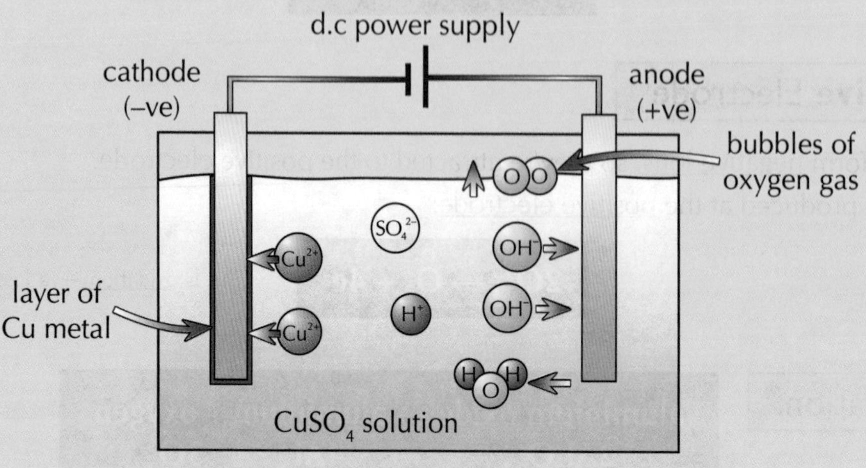

Electrolysis of Aqueous Solutions

Example 2: Electrolysis of Sodium Chloride Solution

A solution of sodium chloride (NaCl) contains four different ions: Na^+, Cl^-, OH^- and H^+.

Sodium metal is more reactive than hydrogen. So at the cathode, hydrogen gas is produced.

$$2H^+ + 2e^- \rightarrow H_2$$

Chloride ions are present in the solution. So at the anode chlorine gas is produced.

$$2Cl^- \rightarrow Cl_2 + 2e^-$$

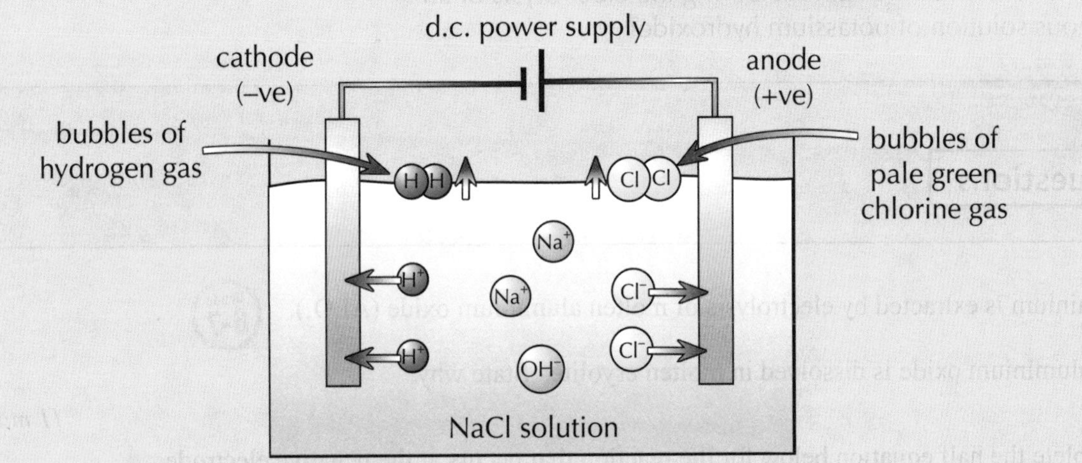

You can do an electrolysis experiment in the lab using a set-up like the one on page 199. Once the experiment is finished you can test any gaseous products to work out what has been produced at the electrodes:

- Chlorine bleaches damp litmus paper, turning it white.
- Hydrogen makes a "squeaky pop" with a lighted splint.
- Oxygen will relight a glowing splint.

For more on tests for gases, turn to page 153.

The Half Equations — Make Sure the Electrons Balance

1) Half equations show the reactions at the electrodes (they show ions or atoms gaining or losing electrons).
2) You can combine the half equations for the reactions at both electrodes to get the ionic equation for the overall reaction. The important thing to remember when you combine half equations is that the number of electrons shown in each half equation must be the same.

For the electrolysis of aqueous sodium chloride solution the half equations are:

Negative Electrode: $2H^+ + 2e^- \rightarrow H_2$

Positive Electrode: $2Cl^- \rightarrow Cl_2 + 2e^-$ (or $2Cl^- - 2e^- \rightarrow Cl_2$)

You can combine these to get the ionic equation: $\mathbf{2H^+ + 2Cl^- \rightarrow H_2 + Cl_2}$

The electrons on each side of the half equations balance, so they cancel out in the full ionic equation.

Remember, if there are no halide ions in an aqueous solution, OH^- ions are discharged at the positive electrode. The half equation for this is: $4OH^- \rightarrow O_2 + 2H_2O + 4e^-$ (or $4OH^- - 4e^- \rightarrow O_2 + 2H_2O$)

Remember — all aqueous solutions contain OH^- and H^+ ions

The main points to remember about the electrolysis of aqueous solutions are: 1) At the cathode, hydrogen gas is made, unless the metal ions are less reactive than hydrogen — then you get a coating of the metal. 2) At the anode, oxygen and water are made, unless halide ions are present — then the halogen is made.

Warm-Up & Exam Questions

Time to test your mettle. Try and get through the following questions. If there's anything you're not quite sure about, have a look at the pages again until you can answer the questions without batting an eyelid.

Warm-Up Questions

1) At which electrode are metals deposited during electrolysis?
2) What products are made when molten zinc chloride is electrolysed?
3) During the electrolysis of molten aluminium oxide are the aluminium ions being reduced or oxidised?
4) What is formed at the cathode during the electrolysis of an aqueous solution of potassium hydroxide?

Exam Questions

1 Aluminium is extracted by electrolysis of molten aluminium oxide (Al_2O_3). Grade 6-7

1.1 The aluminium oxide is dissolved in molten cryolite. State why.

[1 mark]

1.2 Complete the half equation below for the reaction that occurs at the negative electrode.

.... + $3e^-$ → ....

[1 mark]

1.3 Complete the half equation below for the reaction that occurs at the positive electrode.

.... → .... + $4e^-$

[1 mark]

1.4 The positive electrode is made of carbon. Explain why it will need to be replaced over time.

[2 marks]

2 When sodium solution chloride is electrolysed, a gas is produced at each electrode. Grade 6-7

2.1 Name the gas produced at the negative electrode.

[1 mark]

2.2 Give the half equation for the reaction at the negative electrode.

[1 mark]

2.3 Name the gas produced at the positive electrode.

[1 mark]

2.4 Give the half equation for the reaction at the positive electrode.

[1 mark]

2.5 Explain why sodium hydroxide is left in solution at the end of the reaction.

[2 marks]

2.6 If copper chloride solution is electrolysed, copper metal is produced at the negative electrode, instead of the gas named in **2.1**. Explain why.

[1 mark]

Revision Summary for Topic 4

That wraps up Topic 4 — time to put yourself to the test and find out how much you really know.

- Try these questions and tick off each one when you get it right.
- When you've done all the questions under a heading and are completely happy with it, tick it off.

Acids and their Reactions (p.87-92) ☐

1) State whether the following pH values are acidic, alkaline or neutral.
   a) 9 b) 2 c) 7 d) 6 ☐
2) Give the general word equation for the reaction between an acid and a base. ☐
3) What colour does universal indicator turn in a neutral solution? ☐
4) Why should you do a rough titration first when carrying out a titration experiment? ☐
5) Why shouldn't you use universal indicator as an indicator in a titration? ☐
6) Name two suitable indicators for a titration between an acid and an alkali.
   State their colours when in acidic and alkaline solution. ☐
7) What is a strong acid? ☐
8) By what factor does the H^+ concentration increase for a decrease of 1 on the pH scale? ☐
9) Explain the difference between a strong acid and a concentrated one. ☐
10) Write a balanced equation for the reaction between hydrochloric acid and sodium carbonate. ☐

Reactivity of Metals and Redox Reactions (p.94-98) ☐

11) Is zinc more or less reactive than iron? ☐
12) What is the general word equation for the reaction of a metal with an acid? ☐
13) Give the balanced equation for the reaction of calcium with water. ☐
14) Will copper react with water? ☐
15) What product is formed when magnesium is oxidised by oxygen? ☐
16) Explain how you can use a reactivity series to work out whether or not
    it is possible to extract a metal from its oxide by reduction with carbon. ☐
17) Why is gold found as a pure metal in the earth? ☐
18) In terms of electrons, give the definition of oxidation. ☐
19) In a displacement reaction, does the metal atom get reduced or oxidised? ☐

Electrolysis (p.100-103) ☐

20) During electrolysis, which electrode are the positive ions attracted to? ☐
21) Why can ionic solids not undergo electrolysis? ☐
22) During electrolysis, do ions get reduced or oxidised at the anode? ☐
23) During the manufacture of aluminium from bauxite, which electrode is aluminium formed at? ☐
24) In what situation will hydrogen gas be given out during
    the electrolysis of an aqueous solution of an ionic solid? ☐
25) If halide ions are present in an aqueous solution of an ionic solid,
    will oxygen gas be released when the solution is electrolysed? ☐
26) What test could you use to determine if hydrogen gas
    has been produced in an electrolysis reaction? ☐

Exothermic and Endothermic Reactions

Whenever chemical reactions occur, there are changes in energy. This means that when chemicals get together, things either heat up or cool down. I'll give you a heads up — this page is a good 'un.

Energy is **Moved Around** in **Chemical Reactions**

1) Chemicals store a certain amount of energy — and different chemicals store different amounts.
2) If the products of a reaction store more energy than the original reactants, then they must have taken in the difference in energy between the products and reactants from the surroundings during the reaction.
3) But if they store less, then the excess energy was transferred to the surroundings during the reaction.
4) The overall amount of energy doesn't change. This is because energy is conserved in reactions — it can't be created or destroyed, only moved around. This means the amount of energy in the universe always stays the same.

In an **Exothermic** Reaction, Heat is **Given Out**

An exothermic reaction is one which transfers energy to the surroundings, usually by heating. This is shown by a rise in temperature.

1) The best example of an exothermic reaction is burning fuels — also called combustion. This gives out a lot of energy — it's very exothermic.
2) Neutralisation reactions (acid + alkali) are also exothermic.
3) Many oxidation reactions are exothermic. For example, adding sodium to water releases energy, so it must be exothermic — see page 40. The reaction releases energy and the sodium moves about on the surface of the water as it is oxidised.
4) Exothermic reactions have lots of everyday uses. For example:

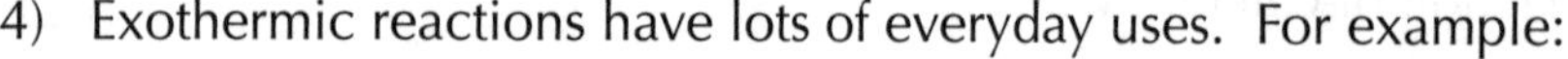

- Some hand warmers use the exothermic oxidation of iron in air (with a salt solution catalyst) to release energy.
- Self heating cans of hot chocolate and coffee also rely on exothermic reactions between chemicals in their bases.

In an **Endothermic** Reaction, Heat is **Taken In**

An endothermic reaction is one which takes in energy from the surroundings. This is shown by a fall in temperature.

Physical processes can also take in or release energy. E.g. freezing is an exothermic process, melting is endothermic.

1) Endothermic reactions are much less common than exothermic reactions, but they include:
   - The reaction between citric acid and sodium hydrogencarbonate.
   - Thermal decomposition — e.g. heating calcium carbonate causes it to decompose into calcium oxide (also called quicklime) and carbon dioxide:

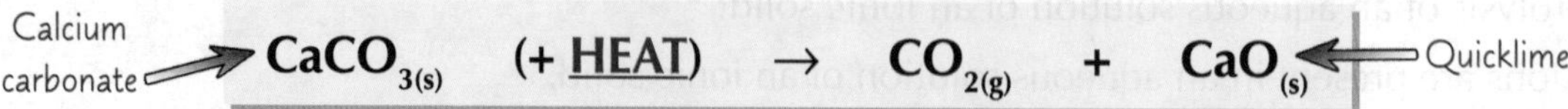

$$CaCO_{3(s)} \quad (+\ \text{HEAT}) \rightarrow CO_{2(g)} + CaO_{(s)}$$

2) Endothermic reactions also have everyday uses. For example:

Endothermic reactions are used in some sports injury packs — the chemical reaction allows the pack to become instantly cooler without having to put it in the freezer.

Exothermic and Endothermic Reactions

Sometimes it's not enough to just know if a reaction is endothermic or exothermic. You may also need to know how much energy is absorbed or released — you can do experiments to find this out. Fun, fun, fun...

Energy Transfer can be Measured

1) You can measure the amount of energy released by a chemical reaction (in solution) by taking the temperature of the reagents (making sure they're the same), mixing them in a polystyrene cup and measuring the temperature of the solution at the end of the reaction (see diagram below).
2) The biggest problem with energy measurements is the amount of energy lost to the surroundings.
3) You can reduce it a bit by putting the polystyrene cup into a beaker of cotton wool to give more insulation, and putting a lid on the cup to reduce energy lost by evaporation.
4) This method works for neutralisation reactions or reactions between metals and acids, or carbonates and acids.
5) You can also use this method to investigate what effect different variables have on the amount of energy transferred — e.g. the mass or concentration of the reactants used.

Here's how you could test the effect of acid concentration on the energy released in a neutralisation reaction between hydrochloric acid (HCl) and sodium hydroxide (NaOH):

1) Put 25 cm^3 of 0.25 mol/dm^3 of hydrochloric acid and sodium hydroxide in separate beakers.
2) Place the beakers in a water bath set to 25 °C until they are both at the same temperature (25 °C).
3) Add the HCl followed by the NaOH to a polystyrene cup with a lid — as in the diagram.
4) Take the temperature of the mixture every 30 seconds, and record the highest temperature.

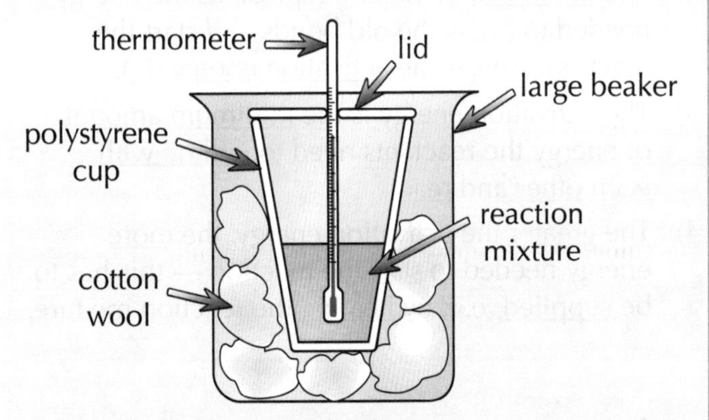

5) Repeat steps 1-4 using 0.5 mol/dm^3 and then 1 mol/dm^3 of hydrochloric acid.

To get a reasonably accurate reading, insulate your reaction

The experiment on this page is an example of a neutralisation reaction, which is an exothermic reaction. Remember, in an exothermic reaction the particles transfer energy to their surroundings. This means that the particles themselves lose energy, but the reaction mixture gets warmer.

Bond Energies

So you know that chemical reactions can take in or release energy — this page is about what causes these energy changes. Hint — it's all to do with making and breaking chemical bonds.

Energy Must Always be Supplied to Break Bonds

There's more on energy transfer on page 106.

1) During a chemical reaction, old bonds are broken and new bonds are formed.
2) Energy must be supplied to break existing bonds — so bond breaking is an endothermic process.
3) Energy is released when new bonds are formed — so bond formation is an exothermic process.

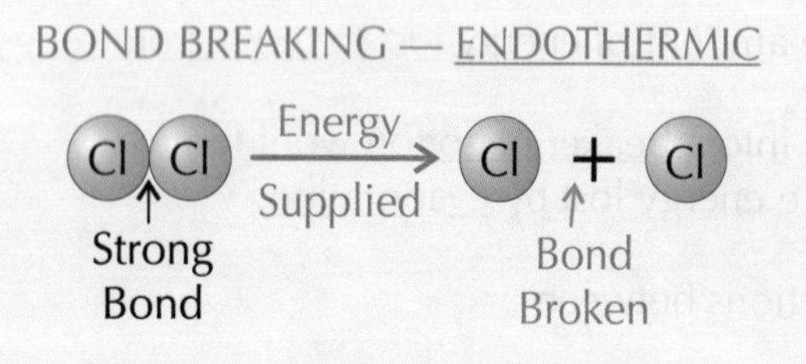

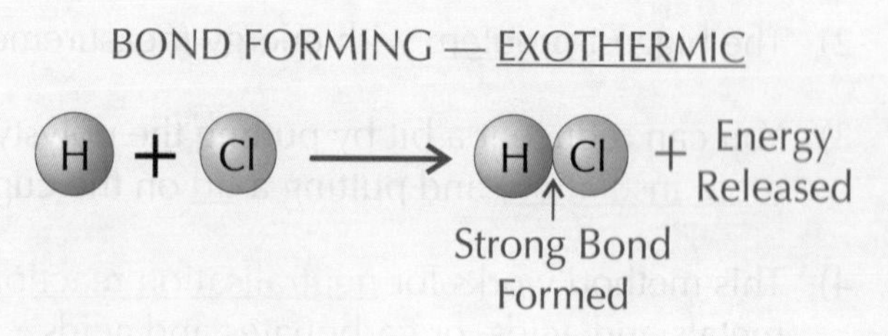

4) In exothermic reactions the energy released by forming bonds is greater than the energy used to break them. In endothermic reactions the energy used to break bonds is greater than the energy released by forming them.

Reaction Profiles Show Energy Changes

Reaction profiles are sometimes called energy level diagrams.

Reaction profiles are diagrams that show the relative energies of the reactants and products in a reaction, and how the energy changes over the course of the reaction.

1) This reaction profile shows an exothermic reaction — the products are at a lower energy than the reactants. The difference in height represents the overall energy change in the reaction (the energy given out) per mole.
2) The initial rise in the line represents the energy needed to break the old bonds and start the reaction. This is the activation energy (E_a).
3) The activation energy is the minimum amount of energy the reactants need to collide with each other and react.
4) The greater the activation energy, the more energy needed to start the reaction — this has to be supplied, e.g. by heating the reaction mixture.

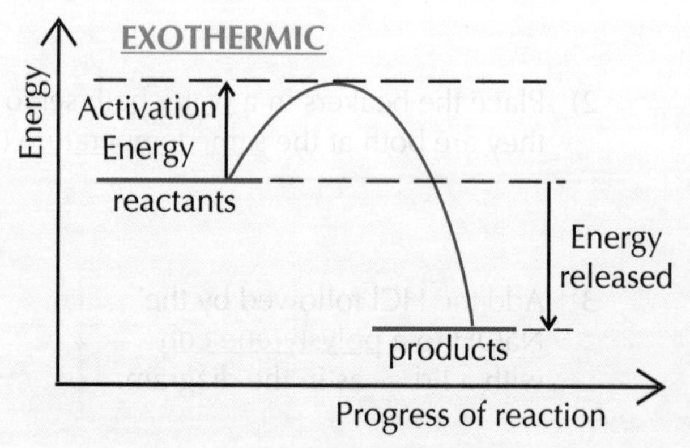

There's more on activation energy and collision theory on pages 118-119.

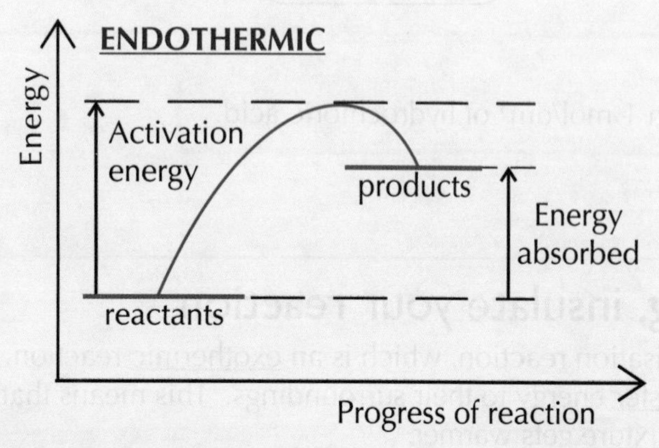

5) This reaction profile shows an endothermic reaction because the products are at a higher energy than the reactants.

6) The difference in height represents the overall energy change during the reaction (the energy taken in) per mole.

Bond Energies

You need to be able to work out the energy change for a particular reaction.
Here's a nice little example, complete with explanation, to get you started.

Bond Energy **Calculations** Need to be **Practised**

1) Every chemical bond has a particular bond energy associated with it. This bond energy varies slightly depending on what compound the bond occurs in.

2) You can use these known bond energies to calculate the overall energy change for a reaction. The overall energy change is the sum of the energies needed to break bonds in the reactants minus the energy released when the new bonds are formed in the products.

3) You need to practise a few of these, but the basic idea is really very simple...

Using the bond energies given below, calculate the energy change for the reaction between H_2 and Cl_2 forming HCl:

$$H—H + Cl—Cl \rightarrow 2H—Cl$$
$$H_2 + Cl_2 \rightarrow 2HCl$$

The bond energies you need are:

- H—H: +436 kJ/mol
- Cl—Cl: +242 kJ/mol
- H—Cl: +431 kJ/mol

1) Find the energy required to break the original bonds:

   (1 × H–H) + (1 × Cl–Cl) = 436 kJ/mol + 242 kJ/mol = 678 kJ/mol

2) Find the energy released by forming the new bonds.

   2 × H–Cl = 2 × 431 kJ/mol = 862 kJ/mol

3) Find the overall energy change for the reaction using this equation:

   Overall energy change = energy required to break bonds – energy released by forming bonds
   = 678 kJ/mol – 862 kJ/mol = –184 kJ/mol

You can't compare the overall energy changes of reactions unless you know the numerical differences in the bond energies.

Chlorine and bromine react with hydrogen in a similar way. Br–Br bonds are weaker than Cl–Cl bonds and H–Br bonds are weaker than H–Cl bonds. So less energy is needed to break the bonds in the reaction with bromine, but less energy is released when the new bonds form. So unless you know the exact difference, you can't say which reaction releases more energy.

You need a balanced chemical equation for these calculations

It's really helpful to lay out all of the information that you're given the way that is shown in the worked example above. That way you won't get all your bond energies mixed up and you'll be able to clearly see what you have do in each part of the calculation.

Warm-Up & Exam Questions

Funny diagrams, bond energy calculations, a practical — there's a lot to get your head around in just a few pages here. Here are some questions so that you can check how you're getting on.

Warm-Up Questions

1) What is an exothermic reaction?
2) What is an endothermic reaction?
3) Is energy released when bonds are formed or when bonds are broken?
4) What is meant by the activation energy of a reaction?

Exam Questions

1 The diagrams in **Figure 1** represent the energy changes in four different chemical reactions.

Figure 1

A
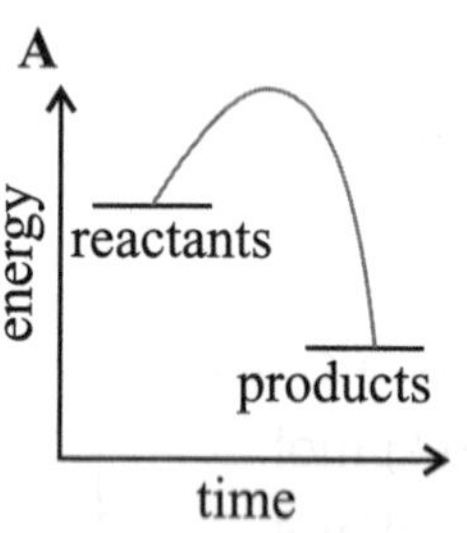

B
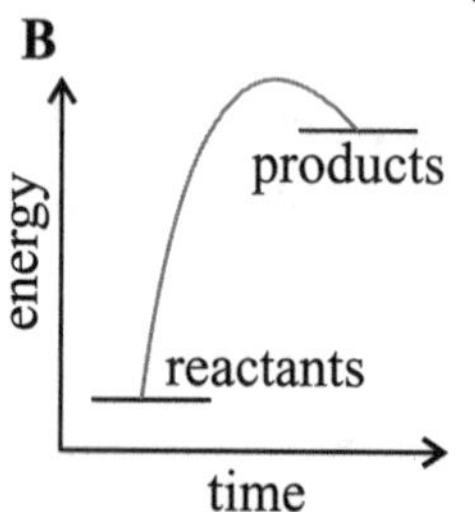

C
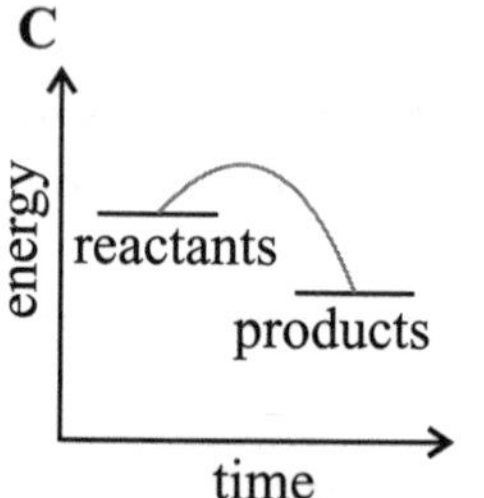

D
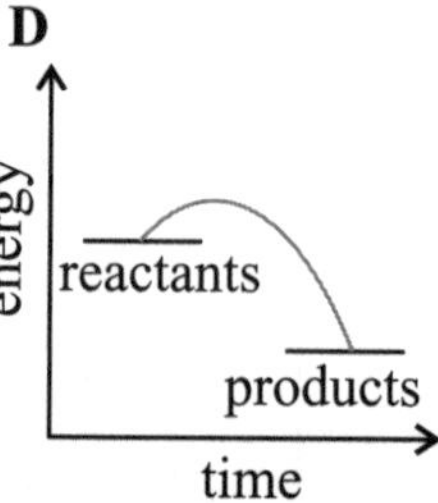

Write the letter of **one** diagram, **A**, **B**, **C** or **D**, which illustrates an endothermic reaction.

[1 mark]

PRACTICAL

2 A student places two beakers of ethanoic acid and dilute potassium hydroxide into a water bath until they are both 25 °C. He adds the ethanoic acid and then the potassium hydroxide to a polystyrene cup. After 1 minute the temperature of the mixture is 28.5 °C.

2.1 Is this reaction endothermic or exothermic? Explain your answer.

[2 marks]

2.2 The student put a lid on the polystyrene cup during the experiment. Suggest why this was done.

[1 mark]

3 When methane burns in air it produces carbon dioxide and water, as shown in **Figure 2**. The bond energies for each bond in the molecules involved are shown in **Figure 3**.

Figure 2

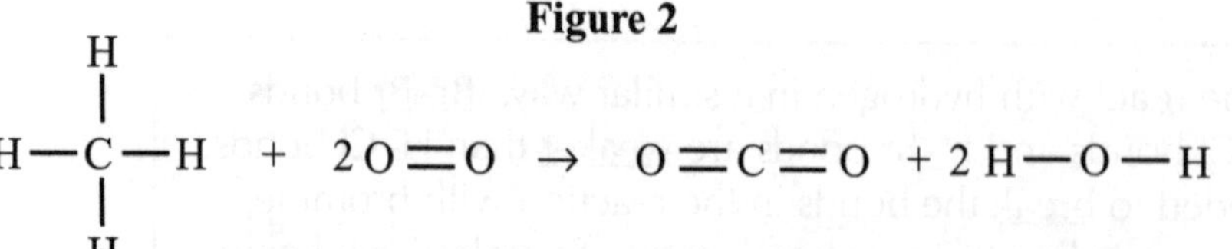

Figure 3

	Bond energies (kJ/mol):
C – H	414
O = O	494
C = O	800
O – H	459

3.1 Which **two** types of bond are broken during the reaction shown in **Figure 2**?

[1 mark]

3.2 Calculate the overall energy change for the reaction shown in **Figure 2**.

[3 marks]

Cells

Cells are really useful, we use them in all sorts of things. You might not be familiar with them, but a simple cell can produce electricity which is a pretty useful for powering a massive variety of things.

Chemical Reactions in a Cell Produce Electricity

An electrochemical cell is a basic system made up of two different electrodes in contact with an electrolyte. An example of a simple electrochemical cell is shown in the diagram below.

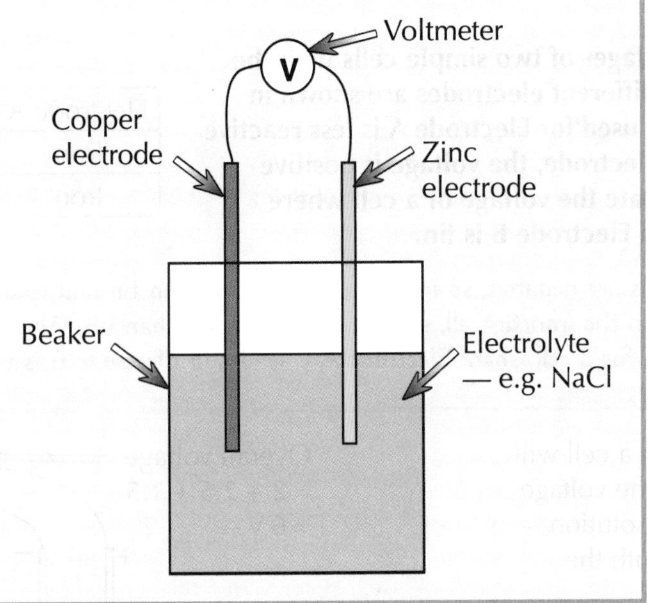

1) The two electrodes must be able to conduct electricity and so are usually metals.

2) The electrolyte is a liquid that contains ions which react with the electrodes.

3) The chemical reactions between the electrodes and the electrolyte set up a charge difference between the electrodes.

4) If the electrodes are then connected by a wire, the charge is able to flow and electricity is produced.

5) A voltmeter can also be connected to the circuit to measure the voltage of the cell.

The two electrodes need to be connected for charge to flow

This stuff can be a bit tricky. Make sure you learn all the parts of a simple cell and what function they have. You could even draw out a chemical cell for yourself to make sure you've learnt it all.

Cells and Batteries

So, now that you know the basics about electrochemical cells, you need to learn a bit more about what affects their voltage, as well as a little bit about batteries.

The **Voltage** of a Cell Depends on Many **Factors**

1) Different metals will react differently with the same electrolyte — this is what causes the charge difference, or the voltage, of the cell. So the type of electrodes used will affect the voltage of the cell.
2) The bigger the difference in reactivity of the electrodes, the bigger the voltage of the cell.

EXAMPLE:

The voltages of two simple cells with the same electrolyte and different electrodes are shown in the table. If the metal used for Electrode A is less reactive than the other metal electrode, the voltage is positive and vice versa. Calculate the voltage of a cell where Electrode A is lead and Electrode B is tin.

Electrode A	Electrode B	Voltage (V)
Iron	Tin	–0.30
Iron	Lead	–0.31

The voltages of both cells are negative, so iron is more reactive than tin and lead. The voltage for the iron/lead cell is more negative than the iron/tin cell, so lead is less reactive than tin. The difference in voltage of these two cells is 0.01 V. So, for a cell where Electrode A is lead and Electrode B is tin, the voltage is +0.01 V.

3) The electrolyte used in a cell will also affect the size of the voltage since different ions in solution will react differently with the metal electrodes used.
4) A battery is formed by connecting two or more cells together in series. The voltages of the cells in the battery are combined so there is a bigger voltage overall.

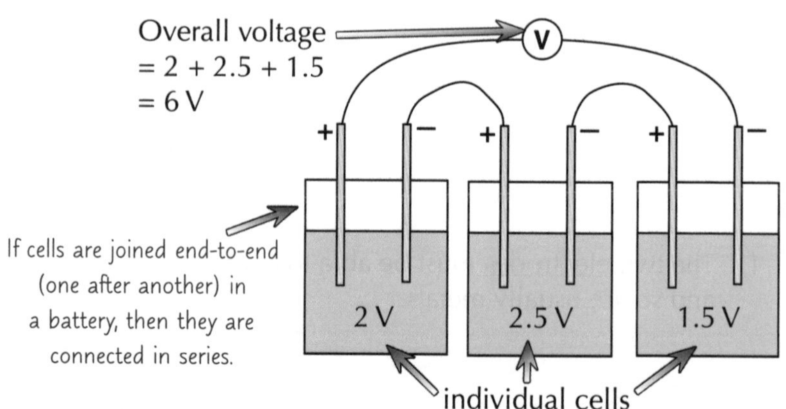

In **Non-Rechargeable** Batteries the **Reactants** Get **Used Up**

1) In some cells, the chemical reactions that happen at the electrodes are irreversible.
2) Over time the reacting particles — the ions in the electrolyte and the metal ions on the electrode — get used up and turned into the products of the reaction.
3) Once any one of the reactants is used up, the reaction can't happen and so no electricity is produced.
4) The products can't be turned back into the reactants, so the cell can't be recharged.
5) A good example of this is what happens with non-rechargeable batteries.
6) In a rechargeable cell, the reaction can be reversed by connecting it to an external electric current.

Non-rechargeable batteries, e.g. alkaline batteries, are made up of cells which use irreversible reactions. Once one of the reactants is used up, they don't produce any more charge and you have to replace them.

Remember, 'voltage' just means the 'charge difference'

Most so-called 'batteries' you use (like the ones in a TV remote) are probably just single cells really. Batteries are made up of several cells joined together, so if you understand how individual cells work, they should be no problem. Remember that the voltage of a battery is the total voltage of its individual cells.

Fuel Cells

We need fuel to move around. Unfortunately, most fuels we use are very polluting — fuel cells might help...

Fuel Cells Use Fuel and Oxygen to Produce Electrical Energy

1) A fuel cell is an electrical cell that's supplied with a fuel and oxygen (or air) and uses energy from the reaction between them to produce electrical energy efficiently.

2) When the fuel enters the cell it becomes oxidised and sets up a potential difference within the cell.

There's an explanation of how these fuel cells work on the next page.

3) There are a few different types of fuel cells, using different fuels and different electrolytes. One of the types is the hydrogen-oxygen fuel cell.

4) This fuel cell combines hydrogen and oxygen to produce nice clean water and energy.

Hydrogen-Oxygen Fuel Cells Can be Used in Vehicles

1) Conventional fuels for vehicles (such as petrol) have a finite supply — they won't last forever, and they're very polluting.
2) This has led to vehicles that use electrical energy becoming more and more popular.
3) Batteries are one way of getting cleaner energy but hydrogen-oxygen fuel cells might be even better:

> 1) Fuel cell vehicles don't produce as many pollutants as other fuels — no greenhouse gases, nitrogen oxides, sulfur dioxide or carbon monoxide.
> 2) The only by-products are water and heat.
> 3) Electric vehicles don't produce many pollutants either — but their batteries are much more polluting to dispose of than fuel cells because they're made from highly toxic metal compounds.
> 4) Batteries in electric vehicles are rechargeable but there's a limit to how many times they can be recharged before they need replacing.
> 5) Batteries are also more expensive to make than fuel cells.
> 6) Batteries store less energy than fuel cells and so would need to be recharged more often — which can take a long time.

Hydrogen-oxygen fuel cells don't produce nasty pollutants

These fuel cells sound great, but don't go thinking they're perfect. Once you've got the hydrogen, they're pretty eco-friendly to use. But producing hydrogen takes a lot of energy (usually from burning fossil fuels). That doesn't mean they won't be more important in the future — but there are some issues to sort out first.

Fuel Cells

You had a nice little intro to fuel cells on the previous page, now it's time to look a bit more in-depth. Roll up your sleeves and get ready to learn how they actually work.

Hydrogen-Oxygen Fuel Cells Involve a Redox Reaction

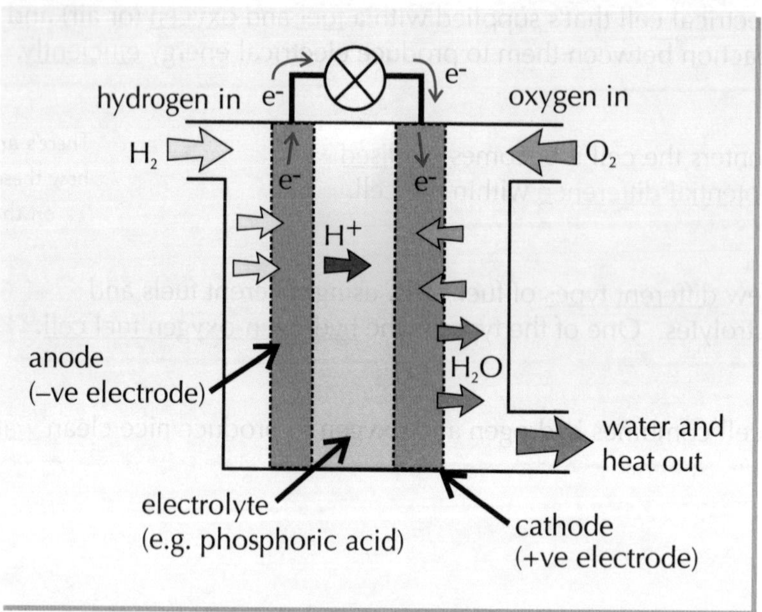

1) The electrolyte is often an acid, such as phosphoric acid, and the electrodes are often porous carbon with a catalyst.

You can also use an alkali (e.g. potassium hydroxide) as the electrolyte. In that case, OH^- ions are produced at the cathode and move through the electrolyte to the anode.

2) Hydrogen goes into the anode compartment and oxygen goes into the cathode compartment.
3) At the negative electrode (the anode), hydrogen loses electrons to produce H^+ ions. This is oxidation.

$$H_2 \rightarrow 2H^+ + 2e^-$$

4) H^+ ions in the electrolyte move to the cathode (+ve).
5) At the positive electrode (the cathode), oxygen gains electrons from the cathode and reacts with H^+ ions (from the acidic electrolyte) to make water. This is reduction.

$$O_2 + 4H^+ + 4e^- \rightarrow 2H_2O$$

These are the half-equations that show what's happening at each electrode.

6) The electrons flow through an external circuit from the anode to the cathode — this is the electric current.
7) The overall reaction is hydrogen plus oxygen, which gives water.

$$2H_2 + O_2 \rightarrow 2H_2O \qquad \text{hydrogen + oxygen} \rightarrow \text{water}$$

There's reduction at the cathode and oxidation at the anode, so the whole thing is a REDOX reaction.

See page 97 for more on redox reactions.

Redox reactions are really important in chemistry, so learn them well

In hydrogen-oxygen fuel cells, different things happen at each electrode — oxidation at the anode and reduction at the cathode. The overall reaction releases electrical energy without a load of pollution.

Warm-Up & Exam Questions

These questions should help you get all of that electrical knowledge into your head.
These Warm-Up questions should get you started and then you can crack on with the Exam questions.

Warm-Up Questions

1) In an electrochemical cell, what is an electrolyte?
2) How do you form a battery from two or more cells?
3) What is a fuel cell?
4) In a hydrogen-oxygen fuel cell, what solution is often used as the electrolyte?

Exam Questions

1 Which of the following can be used to measure the charge difference across a cell?
Tick **one** box.

☐ cathode ☐ electrolyte ☐ anode ☐ voltmeter

[1 mark]

2 Hydrogen-oxygen fuel cells involve redox reactions.
One of the half-equations for the hydrogen-oxygen fuel cell is:

$$O_2 + 4H^+ + 4e^- \rightarrow 2H_2O$$

2.1 Explain why the reaction in a hydrogen-oxygen fuel cell is a redox reaction.

[2 marks]

2.2 Is the half equation shown above an oxidation or a reduction reaction? Explain your choice.

[1 mark]

2.3 What is the overall chemical reaction for a hydrogen-oxygen fuel cell?

[2 marks]

2.4* Explain the advantages of using hydrogen-oxygen fuel cells rather than rechargeable batteries to power cars.

[4 marks]

3 A student investigated the effect of changing one electrode on the voltage produced by a cell.
The other electrode was made of zinc, which is less reactive than the metals tested.
Her results are shown in **Table 1**.

Table 1

Metal used (alongside zinc)	1	2	3
Voltage produced	4 V	2.5 V	3 V

3.1 Write the order of reactivity of the metals used from **lowest** to **highest**.
Explain your answer.

[3 marks]

3.2 Explain how you could construct a 9 V battery using these cells.

[2 marks]

Revision Summary for Topic 5

Right, let's see how much you can remember — don't worry though, you can flick back if you get stuck.

- Try these questions and tick off each one when you get it right.
- When you've done all the questions under a heading and are completely happy with it, tick it off.

Exothermic and Endothermic Reactions (p.106-107) ☑

1) In an exothermic reaction, is energy transferred to or from the surroundings? ☑
2) Name two different types of reaction which are exothermic. ☑
3) Write down the symbol equation for the thermal decomposition of calcium carbonate. ☑

Bond Energies (p.108-109) ☑

4) For the following sentences, use either endothermic or exothermic to fill in the blanks:
   a) Bond breaking is an ________________ process.
   b) Bond forming is an ________________ process.
   c) In an ________________ reaction, the energy released by forming bonds is greater than the energy used to break them. ☑
5) Sketch an energy level diagram for an exothermic reaction. ☑
6) Is the following statement true or false? In an endothermic reaction, the products of the reaction have more energy than the reactants. ☑
7) What three steps would you use to find the overall energy change in a reaction if you were given the known bond energies for the bonds present in the reactants and products? ☑

Cells and Batteries (p.111-112) ☑

8) What can be connected to an electrochemical cell to measure the voltage of the cell? ☑
9) Give two factors that affect the voltage produced by an electrochemical cell. ☑
10) What is a battery? ☑
11) Describe how you would work out the voltage of a battery consisting of three cells with individual voltages of 2 V, 2.5 V and 1.5 V? ☑
12) Explain why non-rechargeable batteries eventually lose their charge. ☑

Fuel Cells (p.113-114) ☑

13) Give one advantage of vehicles which use electrical energy over vehicles which use conventional fuels such as petrol. ☑
14) In a hydrogen-oxygen fuel cell, which molecule becomes oxidised and which becomes reduced? ☑
15) What gas enters the anode compartment in a hydrogen-oxygen fuel cell? ☑
16) Give the half equation for the reaction which takes place at the cathode in a hydrogen-oxygen fuel cell. ☑
17) What are the by-products of a hydrogen-oxygen fuel cell? ☑

Rates of Reaction

Rates of reaction are pretty important. In the chemical industry, the faster you make chemicals, the faster you make money.

Reactions Can Go at All Sorts of **Different Rates**

1) The rate of a chemical reaction is how fast the reactants are changed into products.
2) One of the slowest is the rusting of iron (it's not slow enough though — what about my little Mini).
3) Other slow reactions include chemical weathering — like acid rain damage to limestone buildings.
4) An example of a moderate speed reaction would be the metal magnesium reacting with an acid to produce a gentle stream of bubbles.
5) Burning is a fast reaction, but explosions are even faster and release a lot of gas. Explosive reactions are all over in a fraction of a second.

You Need to Understand **Graphs** for the **Rate of Reaction**

1) You can find the speed of a reaction by recording the amount of product formed, or the amount of reactant used up over time (see page 120).
2) The steeper the line on the graph, the faster the rate of reaction. Over time the line becomes less steep as the reactants are used up.
3) The quickest reactions have the steepest lines and become flat in the least time.
4) The plot below uses the amount of product formed over time to show how the speed of a particular reaction varies under different conditions.

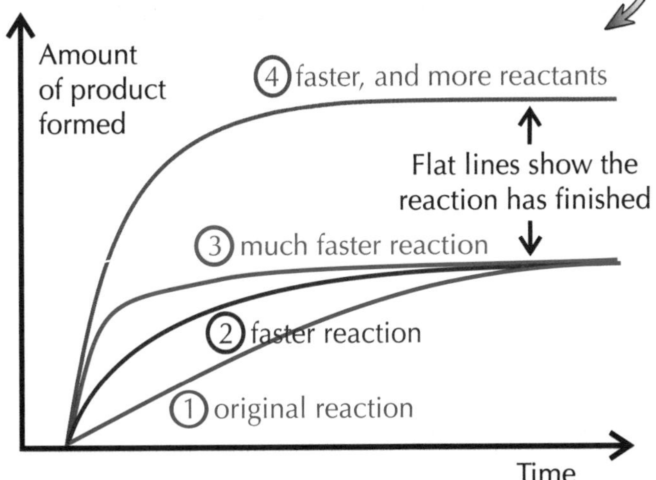

For more on the conditions that affect the rate of reaction — see next page.

- Graph 1 represents the original reaction.
- Graphs 2 and 3 represent the reaction taking place quicker, but with the same initial amounts of reactants. The slopes of the graphs are steeper than for graph 1.
- Graphs 1, 2 and 3 all converge at the same level, showing that they all produce the same amount of product although they take different times to produce it.
- Graph 4 shows more product and a faster reaction. This can only happen if more reactant(s) are added at the start.

Factors Affecting Rates of Reaction

I'd ask you to guess what these two pages are about, but the title pretty much says it all really. Read on...

Particles Must **Collide** with **Enough Energy** in Order to **React**

1) Reaction rates are explained perfectly by collision theory. The rate of a chemical reaction depends on:
    - The collision frequency of reacting particles (how often they collide). The more collisions there are the faster the reaction is. E.g. doubling the frequency of collisions doubles the rate.
    - The energy transferred during a collision. Particles have to collide with enough energy for the collision to be successful.

A successful collision is a collision that ends in the particles reacting to form products.

2) You might remember from page 108 that the minimum amount of energy that particles need to react is called the activation energy. Particles need this much energy to break the bonds in the reactants and start the reaction.
3) Factors that increase the number of collisions (so that a greater proportion of reacting particles collide) or the amount of energy particles collide with will increase the rate of the reaction.

The **Rate of Reaction** Depends on **Four Things**

1) Temperature.
2) The concentration of a solution or the pressure of gas.
3) Surface area — this changes depending on the size of the lumps of a solid.
4) The presence of a catalyst.

More Collisions Increases the **Rate of Reaction**

All four methods of increasing the rate of a reaction can be explained in terms of increasing the number of successful collisions between the reacting particles:

Increasing the **Temperature** Increases the Rate

1) When the temperature is increased, the particles all move faster.
2) If they're moving faster, they're going to collide more frequently.
3) Also the faster they move the more energy they have, so more of the collisions will have enough energy to make the reaction happen.

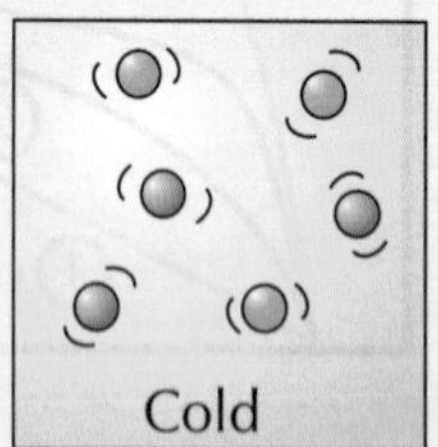

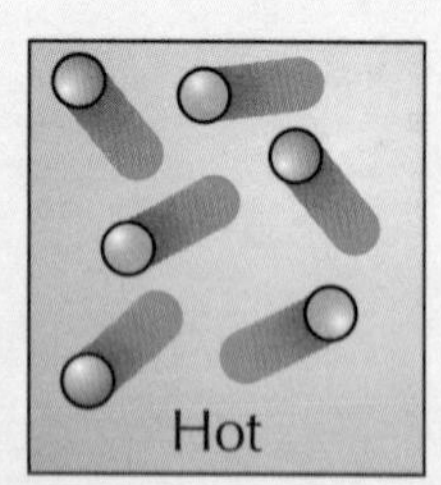

Increasing the **Concentration** or **Pressure** Increases the Rate

1) If a solution is made more concentrated, it means there are more particles knocking about in the same volume of water (or other solvent).
2) Similarly, when the pressure of a gas is increased, it means that the same number of particles occupies a smaller space.
3) This makes collisions between the reactant particles more frequent.

Low concentration/ pressure

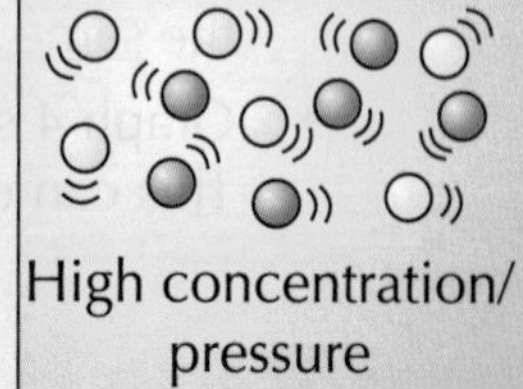

Factors Affecting Rates of Reaction

Increasing the **Surface Area** Increases the Rate

1) If one of the reactants is a solid, then breaking it up into smaller pieces will increase its surface area to volume ratio.
2) This means that for the same volume of the solid, the particles around it will have more area to work on — so there will be collisions more frequently.

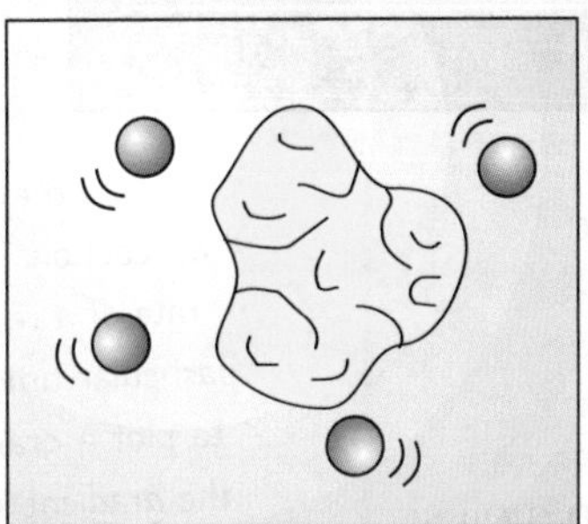
Small surface area

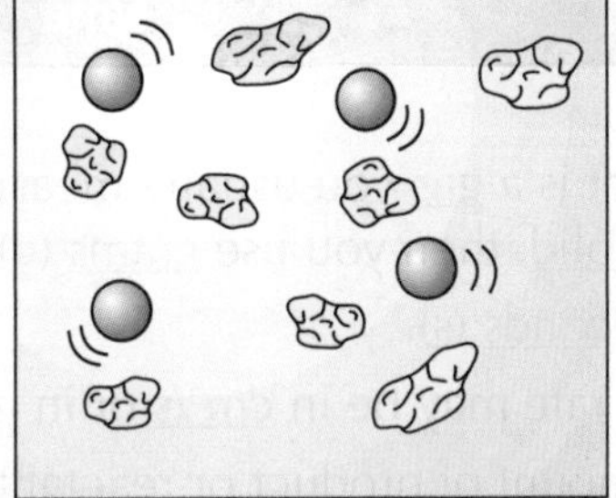
Big surface area

Using a **Catalyst** Increases the Rate

1) A catalyst is a substance that speeds up a reaction, without being used up in the reaction itself. This means it's not part of the overall reaction equation.
2) Different catalysts are needed for different reactions, but they all work by decreasing the activation energy needed for the reaction to occur. They do this by providing an alternative reaction pathway with a lower activation energy.

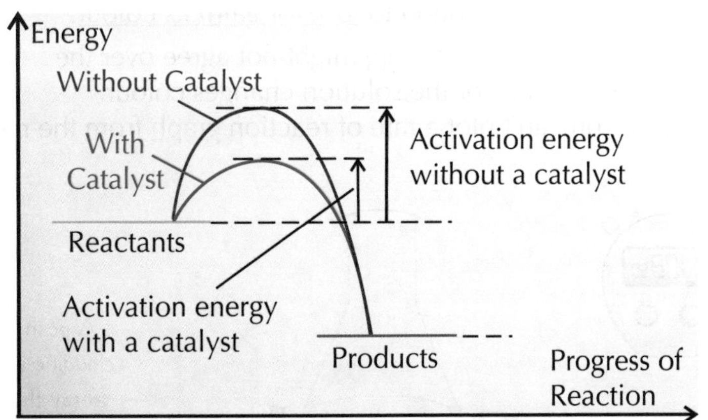

This is a reaction profile. There's more on these on p.108.

3) Enzymes are biological catalysts — they catalyse reactions in living things.

It's easier to learn stuff when you know the reasons for it

Once you've learnt everything off these two pages, the rates of reaction stuff should start making a lot more sense to you. The concept's fairly simple — the more often particles bump into each other, and the harder they hit when they do, the faster the reaction happens.

Measuring Rates of Reaction

All this talk about rates of reactions is fine and dandy, but it's no good if you can't measure it.

You can **Calculate** the Rate of Reaction from **Experimental Results**

1) The rate of a reaction can be observed either by how quickly the reactants are used up or how quickly the products are formed:

$$\text{Rate of Reaction} = \frac{\text{Amount of reactant used or amount of product formed}}{\text{Time}}$$

2) When the product or reactant is a gas you usually measure the amount in cm^3. If it's a solid, then you use grams (g).
3) Time is often measured in seconds (s).
4) This means that the units for rate may be in cm^3/s or in g/s.
5) You can also measure the amount of product or reactant in moles — so the units of rate could also be mol/s.
6) There are three different ways of measuring the rate of a reaction:

This is the mean rate of reaction. To find the rate of a reaction at a particular time, you'll need to plot a graph and find the gradient at that time (see page 124).

1) **Precipitation** and **Colour Change**

1) You can record the visual change in a reaction if the initial solution is transparent and the product is a precipitate which clouds the solution (it becomes opaque).
2) You can observe a mark through the solution and measure how long it takes for it to disappear — the faster the mark disappears, the quicker the reaction.
3) If the reactants are coloured and the products are colourless (or vice versa), you can time how long it takes for the solution to lose (or gain) its colour.
4) The results are very subjective — different people might not agree over the exact point when the mark 'disappears' or the solution changes colour. Also, if you use this method, you can't plot a rate of reaction graph from the results.

A posh way of saying that the cloudiness of a solution changes is to say that its 'turbidity' changes.

Make sure you use a method appropriate to your experiment

The method shown on this page only works if there's a really obvious change in the solution. If there's only a small change in colour, it might not be possible to observe and time the change.

Measuring Rates of Reaction

PRACTICAL

2) Change in Mass (Usually Gas Given Off)

Putting cotton wool in the top of the flask lets the gas escape but stops the acid spitting out.

1) Measuring the speed of a reaction that produces a gas can be carried out using a mass balance.
2) As the gas is released, the mass disappearing is measured on the balance.
3) The quicker the reading on the balance drops, the faster the reaction.
4) If you take measurements at regular intervals, you can plot a rate of reaction graph and find the rate quite easily (see page 124 for more).
5) This is the most accurate of the three methods described because the mass balance is very accurate. But it has the disadvantage of releasing the gas straight into the room.

3) The Volume of Gas Given Off

1) This involves the use of a gas syringe to measure the volume of gas given off.
2) The more gas given off during a given time interval, the faster the reaction.
3) Gas syringes usually give volumes accurate to the nearest cm^3, so they're quite accurate. You can take measurements at regular intervals and plot a rate of reaction graph using this method too. You have to be quite careful though — if the reaction is too vigorous, you can easily blow the plunger out of the end of the syringe.

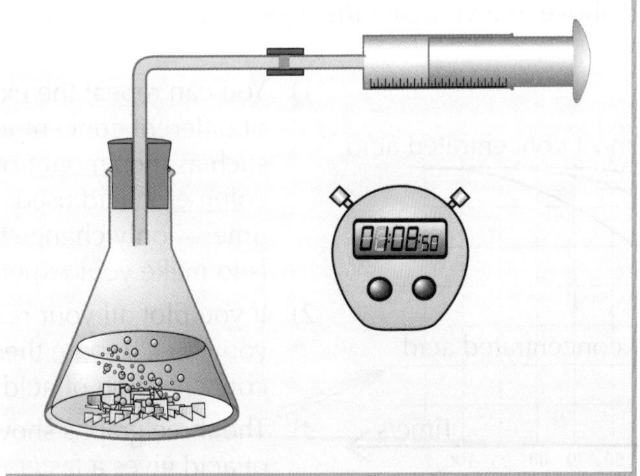

Each of these three methods has pros and cons

The mass balance method is only accurate as long as the flask isn't too hot, otherwise the loss in mass that you see might be partly due to evaporation of liquid as well as being due to the loss of gas formed during the reaction. The first method (on the previous page) is subjective so it isn't very accurate, but if you're not producing a gas you can't use either of the other two.

Rate Experiments

This page shows how you can use the method on the previous page in a real investigation. Get your safety goggles on and let's go...

Magnesium and HCl React to Produce H_2 Gas

1) Start by adding a set volume of dilute hydrochloric acid to a conical flask.
2) Now add some magnesium ribbon to the acid and quickly attach an empty gas syringe to the flask.
3) Start the stopwatch. Take readings of the volume of gas in the gas syringe at regular intervals.

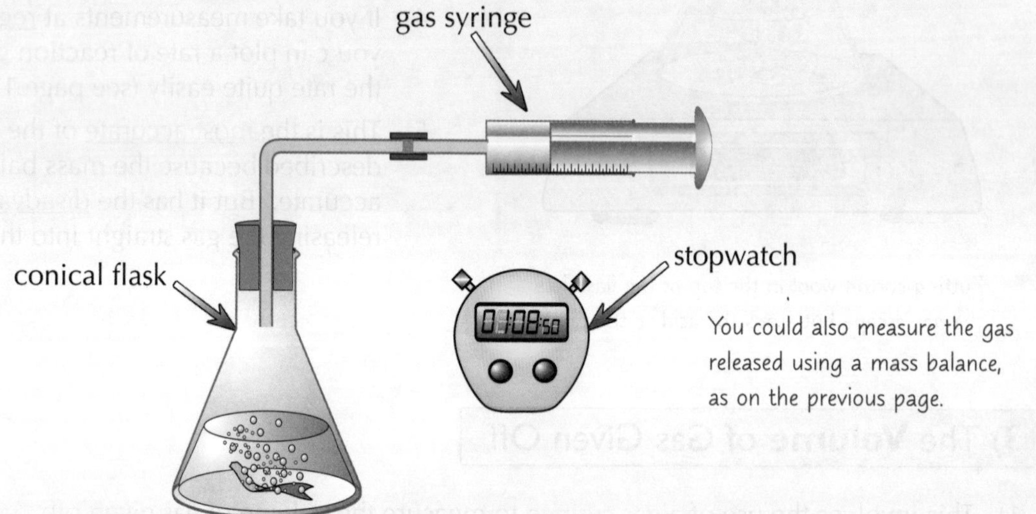

You could also measure the gas released using a mass balance, as on the previous page.

4) Plot the results in a table.
5) Now you can plot a graph with time on the *x*-axis and volume of gas produced on the *y*-axis.

You Can Investigate the Effect of Using Different Acid Concentrations

You can use the method above to investigate the effect of acid concentration on the rate of reaction:

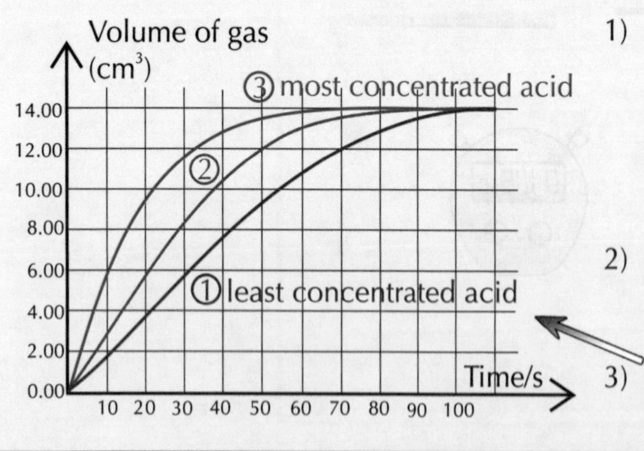

1) You can repeat the experiment above with a number of different concentrations of acid. Variables such as the amount of magnesium ribbon and the volume of acid used should be kept the same each time — only change the acid's concentration. This is to make your experiment a fair test — see p.5.
2) If you plot all your results on the same graph, you can compare them to see how the concentration of acid affects the rate of reaction.
3) The three graphs show that a higher concentration of acid gives a faster rate of reaction.

The hypothesis for this experiment could be something like 'As the concentration of acid increases, the rate of reaction will increase.'

Don't forget about any safety precautions you need to take

The graph above shows how to compare rates of reactions when using different concentrations of acid, but if you want to calculate a numerical value for the rate you need to use a calculation. Take a look at p.124 for how to find the rate of reaction from a graph.

Rate Experiments

Here's how to use the method that you saw on page 120. It's a bit less accurate than using a mass balance, but you need to know how to do it in case you want to investigate a reaction that doesn't produce a gas.

Sodium Thiosulfate and HCl Produce a Cloudy Precipitate

1) These two chemicals are both clear solutions.
They react together to form a yellow precipitate of sulfur.
2) Start by adding a set volume of dilute sodium thiosulfate to a conical flask.
3) Place the flask on a piece of paper with a black cross drawn on it.
4) Add some dilute HCl to the flask and start the stopwatch.
5) Now watch the black cross disappear through the cloudy sulfur and time how long it takes to go.

This reaction releases sulfur dioxide, so the experiment should be carried out in a well-ventilated place.

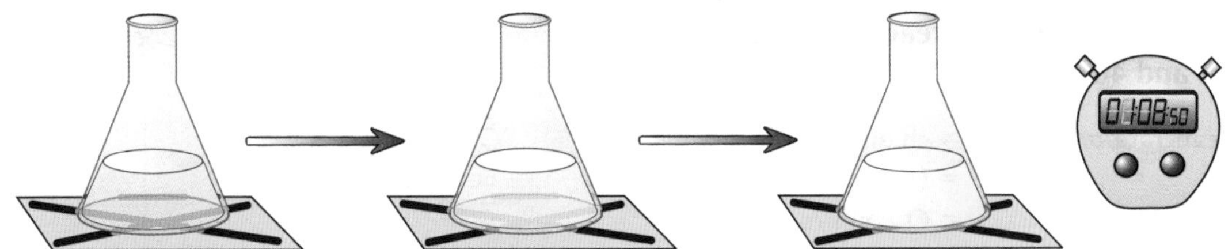

The hypothesis for this experiment could be something like 'As the concentration of acid increases, the solution will go cloudy more quickly.'

You Can Investigate How the Concentration of Acid Affects the Rate

1) The reaction can be repeated with solutions of either reactant at different concentrations. (Only change the concentration of one reactant at a time though). The depth of the liquid must be kept the same each time.
2) These results show the effect of increasing the concentration of HCl on the rate of reaction, when added to an excess of sodium thiosulfate.

Concentration of HCl (mol/dm^3)	0.5	1	1.5	2	2.5
Time taken for mark to disappear (s)	193	184	178	171	164

3) The higher the concentration, the quicker the reaction and therefore the less time it takes for the mark to disappear.
4) One sad thing about this reaction is that it doesn't give a set of graphs. Well I think it's sad. All you get is a set of readings of how long it took till the mark disappeared for each concentration. Boring.

Although you could draw a graph of concentration against 1/time which will give you an approximate rate.

Make sure you can clearly see the cross through the flask at the start

Makes sure you learn both the methods on these two pages for investigating rate, but don't forget that other methods (e.g. the one using a balance that you saw earlier in the topic) can also be used. It's not just the effect of concentration you can investigate with these experiments, either — you can also use them to see how other factors (such as temperature, surface area or the presence of a catalyst) affect rate.

Finding Reaction Rates from Graphs

You might remember a bit about how to interpret graphs on reaction rate from page 117 — well this page shows you how to use them to calculate rates.

You can Calculate the **Mean Reaction Rate** from a **Graph**

1) Remember, a rate of reaction graph shows the amount of product formed or amount of reactant used up on the y-axis and time on the x-axis.
2) So to find the mean rate for the whole reaction, you just work out the overall change in the y-value and then divide this by the total time taken for the reaction.
3) You can also use the graph to find the mean rate of reaction between any two points in time:

The graph shows the volume of gas released by a reaction, measured at regular intervals. Find the mean rate of reaction between 20 s and 40 s.

Mean rate of reaction = change in y ÷ change in x
= ($19\ cm^3 - 15\ cm^3$) ÷ 20 s
= $0.2\ cm^3/s$

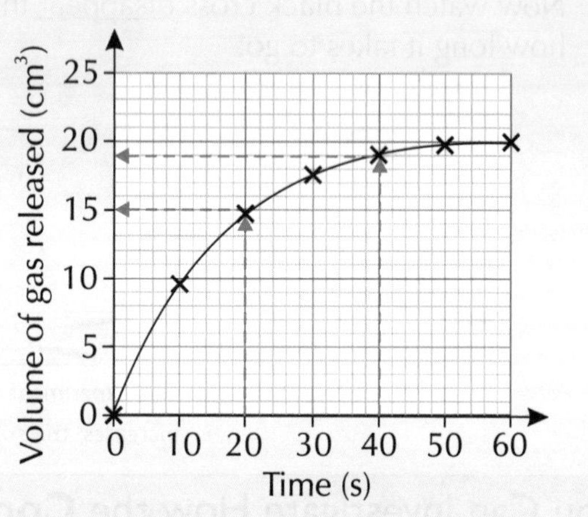

If you're asked to find the mean rate of reaction for the whole reaction, remember that the reaction finishes as soon as the line on the graph goes flat.

Draw a **Tangent** to Find the **Reaction Rate** at a **Particular Point**

If you want to find the rate of the reaction at a particular point in time, you need to find the gradient (slope) of the curve at that point. The easiest way to do this is to draw a tangent to the curve — a straight line that touches the curve at one point and doesn't cross it. You then work out the gradient of the tangent. It's simpler than it sounds, honest...

EXAMPLE:

The graph below shows the mass of reactant used up measured at regular intervals during a chemical reaction. What is the rate of reaction at 3 minutes?

1) Position a ruler on the graph at the point where you want to know the rate — here it's 3 minutes.
2) Adjust the ruler until the space between the ruler and the curve is equal on both sides of the point.
3) Draw a line along the ruler to make the tangent. Extend the line right across the graph.

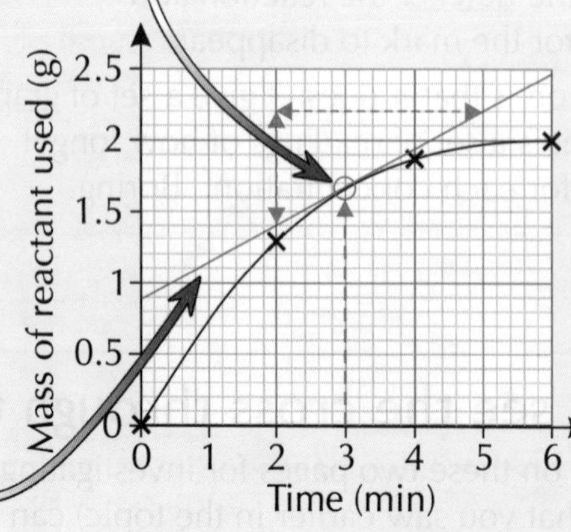

4) Pick two points on the line that are easy to read. Use them to calculate the gradient of the tangent in order to find the rate:

gradient = change in y ÷ change in x
= (2.2 – 1.4) ÷ (5.0 – 2.0)
= 0.8 ÷ 3.0
= 0.27

So, the rate of reaction at 3 minutes was 0.27 g/min.

Warm-Up & Exam Questions

It's easy to think that you've understood something when you've just read through it. These questions should test whether you really understand the previous chunk of pages, and get you set for the next bit.

Warm-Up Questions

1) Give an example of a reaction that happens very slowly, and one that is very fast.
2) According to collision theory, what must happen in order for two particles to react?
3) Why does increasing the concentration of solutions increase the rate of a reaction?
4) What units of rate would be used for a reaction where the change in mass of a reaction vessel (measured in grams) has been recorded over time (measured in seconds)?
5) Describe how you could measure the rate of a precipitation reaction.

Exam Questions

1 **Figure 1** shows one method of measuring the rate of a reaction which produces a gas.

Figure 1

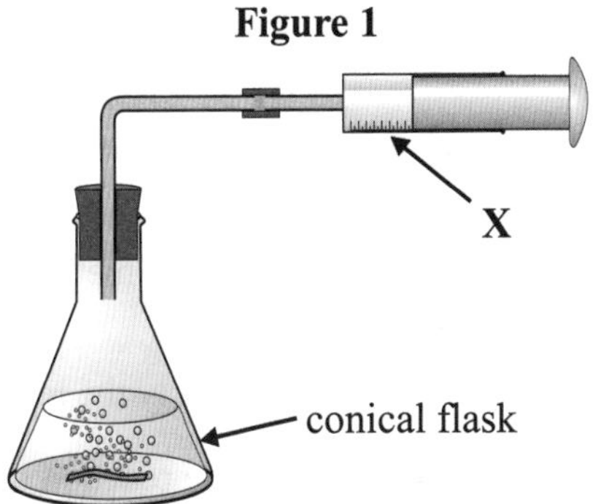

1.1 What piece of apparatus necessary for measuring the rate of this reaction, is missing from **Figure 1**?

[1 mark]

1.2 Name of the piece of apparatus in **Figure 1** labelled **X**.

[1 mark]

1.3 Describe **one** other method of measuring the rate of a reaction which produces a gas.

[2 marks]

2 Set volumes of sodium thiosulfate and hydrochloric acid were reacted at different temperatures. The time taken for a black cross to be obscured by the sulfur precipitate was measured at each temperature. The results are shown in **Table 1**.

Table 1

Temperature (°C)	Time (s)
55	6
36	11
24	17
16	27
9	40
5	51

2.1 Give **two** variables that should be kept constant in this experiment.

[2 marks]

2.2 Plot the results on a graph (with time on the *x*-axis) and draw a line of best fit.

[4 marks]

2.3 Describe the relationship illustrated by your graph.

[1 mark]

2.4 Describe how the results would change if the experiment was repeated with a **lower** concentration of sodium thiosulfate.

[1 mark]

2.5 Explain how using a lower reactant concentration affects the rate of a reaction.

[3 marks]

2.6 Suggest how you could assess if the results of the experiment are repeatable.

[2 marks]

PRACTICAL

Exam Questions

3 A teacher demonstrated an experiment to investigate the effect of temperature on the rate of a reaction. The teacher added dilute hydrochloric acid at 20 °C to marble chips and measured the volume of gas produced at regular time intervals. The teacher then repeated the experiment at 30 °C using the same mass of marble chips of the same size. The results are shown in **Figure 2**.

Figure 2

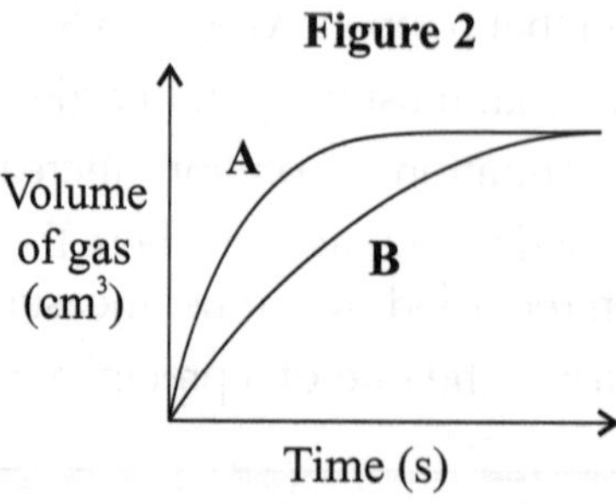

3.1 On **Figure 2**, which curve, **A** or **B**, shows the result of the experiment at 30 °C?

[1 mark]

3.2 Sketch the curve you would expect if you repeated the experiment at 25 °C onto **Figure 2**. Label it **C**.

[1 mark]

3.3 The teacher made sure that the same volume and concentration of acid was used in each repeat. Explain why these variables needed to be controlled.

[1 mark]

3.4 Does **Figure 2** suggest that the teacher successfully controlled these variables? Explain your answer.

[1 mark]

3.5 Which of the following methods could also be used to measure the rate of this reaction? Tick **one** box.

Measuring how quickly the reaction loses mass. ☐

Timing how long the reaction takes to go cloudy. ☐

Timing how long the reaction takes to start. ☐

[1 mark]

4 Calcium carbonate powder was added to a conical flask containing dilute HCl. CO_2 was produced and collected in a gas syringe. The volume of gas released was recorded at 10 second intervals in **Table 2**:

Table 2

Time (s)	0	10	20	30	40	50	60
Volume of CO_2 (cm^3)	0	24	32	36	38	39	40

4.1 Calculate the mean rate of reaction between 0 and 60 seconds.

[2 marks]

4.2 Plot these results on a graph and draw a line of best fit.

[4 marks]

4.3 Find the rate of the reaction 25 seconds after starting the experiment.

[4 marks]

5* Hydrogen gas and ethene gas react to form ethane. Nickel can be used as a catalyst for this reaction.

Using your knowledge of collision theory, explain how the rate of this reaction can be increased.

[6 marks]

Reversible Reactions

Some reactions can go backwards. Honestly, that's all you need...

Reversible Reactions Will Reach Equilibrium

This equation shows a reversible reaction — the products (C and D) can react to form the reactants (A and B) again:

The '⇌' shows the reaction goes both ways.

1) As the reactants react, their concentrations fall — so the forward reaction will slow down (see page 118). But as more and more products are made and their concentrations rise, the backward reaction will speed up.
2) After a while the forward reaction will be going at exactly the same rate as the backward one — the system is at equilibrium.
3) At equilibrium, both reactions are still happening, but there's no overall effect (it's a dynamic equilibrium). This means the concentrations of reactants and products have reached a balance and won't change.
4) Equilibrium is only reached if the reversible reaction takes place in a 'closed system'. A closed system just means that none of the reactants or products can escape and nothing else can get in.

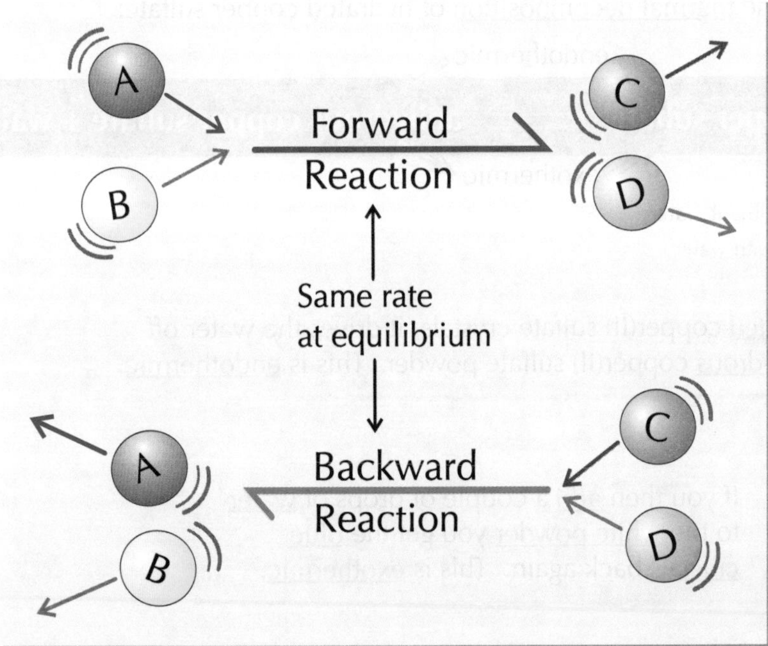

Dynamic equilibrium — lots of activity, but not to any great effect

The idea of dynamic equilibrium is something that you need to get to grips with, as things will get more complicated on the next couple of pages. Have another read and make sure you've got the basics sorted.

Reversible Reactions

In a reversible reaction, both the forward and the reverse reactions are happening at the same time. Certain conditions can be more favourable to one reaction than the other, making it happen faster.

The Position of Equilibrium Can be on the Right or the Left

1) When a reaction's at equilibrium it doesn't mean the amounts of reactants and products are equal.
2) If the equilibrium lies to the right, the concentration of products is greater than that of the reactants.
3) If the equilibrium lies to the left, the concentration of reactants is greater than that of the products.
4) The position of equilibrium depends on the following conditions (as well as the reaction itself):

    1) the temperature,
    2) the pressure (this only affects equilibria involving gases),
    3) the concentration of the reactants and products.

E.g. ammonium chloride ⇌ ammonia + hydrogen chloride
Heating this reaction moves the equilibrium to the right (more ammonia and hydrogen chloride) and cooling it moves it to the left (more ammonium chloride).

The next page tells you why these things affect equilibrium position.

Reversible Reactions Can Be Endothermic and Exothermic

1) In reversible reactions, if the reaction is endothermic in one direction, it will be exothermic in the other.
2) The energy transferred from the surroundings by the endothermic reaction is equal to the energy transferred to the surroundings during the exothermic reaction.
3) A good example is the thermal decomposition of hydrated copper sulfate:

See page 106 for more on endothermic and exothermic reactions.

endothermic

hydrated copper sulfate ⇌ anhydrous copper sulfate + water

exothermic

'Anhydrous' just means 'without water', and 'hydrated' means 'with water'.

If you heat blue hydrated copper(II) sulfate crystals, it drives the water off and leaves white anhydrous copper(II) sulfate powder. This is endothermic.

If you then add a couple of drops of water to the white powder you get the blue crystals back again. This is exothermic.

More of the products = equilibrium lies to the right

This whole energy transfer thing is a fairly simple idea — don't be put off by the long words. Remember, "exo-" = external, "-thermic" = heat, so an exothermic reaction is one that gives out heat. And "endo-" = erm... the other one. OK, there's no easy way to remember that one. Tough.

Le Chatelier's Principle

Reversible reactions don't like being messed around — so if you change something, the system will respond to undo the change.

Reversible Reactions Try to Counteract Changes...

1) Le Chatelier's Principle is the idea that if you change the conditions of a reversible reaction at equilibrium, the system will try to counteract that change.
2) It can be used to predict the effect of any changes you make to a reaction system.

...Such as Changes to the Temperature...

1) All reactions are exothermic in one direction and endothermic in the other (see previous page).
2) If you decrease the temperature, the equilibrium will move in the exothermic direction to produce more heat. This means you'll get more products for the exothermic reaction and fewer products for the endothermic reaction.
3) If you raise the temperature, the equilibrium will move in the endothermic direction to try and decrease it. You'll now get more products for the endothermic reaction and fewer products for the exothermic reaction.

$N_2 + 3H_2 \rightleftharpoons 2NH_3$

Here the forward reaction is exothermic — a decrease in temperature moves equilibrium to the right (more NH_3).

...Pressure...

1) Changing the pressure only affects an equilibrium involving gases.
2) If you increase the pressure, the equilibrium tries to reduce it — it moves in the direction where there are fewer molecules of gas.
3) If you decrease the pressure, the equilibrium tries to increase it — it moves in the direction where there are more molecules of gas.
4) You can use the balanced symbol equation for a reaction to see which side has more molecules of gas.

$N_2 + 3H_2 \rightleftharpoons 2NH_3$

There are 4 moles on the left (1 of N_2 and 3 of H_2) but only 2 on the right. So, if you increase the pressure, the equilibrium shifts to the right (more NH_3).

...or Concentration

1) If you change the concentration of either the reactants or the products, the system will no longer be at equilibrium.
2) So the system responds to bring itself back to equilibrium again.
3) If you increase the concentration of the reactants the system tries to decrease it by making more products.
4) If you decrease the concentration of products the system tries to increase it again by reducing the amount of reactants.

$N_2 + 3H_2 \rightleftharpoons 2NH_3$

If more N_2 or H_2 is added, the forward reaction increases to produce more NH_3.

So, you do one thing, and the reaction does the other...

The best way to get your head around all this is to practise it. So find a reversible reaction, and then think about how changing each condition will affect the position of equilibrium.

Warm-Up & Exam Questions

Not long now 'til this section's over, but first there are some questions for you to tackle.

Warm-Up Questions

1) What can you say about forward and backward reaction rates at equilibrium?
2) For a reversible reaction, what is the effect on equilibrium of removing some of the reactants from the reaction mixture?
3) For the reaction, $N_{2(g)} + O_{2(g)} \rightleftharpoons 2NO_{(g)}$, what would be the effect on the equilibrium of changing the gas pressure?

Exam Questions

1 In the reaction below, substances A and B react to form substances C and D.

$$2A + B \rightleftharpoons 2C + D$$

1.1 What can you deduce about this reaction from the symbol $\rightleftharpoons$?

[1 mark]

1.2 What is meant by the term **dynamic equilibrium**?

[1 mark]

2 This question is about how pressure affects the position of equilibrium.

Reaction 1: $N_2O_{4(g)} \rightleftharpoons 2NO_{2(g)}$

Reaction 2: $ClNO_{2(g)} + NO_{(g)} \rightleftharpoons NO_{2(g)} + ClNO_{(g)}$

2.1 For Reaction 1, explain the effect of an **increase** in pressure on the amount of products at equilibrium.

[2 marks]

2.2 For Reaction 2, explain the effect of a **decrease** in pressure on the amount of products at equilibrium.

[2 marks]

3 When calcium carbonate is heated to a high temperature in a closed system, an equilibrium is reached:

$$CaCO_{3(s)} \rightleftharpoons CaO_{(s)} + CO_{2(g)}$$

The forward reaction is endothermic.

3.1 Does the reverse reaction take in or give out energy? Explain your answer.

[2 marks]

3.2 Explain why changing the temperature of a reversible reaction always affects the position of the equilibrium.

[2 marks]

3.3 For the reaction shown above, describe what would happen to the equilibrium position if the temperature was raised.

[1 mark]

Revision Summary for Topic 6

Well you've almost made it — you're just one more page away from a lovely cup of tea and a biscuit...

- Try these questions and tick off each one when you get it right.
- When you've done all the questions under a heading and are completely happy with it, tick it off.

Rates of Reaction and Factors Affecting Them (p.117-119) ☐

1) On a rate of reaction graph, what does a steep line show? ☑
2) On a rate of reaction graph, what does a flat line show? ☑
3) What two factors relating to the collisions between particles influence the rate of a reaction? ☑
4) What are the four factors that affect the rate of a reaction? ☑
5) Why does increasing the temperature of a reaction mixture increase the rate of a reaction? ☑
6) What is a catalyst? ☑
7) How does a catalyst increase the rate of a reaction? ☑

Measuring and Calculating Rates of Reaction (p.120-124) ☑

8) State the equation that could be used to calculate the mean rate of a reaction. ☑
9) Give three possible units for the rate of a chemical reaction. ☑
10) Explain why measuring a mass change during a reaction is usually an accurate method of measuring rate. ☑
11) Describe how you could investigate the effect of increasing HCl concentration on the rate of reaction between HCl and Mg. ☑
12) Describe how you could use a graph to find the mean rate of a reaction between two points in time. ☑
13) What is a tangent? ☑
14) How would you use a tangent to find the gradient of a curve at a particular point? ☑

Reversible Reactions and Le Chatelier's Principle (p.127-129) ☑

15) Which one of the following statements is true?
    a) In a reaction at equilibrium, there is the same amount of products as reactants.
    b) If the forward reaction in a reversible reaction is exothermic, then the reverse reaction is endothermic.
    c) If the equilibrium of a system lies to the right, then the concentration of products is less than the concentration of reactants. ☑
16) What effect will decreasing the temperature have on a reversible reaction in which the forward reaction is exothermic? ☑
17) In a reversible reaction between gases, how will the equilibrium position change if the pressure is increased? ☑
18) In a reversible reaction, what effect will decreasing the concentration of the products for the forward reaction have on the equilibrium position? ☑

Hydrocarbons

Organic chemistry is about compounds that contain carbon. Hydrocarbons are the simplest organic compounds. As you're about to discover, their properties are affected by their structure.

Hydrocarbons Only Contain Hydrogen and Carbon Atoms

1) A hydrocarbon is any compound that is formed from carbon and hydrogen atoms only.
2) So $C_{10}H_{22}$ (decane, an alkane) is a hydrocarbon, but $CH_3COOC_3H_7$ (an ester) is not — it contains oxygen.

Alkanes Have All C–C Single Bonds

1) Alkanes are the simplest type of hydrocarbon you can get. Their general formula is:

$$C_nH_{2n+2}$$

2) The alkanes are a homologous series — a group of organic compounds that react in a similar way.
3) Alkanes are saturated compounds — each carbon atom forms four single covalent bonds.
4) The first four alkanes are methane, ethane, propane and butane.

A drawing showing all the atoms and bonds in a molecule is called a displayed formula.

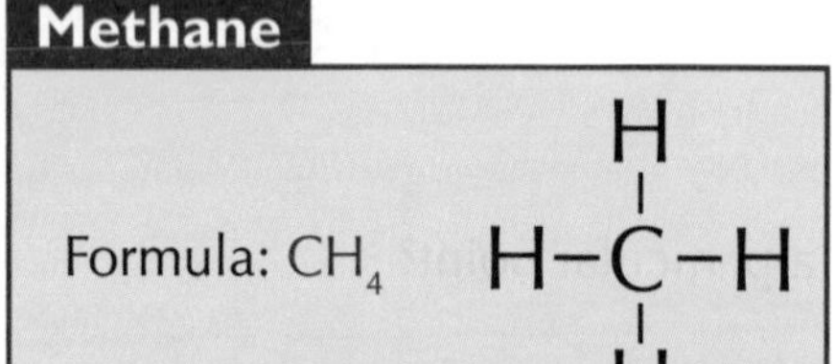

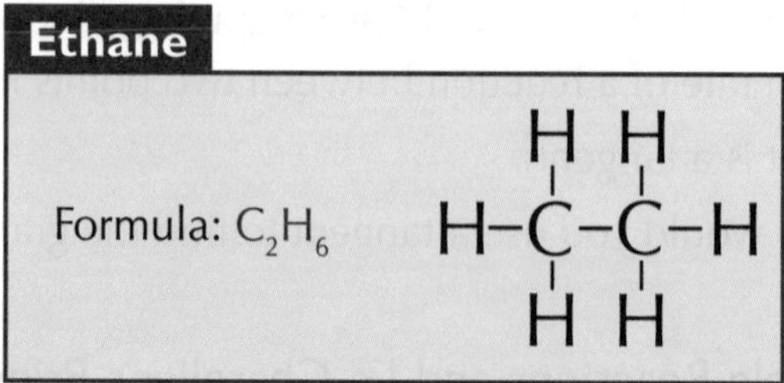

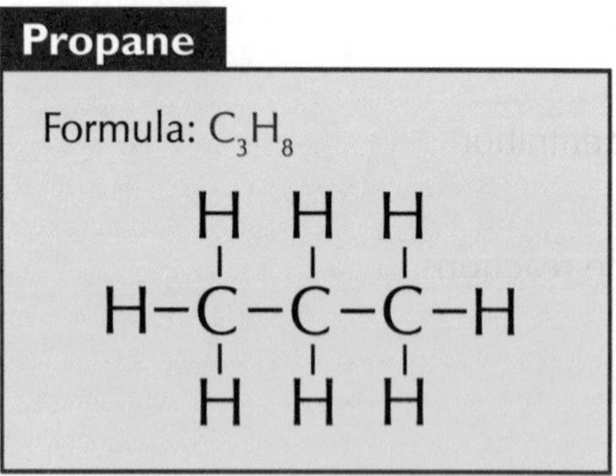

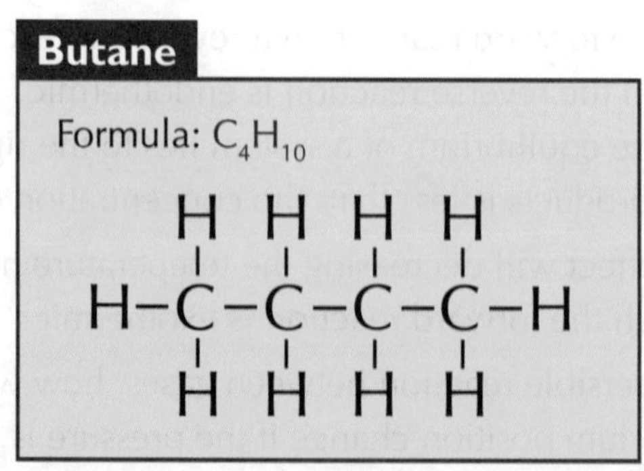

Hydrocarbons only contain hydrogen and carbon

To help remember the names of the first four alkanes just remember: Mice Eat Peanut Butter. Practice drawing the structures and get to grips with their formulas using the general formula.

Hydrocarbons

Hydrocarbon **Properties Change** as the Chain Gets **Longer**

As the length of the carbon chain changes, the properties of the hydrocarbon change:

1) The shorter the carbon chain, the more runny the hydrocarbon is — that is, the less viscous (gloopy) it is.

2) The shorter the carbon chain, the more volatile the hydrocarbon is. "More volatile" means it turns into a gas at a lower temperature. So, the shorter the carbon chain, the lower the temperature at which that hydrocarbon vaporises or condenses — and the lower its boiling point.

3) Also, the shorter the carbon chain, the more flammable (easier to ignite) the hydrocarbon is.

The properties of hydrocarbons affect how they're used for fuels. E.g. short chain hydrocarbons with lower boiling points are used as 'bottled gases' — stored under pressure as liquids in bottles.

Complete Combustion Occurs When There's Plenty of **Oxygen**

1) The complete combustion of any hydrocarbon in oxygen releases lots of energy. The only waste products are carbon dioxide and water vapour.

hydrocarbon + oxygen → carbon dioxide + water (+ energy)

2) During combustion, both carbon and hydrogen from the hydrocarbon are oxidised. Oxidation can be defined as the gain of oxygen.
3) Hydrocarbons are used as fuels due to the amount of energy released when they combust completely.
4) You need to be able to give a balanced symbol equation for the complete combustion of a simple hydrocarbon fuel when you're given its molecular formula. It's pretty easy — here's an example:

EXAMPLE:

Write a balanced equation for the complete combustion of methane (CH_4).

1) On the left hand side, there's one carbon atom, so only one molecule of CO_2 is needed to balance this. $CH_4 + ?O_2 \rightarrow CO_2 + ?H_2O$

2) On the left hand side, there are four hydrogen atoms, so two water molecules are needed to balance them. $CH_4 + ?O_2 \rightarrow CO_2 + 2H_2O$

3) There are four oxygen atoms on the right hand side of the equation. Two oxygen molecules are needed on the left to balance them. $CH_4 + 2O_2 \rightarrow CO_2 + 2H_2O$

Alkanes are useful fuels as they release energy when burnt

So shorter hydrocarbons are less viscous, more volatile and easier to ignite than longer hydrocarbons.

Fractional Distillation

Crude oil can be used to make loads of useful things, such as fuels. But you can't just put crude oil in your car. First, the different hydrocarbons have to be separated. That's where fractional distillation comes in.

Crude Oil is Made Over a Long Period of Time

1) Crude oil is a fossil fuel. It's formed from the remains of plants and animals, mainly plankton, that died millions of years ago and were buried in mud. Over millions of years, with high temperature and pressure, the remains turn to crude oil, which can be drilled up from the rocks where it's found.
2) Fossil fuels like coal, oil and gas are called non-renewable fuels as they take so long to make that they're being used up much faster than they're being formed. They're finite resources (see p.178) — one day they'll run out.

Fractional Distillation is Used to Separate Hydrocarbon Fractions

1) Crude oil is a mixture of lots of different hydrocarbons, most of which are alkanes.
2) The different compounds in crude oil are separated by fractional distillation.

Hydrocarbons are molecules containing only hydrogen and carbon.

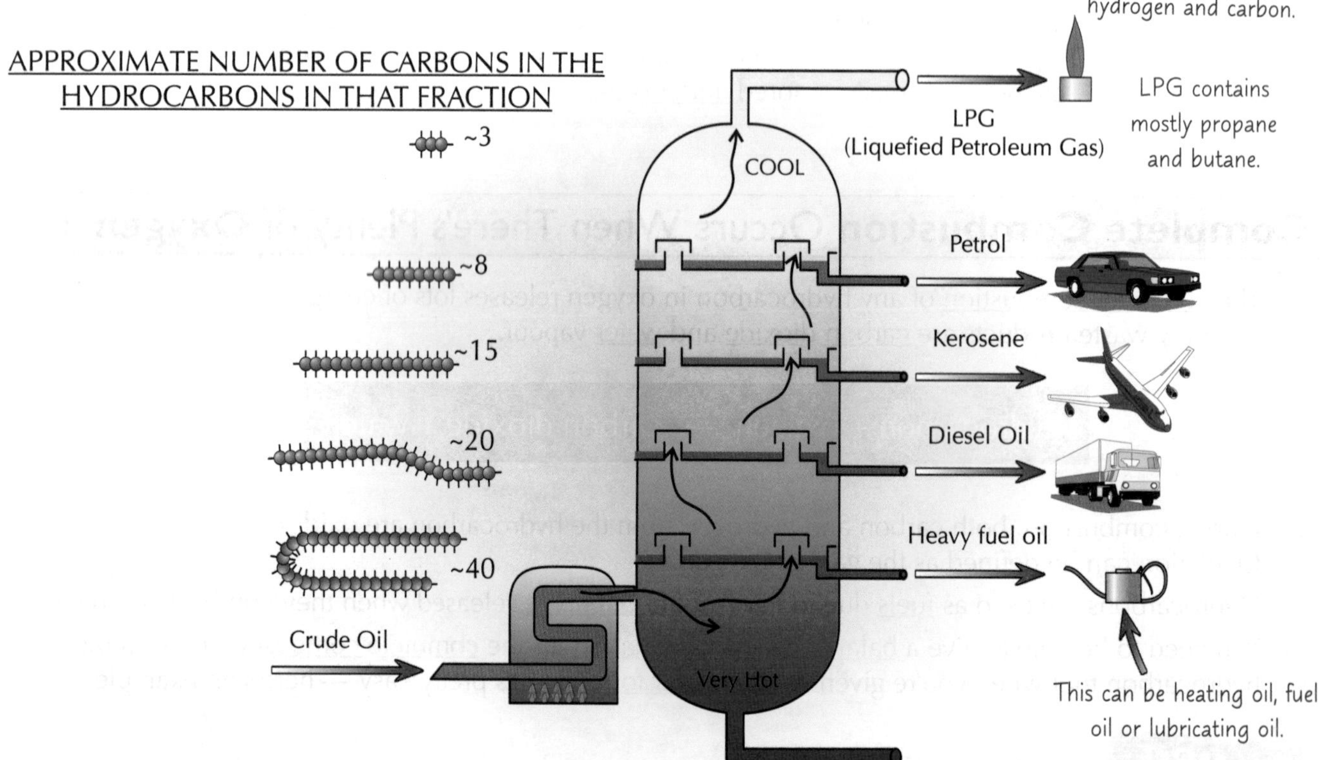

Here's how it works:

- The oil is heated until most of it has turned into gas. The gases enter a fractionating column (and the liquid bit is drained off).
- In the column there's a temperature gradient (it's hot at the bottom and gets cooler as you go up).
- The longer hydrocarbons have high boiling points. They condense back into liquids and drain out of the column early on, when they're near the bottom. The shorter hydrocarbons have lower boiling points. They condense and drain out much later on, near to the top of the column where it's cooler.
- You end up with the crude oil mixture separated out into different fractions. Each fraction contains a mixture of hydrocarbons that all contain a similar number of carbon atoms, so have similar boiling points.

Uses and Cracking of Crude Oil

Crude oil has fuelled modern civilisation — it would be a very different world if we hadn't discovered oil.

Crude Oil has Various Uses Important in Modern Life

1) Oil provides the fuel for most modern transport — cars, trains, planes, the lot. Diesel oil, kerosene, heavy fuel oil and LPG (liquid petroleum gas) all come from crude oil.
2) The petrochemical industry uses some of the hydrocarbons from crude oil as a feedstock to make new compounds for use in things like polymers, solvents, lubricants, and detergents.
3) All the products you get from crude oil are examples of organic compounds (compounds containing carbon atoms). The reason you get such a large variety of products is because carbon atoms can bond together to form different groups called homologous series. These groups contain similar compounds with many properties in common. Alkanes, alkenes, as well as the other families you'll meet in this topic, are all examples of different homologous series.

Cracking Means Splitting Up Long-Chain Hydrocarbons

1) Short-chain hydrocarbons are flammable so make good fuels and are in high demand. However, long-chain hydrocarbons form thick gloopy liquids like tar which aren't all that useful.
2) As a result of this a lot of the longer alkane molecules produced from fractional distillation are turned into smaller, more useful ones by a process called cracking.
3) As well as alkanes, cracking also produces another type of hydrocarbon called alkenes. Alkenes are used as a starting material when making lots of other compounds and can be used to make polymers (see p.141).
4) Some of the products of cracking are useful as fuels, e.g. petrol for cars and paraffin for jet fuel.

You can test for alkenes using bromine water. See p.140.

There are Different Methods of Cracking

1) Cracking is a thermal decomposition reaction — breaking molecules down by heating them.
2) The first step is to heat long-chain hydrocarbons to vaporise them (turn them into a gas).
3) Then the vapour can be passed over a hot powdered aluminium oxide catalyst.
4) The long-chain molecules split apart on the surface of the specks of catalyst — this is catalytic cracking.
5) You can also crack hydrocarbons if you vaporise them, mix them with steam and then heat them to a very high temperature. This is known as steam cracking.

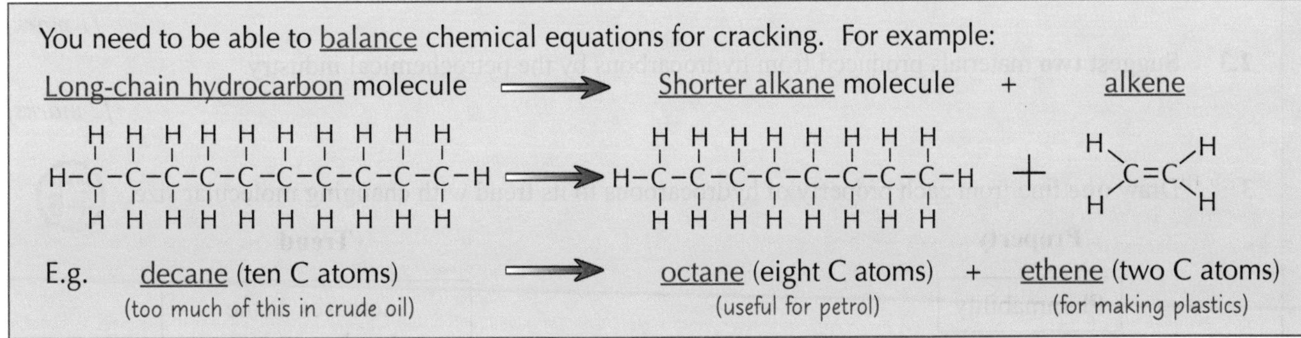

You need to be able to balance chemical equations for cracking. For example:
Long-chain hydrocarbon molecule ⟹ Shorter alkane molecule + alkene
E.g. decane (ten C atoms) ⟹ octane (eight C atoms) + ethene (two C atoms)
(too much of this in crude oil)
(useful for petrol)
(for making plastics)

Make sure that, when writing equations for cracking, there are the same number of carbon and hydrogen atoms on both sides of the equation.

Cracking breaks down long hydrocarbons into shorter ones

Hydrocarbons are used in all areas of industry. Unfortunately it's really only the smaller molecules which are really useful. Luckily, we have cracking to get as many useful products from crude oil as possible.

Warm-Up & Exam Questions

Hydrocarbons contain only hydrogen and carbon atoms. This page contains only Warm-Up and Exam Questions. Time to get thinking.

Warm-Up Questions

1) Name the first four alkanes.
2) What kind of bonds are present in alkanes?
3) Why are alkanes often used as fuels?
4) Suggest a use of crude oil fractions that modern society depends on.
5) What sort of hydrocarbon molecules are cracked?

Exam Questions

1 Choose the statement which best describes the bonding in alkanes. (Grade 4-6)
Tick **one** box.

Carbon-carbon single bonds and carbon-hydrogen single bonds. ☐

Carbon-carbon single bonds and carbon-hydrogen double bonds. ☐

Carbon-carbon double bonds and carbon-hydrogen single bonds. ☐

Carbon-carbon double bonds and carbon-hydrogen double bonds. ☐

[1 mark]

2 Alkanes are a homologous series of hydrocarbons made up of chains of carbon atoms surrounded by hydrogen atoms. (Grade 4-6)

Pentane is an alkane with the formula C_5H_{12}.

2.1 State the name of the alkane that comes before pentane in the homologous series.

[1 mark]

2.2 State the formula of the alkane that comes before pentane in the homologous series.

[1 mark]

2.3 Suggest **two** materials produced from hydrocarbons by the petrochemical industry.

[2 marks]

3 Draw **one** line from each property of hydrocarbons to its trend with changing molecular size. (Grade 4-6)

Property	Trend
Flammability	Increases as the molecules get bigger
Viscosity	
Boiling point	Decreases as the molecules get bigger

[3 marks]

Exam Questions

4 Crude oil can be separated into a number of different compounds in a fractional distillation column. **Figure 1** shows a fractional distillation column. 

Figure 1

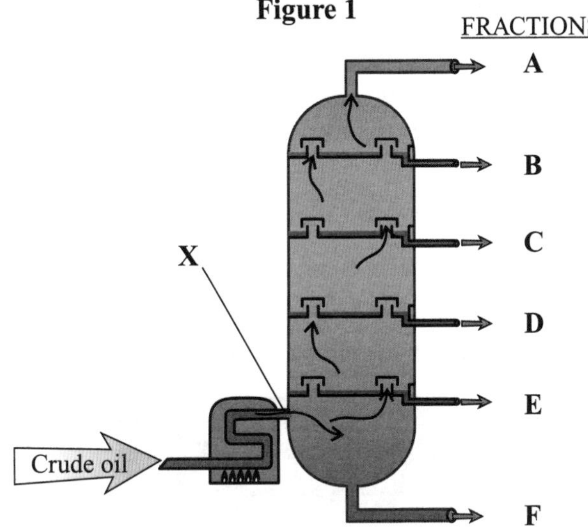

4.1 Which letter, **A-F**, represents the fraction with the longest hydrocarbon molecules?

[1 mark]

4.2 Which letter, **A-F**, represents the fraction with the lowest boiling point?

[1 mark]

4.3 Gaseous crude oil enters near the bottom of the fractional distillation column (point **X** in **Figure 1**). Explain why different fractions exit the column at different points and how this relates to their structure.

[3 marks]

5 Alkanes burn in excess oxygen by complete combustion. 

5.1 What two products are made by the complete combustion of an alkane?

[2 marks]

5.2 Complete the equation below for the complete combustion of propane (C_3H_8):

$$C_3H_8 + \ldots\ldots O_2 \rightarrow \ldots\ldots\ldots\ldots + \ldots\ldots\ldots\ldots$$

[2 marks]

5.3 Carbon and hydrogen atoms gain oxygen during the combustion of hydrocarbons. What type of reaction is this?

[1 mark]

6 Cracking is a process used to break longer chain molecules in crude oil down to shorter ones. 

6.1 Why is cracking an important process in the petrochemical industry?

[1 mark]

6.2 Octane (C_8H_{18}) can be cracked to form two products, ethane and propene (C_3H_6). Complete the equation below for this reaction:

$$C_8H_{18} \rightarrow \ldots\ldots\ldots\ldots\ldots + \ldots\ldots C_3H_6$$

[2 marks]

6.3 Describe the **two** methods used by the petrochemical industry for cracking hydrocarbons.

[4 marks]

Alkenes and their Reactions

Alkenes are unsaturated because they have a double carbon-carbon bond. They're hydrocarbons, like alkanes.

Alkenes Have a C=C Double Bond

1) Alkenes are hydrocarbons which have a double bond between two of the carbon atoms in their chain.
2) The C=C double bond means that alkenes have two fewer hydrogens compared with alkanes containing the same number of carbon atoms. This makes them unsaturated.
3) The C=C double bond can open up to make a single bond, allowing the two carbon atoms to bond with other atoms (see the next page). This makes alkenes reactive — far more reactive than alkanes.
4) The first four alkenes are ethene (with two carbon atoms), propene (three Cs), butene (four Cs) and pentene (five Cs).
5) Straight-chain alkenes have twice as many hydrogen atoms as carbon:

6) Here are the structures of the first four alkenes:

This is a double bond.

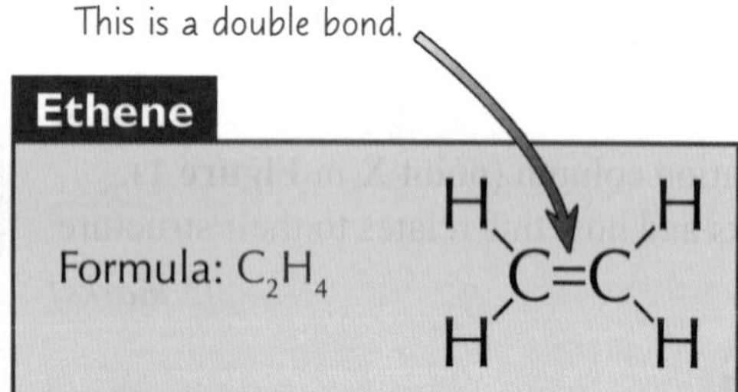

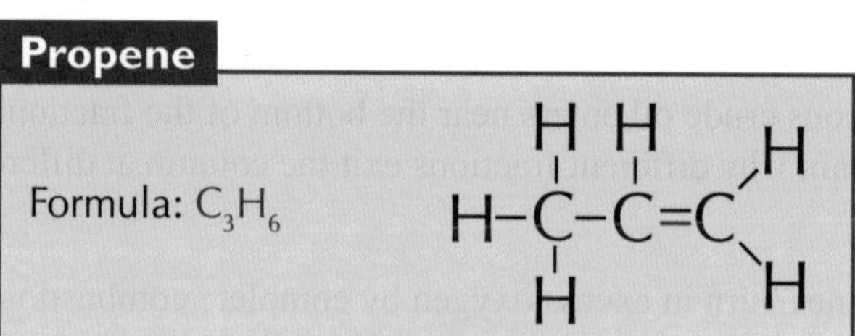

Butene

Formula: C_4H_8

H-C-C=C-C-H Or C=C-C-C-H

Pentene

Formula: C_5H_{10}

C=C-C-C-C-H Or H-C-C=C-C-C-H

There are two different structures for butene (C_4H_8) and pentene (C_5H_{10}) as the double bond can be in two different places.

That double bond makes all the difference...

Don't get alkenes confused with alkanes. Alkenes have a C=C bond, alkanes don't.

Alkenes and their Reactions

The double bond means that alkenes have reactions with lots of different compounds.

Alkenes Burn With a Smoky Flame

1) In a large amount of oxygen, alkenes combust completely to produce only water and carbon dioxide (see p.133).
2) However, when you burn them in air they tend to undergo incomplete combustion. Carbon dioxide and water are still produced, but you can also get carbon and carbon monoxide (CO) which is a poisonous gas.
3) Here's the standard equation for incomplete combustion of alkenes:

$$\text{alkene} + \text{oxygen} \rightarrow \text{carbon} + \text{carbon monoxide} + \text{carbon dioxide} + \text{water} \quad (+\ \text{energy})$$

4) Incomplete combustion results in a smoky yellow flame, and less energy being released compared to complete combustion of the same compound.
5) Here's an example of an equation for incomplete combustion of butene:

$$C_4H_8 + 5O_2 \rightarrow 2CO + 2CO_2 + 4H_2O$$

6) The example above is just one possibility. The products depend on how much oxygen is present. For example, you could also have: $C_4H_8 + 3O_2 \rightarrow 2C + 2CO + 4H_2O$ — but the equation always has to be balanced.

Alkenes React via Addition Reactions

1) A functional group is a group of atoms in a molecule that determines how that molecule typically reacts.
2) All alkenes have the functional group 'C=C', so they all react in similar ways. So you can suggest the products of a reaction based on your knowledge of how alkenes react in general.
3) Most of the time, alkenes react via addition reactions. The carbon-carbon double bond will open up to leave a single bond and a new atom is added to each carbon:

R = Carbon chain

$$\text{R(H)C=CH}_2 + \text{X–Y} \longrightarrow \text{R–C(Y)(H)–C(H)(X)–H}$$

Alkenes are a homologous series because they all have the same functional group and react in similar ways.

Addition of Hydrogen is Known as Hydrogenation

Hydrogen can react with the double-bonded carbons to open up the double bond and form the equivalent, saturated, alkane. The alkene is reacted with hydrogen in the presence of a catalyst:

$$\text{R(H)C=CH}_2 + H_2 \xrightarrow{\text{catalyst}} \text{R–C(H)(H)–C(H)(H)–H}$$

Alkenes and their Reactions

Halogens can React with Alkenes

1) Alkenes will also react in addition reactions with halogens such as bromine, chlorine and iodine. The molecules formed are saturated, with the C=C carbons each becoming bonded to a halogen atom.
2) For example, bromine and ethene react together to form dibromoethane:

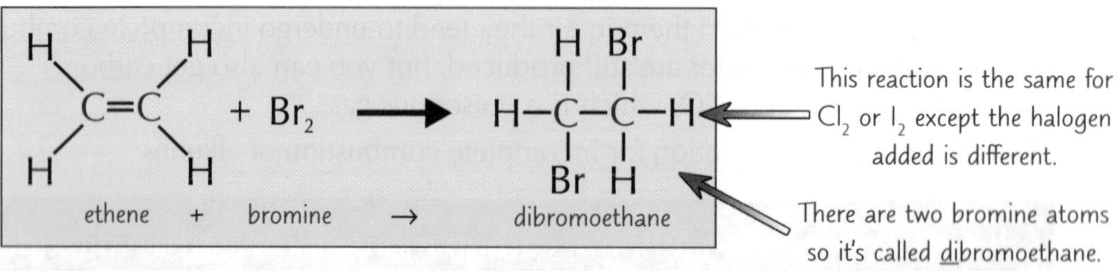

+ Br_2
ethene + bromine → dibromoethane
This reaction is the same for Cl_2 or I_2 except the halogen added is different.
There are two bromine atoms so it's called dibromoethane.

Alkenes Turn Bromine Water Colourless

The addition of bromine to a double bond can be used to test for alkenes:

1) When orange bromine water is added to a saturated compound, like an alkane, no reaction will happen and it'll stay bright orange.
2) If it's added to an alkene the bromine will add across the double bond, making a colourless dibromo-compound — so the bromine water is decolourised.

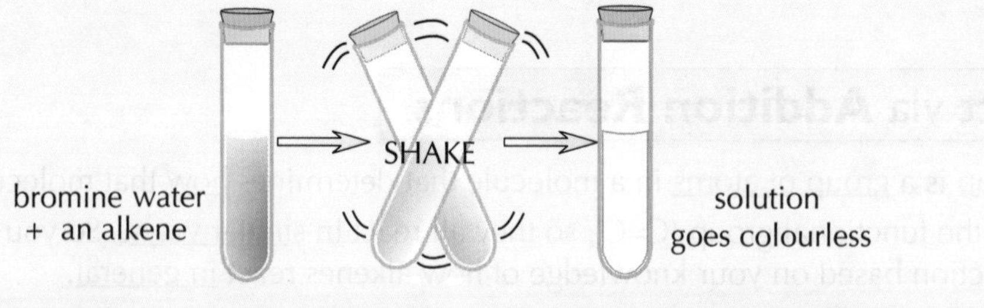

Steam can React with Alkenes to Form Alcohols

1) When alkenes react with steam, water is added across the double bond and an alcohol is formed.
2) For example, ethanol can be made by mixing ethene with steam and then passing it over a catalyst:

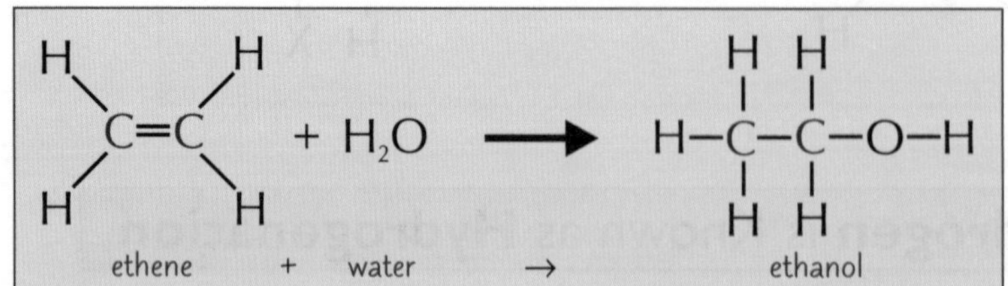

Have a look at p.144-145 for more on alcohols

3) The conversion of ethene to ethanol is one way of making ethanol industrially. After the reaction has taken place, the reaction mixture is passed from the reactor into a condenser. Ethanol and water have a higher boiling point than ethene, so both condense whilst any unreacted ethene gas is recycled back into the reactor. The alcohol can then be purified from the mixture by fractional distillation.

A few more reactions to learn here

A really important test is using bromine water to see if you've got an alkane or an alkene. With the alkane, the bromine water will stay orange whereas with an alkene the solution will turn colourless.

Addition Polymers

Polymers are made up of lots of the same molecule joined together in one long chain. They're what make up plastics. They have lots of weird and wonderful properties which make them very useful to modern society.

Plastics are Made Up of Long-Chain Molecules Called Polymers

1) Polymers are long molecules formed when lots of small molecules called monomers join together. This reaction is called polymerisation — and it usually needs high pressure and a catalyst.
2) Plastics are made up of polymers. They're usually carbon based and their monomers are often alkenes (see page 138).

Addition Polymers are Made From Unsaturated Monomers

1) The monomers that make up addition polymers have a double covalent bond.

2) Lots of unsaturated monomer molecules (alkenes) can open up their double bonds and join together to form polymer chains. This is called addition polymerisation.

For example, lots of ethene molecules can react together to form poly(ethene) (or polythene):

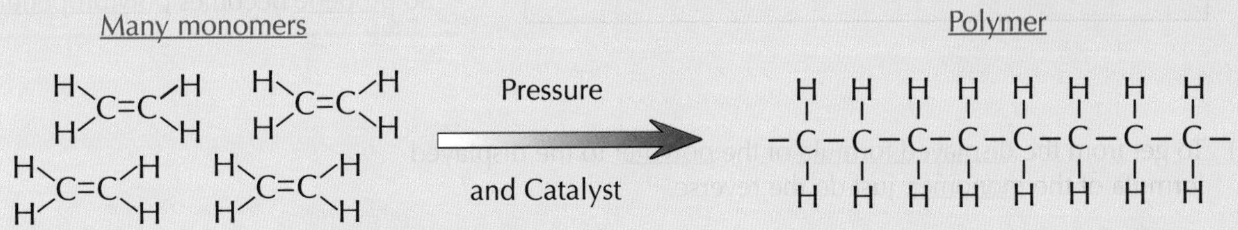

Many monomers

Pressure and Catalyst

Polymer

The reaction can also be shown like this:

The 'n' just means there can be any number of monomers.

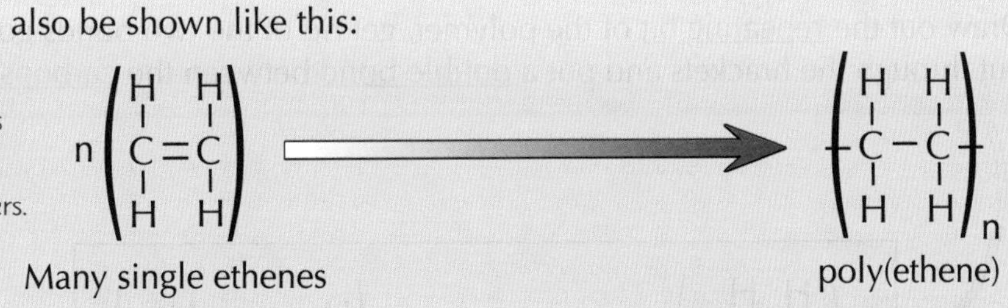

Many single ethenes

poly(ethene)

This is a repeating unit. It's the standard way of representing the structure of a polymer (see page 57).

3) When the monomers react in addition polymerisation reactions, the only product is the polymer, so an addition polymer contains exactly the same type and number of atoms as the monomers that formed it.

Addition polymerisation usually needs high pressure and a catalyst

The monomer a polymer is made from affects the properties of the plastic — as there are lots of monomers, there are many different plastics. Make sure you know the difference between a monomer and a polymer.

Addition Polymers

Polymers aren't usually drawn in diagrams as long chains — they're represented using repeating units.

You May Need to Draw the **Repeating Unit** of a Polymer

1) Drawing the displayed formula of an addition polymer from the displayed formula of its monomer is easy.

- Start by drawing the two alkene carbons, replace the double bond with a single bond and add an extra single bond to each of the carbons.
- Then fill in the rest of the groups in the same way that they surrounded the double bond in the monomer.
- Finally, stick a pair of brackets around the repeating bit, and put an 'n' after it (to show that there are lots of monomers).

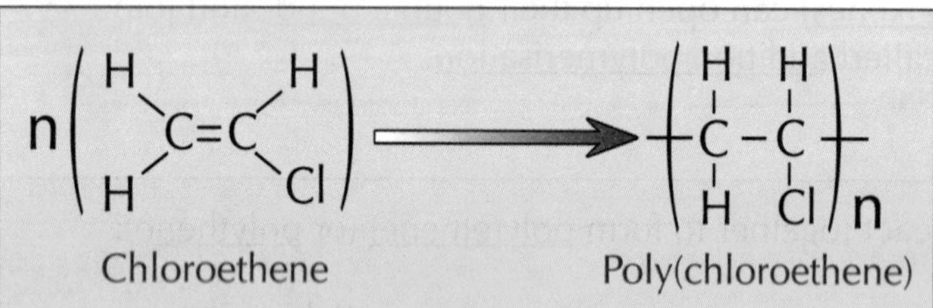

Chloroethene
Poly(chloroethene)

The name of the polymer comes from the type of monomer it's made from — you just stick the word 'poly' in front of it and put the monomer name in brackets. So propene becomes poly(propene).

2) To get from the displayed formula of the polymer to the displayed formula of the monomer, just do the reverse.

Draw out the repeating bit of the polymer, get rid of the two bonds going out through the brackets and put a double bond between the carbons.

H H
C–C
H CH_3
n
n
H
H
C=C
H
CH_3

Poly(propene)
Propene

Addition polymer carbon chains have no C=C bonds

REVISION TIP

Make sure you can recognise and draw polymers and their monomers by practising lots of times.

Warm-Up & Exam Questions

Alkenes and addition polymers are really useful compounds — you need to know how they are made, some of their reactions and how to draw them. Try these questions to test yourself out on.

Warm-Up Questions

1) What are alkenes?
2) Why is an alkene described as unsaturated?
3) What happens to alkenes when they react with halogens?
4) What will you observe if you add propene to bromine water?

Exam Questions

1 Complete the table to show the missing information for the two alkenes given. 

Name of alkene	Formula	Displayed formula
ethene		
	C_3H_6	

[4 marks]

2 Ethene can be used to make ethanol. 

2.1 Write a chemical equation for the conversion of ethene into ethanol. Include state symbols.

[2 marks]

2.2 State the name for this type of reaction.

[1 mark]

2.3 Ethene can also be used to make ethane.
Write a chemical equation for the conversion of ethene into ethane.

[1 mark]

2.4 State why alkenes tend to burn with a smoky yellow flame in air.

[1 mark]

3 Propene can undergo an addition polymerisation reaction to form a polymer. 

3.1 Draw the bonds to complete the displayed formula of the polymer formed in this addition polymerisation reaction.

$$\left(\begin{array}{ll} H & H \\ C & C \\ CH_3 & H \end{array}\right)_n$$

[1 mark]

3.2 Name the polymer formed in this reaction.

[1 mark]

Alcohols

Alcohols are yet another homologous series of organic compounds. They're handy for lots of things, and you need to know how they're used and their properties.

Alcohols Have an '-OH' Functional Group and End in '-ol'

1) The general formula of an alcohol is:

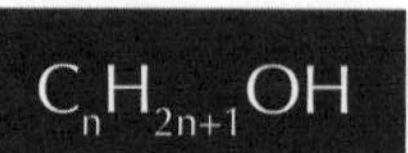

So an alcohol with 2 carbons has the formula C_2H_5OH.

2) All alcohols contain an -OH group.
3) Here are the first 4 alcohols in the homologous series:

Remember — a homologous series is a group of chemicals that react in a similar way because they have the same functional group (in alcohols it's the -OH group).

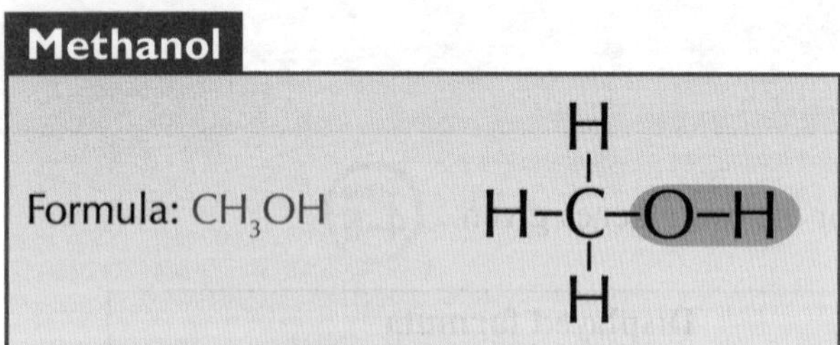

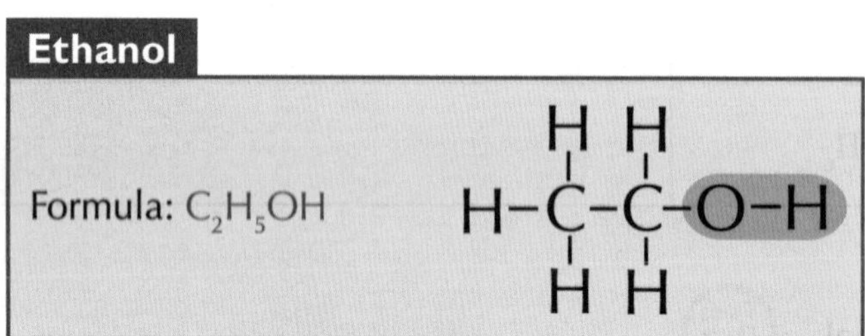

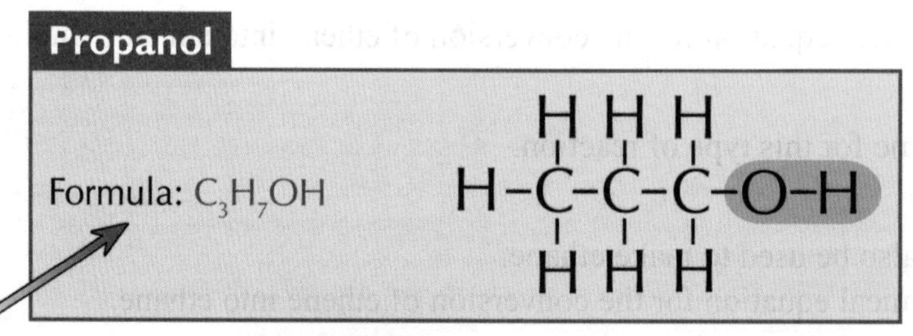

The formula of each alcohol can also be shown with the atoms bonded to each carbon in the molecule shown separately. For example, C_2H_5OH can be shown as CH_3CH_2OH.

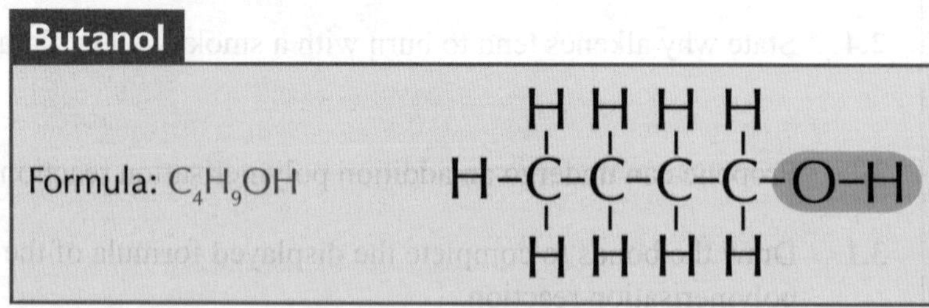

4) The basic naming system is the same as for alkanes — but replace the final '-e' with '-ol'.
5) Don't write CH_4O instead of CH_3OH — it doesn't show the -OH functional group.

The general formula of alcohols is $C_nH_{2n+1}OH$

For the exam, make sure you learn the structures of the first four alcohols.
You also need to be able to recognise alcohols from their names or formulae.

Alcohols

The First Four Alcohols Have Similar Properties

1) Alcohols are flammable. They undergo complete combustion in air to produce carbon dioxide and water. For example:

$$2CH_3OH_{(l)} + 3O_{2(g)} \rightarrow 2CO_{2(g)} + 4H_2O_{(g)}$$

Make sure you can write balanced equations for the combustion of alcohols.

2) Methanol, ethanol, propanol and butanol are all soluble in water. Their solutions have a neutral pH.
3) They also react with sodium. One of the products of this reaction is hydrogen.
4) Alcohols can be oxidised by reacting with oxygen (e.g. from the air) to produce a carboxylic acid (see the next page):
5) Different alcohols form different carboxylic acids. For example, methanol is oxidised to methanoic acid, while ethanol is oxidised to ethanoic acid.

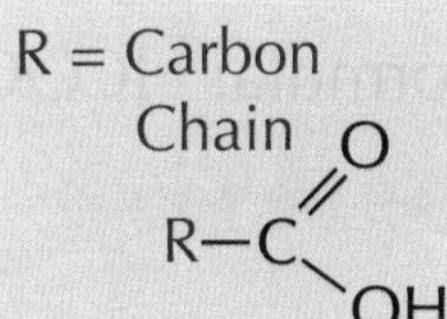

Alcohols are Used as Solvents and Fuels

1) Alcohols such as methanol and ethanol are used as solvents in industry. This is because they can dissolve most things water can dissolve, but they can also dissolve substances that water can't dissolve — e.g. hydrocarbons, oils and fats.

2) The first four alcohols are used as fuels. For example, ethanol is used as a fuel in spirit burners — it burns fairly cleanly and it's non-smelly.

Ethanol can be Made by Fermentation

Ethanol is the alcohol found in alcoholic drinks such as wine or beer. It's usually made using fermentation.

1) Fermentation uses an enzyme in yeast to convert sugars into ethanol. Carbon dioxide is also produced. The reaction occurs in solution so the ethanol produced is aqueous.

$$\text{sugar} \xrightarrow{\text{yeast}} \text{ethanol} + \text{carbon dioxide}$$

Ethanol can also be produced from ethene. See p.140.

2) Fermentation happens fastest at a temperature of around 37 °C, in a slightly acidic solution and under anaerobic conditions (no oxygen).
3) Under these conditions the enzyme in yeast works best to convert the sugar to alcohol. If the conditions were different, for example a lower pH/higher temperature or higher pH/lower temperature, the enzyme could be denatured (destroyed) or could work at a much slower rate.

The first four alcohols behave in similar ways

Alcohols can burn in air, are soluble in water, react with sodium and are oxidised to form carboxylic acids. Make sure you can describe the reactions and be able to write equations for the combustion of alcohols.

Carboxylic Acids

You may have seen carboxylic acids mentioned on the previous page. Here's some more about them.

Carboxylic Acids Have the Functional Group **-COOH**

1) Carboxylic acids are a homologous series of compounds that all have '-COOH' as a functional group.
2) Their names end in '-anoic acid' (and start with the normal 'meth/eth/prop/but')

Methanoic Acid

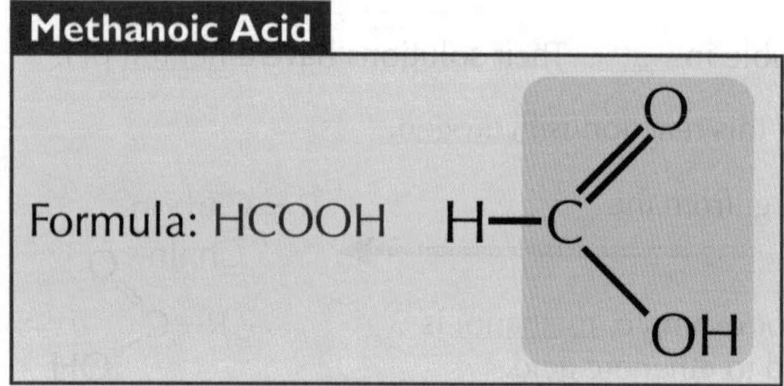

Make sure you know the names and the structures of these four carboxylic acids. You could be asked about them in the exams.

Ethanoic Acid

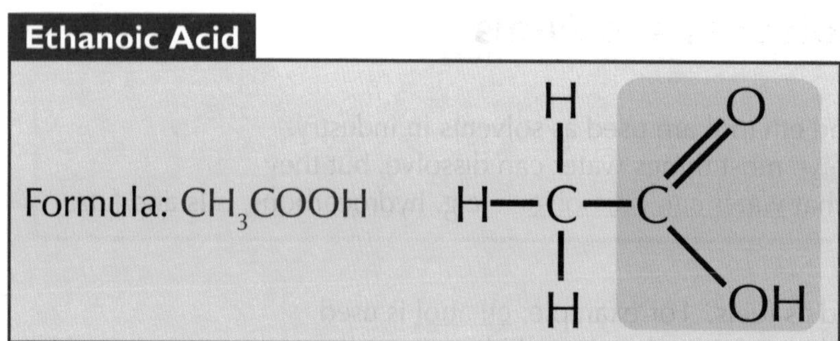

Propanoic Acid

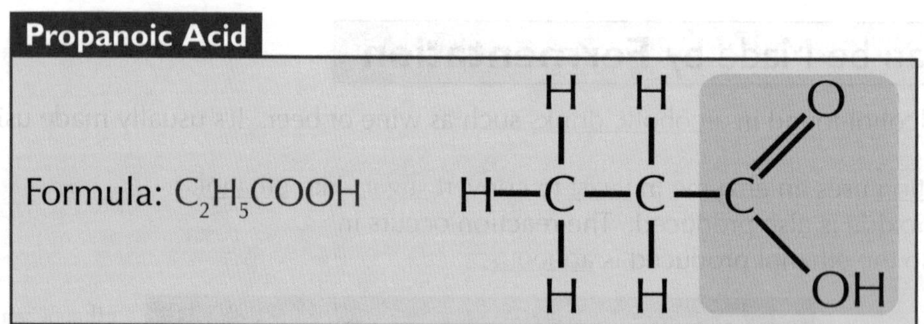

Butanoic Acid

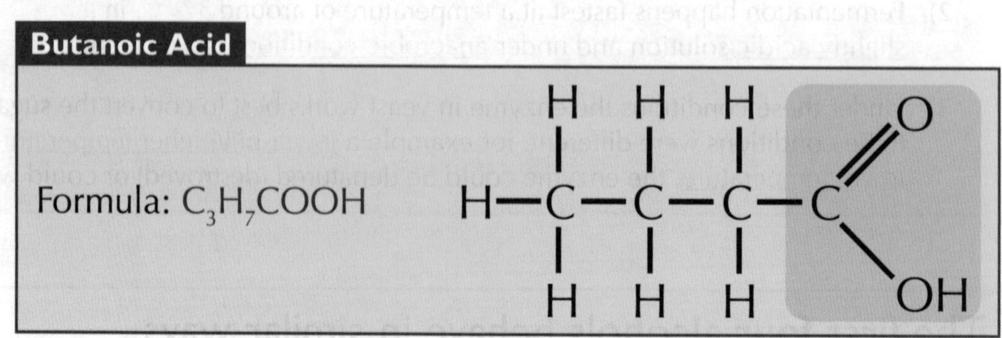

Carboxylic Acids

Carboxylic Acids React Like Other **Acids**

1) They react (like any other acid) with carbonates to produce a salt, water and carbon dioxide.
2) The salts formed in these reactions end in -anoate — e.g. methanoic acid will form a methanoate, ethanoic acid an ethanoate, etc. For example:

ethanoic acid + sodium carbonate → sodium ethanoate + water + carbon dioxide

3) Carboxylic acids can dissolve in water. When they dissolve, they ionise and release H^+ ions resulting in an acidic solution. But, because they don't ionise completely (not all the acid molecules release their H^+ ions), they just form weak acidic solutions. This means that they have a higher pH (are less acidic) than aqueous solutions of strong acids with the same concentration. There's more about strong and weak acids on page 90.

Esters can be Made from **Carboxylic Acids**

1) Esters have the functional group '-COO-'.
2) Esters are formed from an alcohol and a carboxylic acid.
3) An acid catalyst is usually used (e.g. concentrated sulfuric acid).

alcohol + carboxylic acid $\xrightarrow{\text{acid catalyst}}$ ester + water

Ethyl ethanoate can be made from ethanoic acid and ethanol with an acid catalyst:

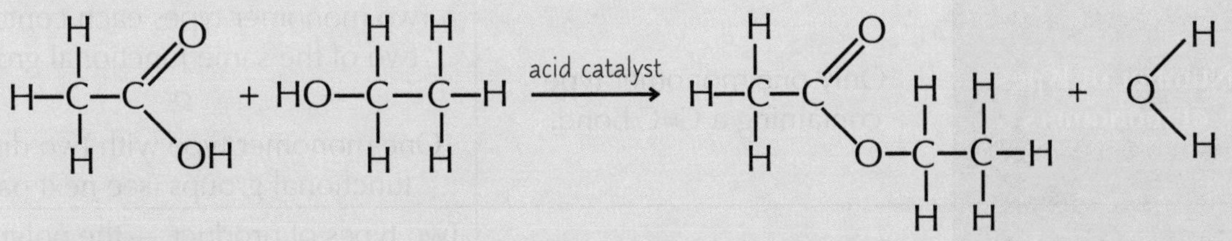

```
  H     O          H   H                   H     O                H
  |    //          |   |   acid catalyst   |    //               /
H-C-C       + HO-C-C-H  ───────────→  H-C-C      H   H    +  O
  |    \           |   |                   |    \   |   |        \
  H     OH         H   H                   H     O-C-C-H         H
                                                    |   |
                                                    H   H
```

CH_3COOH	C_2H_5OH	$CH_3COOC_2H_5$	H_2O
Ethanoic acid	Ethanol	Ethyl ethanoate	Water

The names of esters can be a bit complicated and are often tongue twisters. The one you need to make sure you learn is ethyl ethanoate — the examiners could ask you about this one.

Carboxylic acids — just like other acids...

...they don't ionise completely, so are weak acids. Also, don't forget they can form esters with alcohols.

Condensation Polymers

You might remember addition polymerisation from back on page 141, but there is another type of polymerisation that you could be asked about. This time the monomers need to have TWO functional groups.

Polymers can be Made by Condensation Polymerisation

1) Condensation polymerisation involves monomers which contain different functional groups.
2) The monomers react together and bonds form between them, making polymer chains.
3) For each new bond that forms, a small molecule (for example, water) is lost. This is why it's called condensation polymerisation.
4) The simplest types of condensation polymers contain two different types of monomer, each with two of the same functional groups.
5) For example, here's how a polyester can be made by condensation polymerisation:

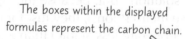

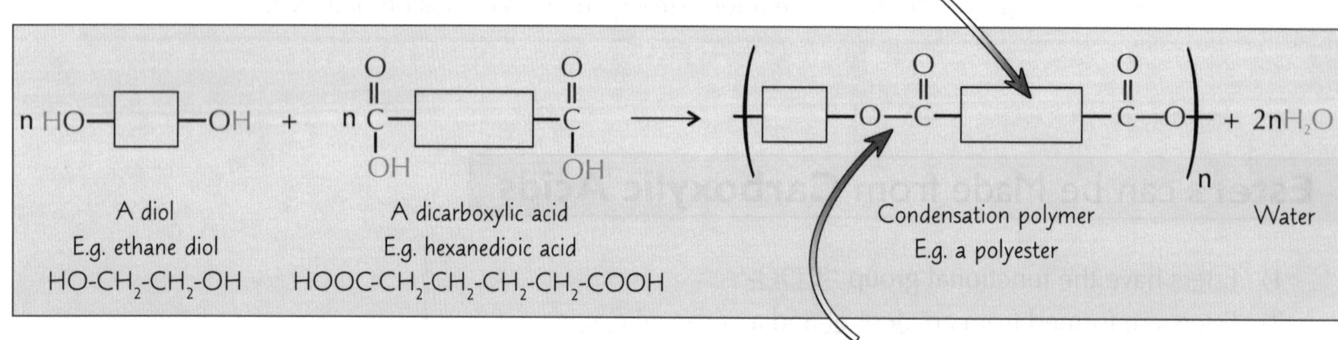

Addition and Condensation Polymerisation are Different

Both addition and condensation polymerisation produce polymers.
However, the products and reactants are very different:

	Addition Polymerisation	Condensation Polymerisation
Number of types of monomers	Only one monomer type containing a C=C bond.	Two monomer types each containing two of the same functional groups. or One monomer type with two different functional groups (see next page).
Number of products	Only one product formed.	Two types of product — the polymer and a small molecule (e.g. water).
Functional groups involved in polymerisation	Carbon-carbon double bond in monomer.	Two reactive groups on each monomer.

Condensation polymerisation results in the loss of a small molecule

So, addition polymers are formed when identical monomers containing C=C bonds are joined together in a polymer chain. Condensation polymers are made when different monomers, or monomers with two different reactive functional groups, react together with the loss of a small molecule such as water.

Naturally Occurring Polymers

Polymers aren't just something created in the lab — they are also found in the natural world.

Amino Acids have an Amino Group and a Carboxyl Group

1) An amino acid contains two different functional groups — a basic amino group (NH_2) and an acidic carboxyl group (COOH).
2) An example of an amino acid is glycine — the smallest and simplest amino acid possible.

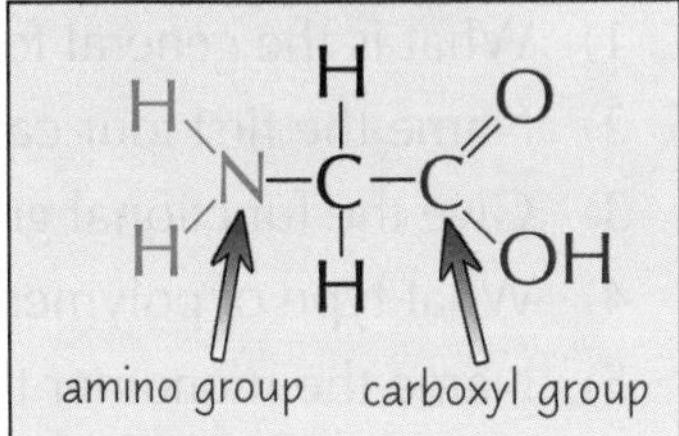

Proteins are Polymers of Amino Acids

1) Amino acids can form polymers known as polypeptides via condensation polymerisation.
2) The amino group of an amino acid can react with the acid group of another, and so on, to form a polymer chain. For every new bond that is formed a molecule of water is lost.

Glycine

$$n\,H_2N{-}CH_2{-}COOH \longrightarrow {-}\!\left[NH{-}CH_2{-}CO\right]_n\!{-} + nH_2O$$

3) One or more long-chains of polypeptides are known as proteins. Proteins have loads of important uses in the human body. For example, enzymes work as catalysts, haemoglobin transports oxygen, antibodies form part of the immune system, and the majority of body tissue is made from proteins.
4) Polypeptides and proteins can contain different amino acids in their polymer chains. The order of the amino acids is what gives proteins their different properties and shapes.

DNA Molecules are Made From Nucleotide Polymers

DNA (deoxyribonucleic acid) is found in every living thing and many viruses. It contains genetic instructions that allow the organism to develop and operate. It's a large molecule that takes a double helix structure.

1) DNA is made of two polymer chains of monomers called 'nucleotides'. The nucleotides each contain a small molecule known as a 'base'. There are four different bases, known by their initials — A, C, G, T.
2) The bases on the different polymer chains pair up with each other and form cross links keeping the two strands of nucleotides together and giving the double helix structure.
3) The order of the bases acts as a code for an organism's genes.

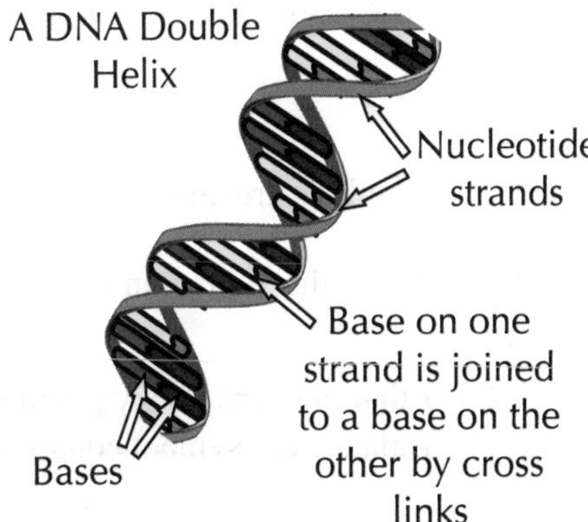

Simple Sugars Can Form Polymers

1) Sugars are small molecules that contain carbon, oxygen and hydrogen.
2) Sugars can react together through polymerisation reactions to form larger polymers, e.g. starch, which living things use to store energy, and cellulose, which is found in plant cell walls.

Warm-Up & Exam Questions

I'm sure you have everything on these pages learnt and understood. But just to check, try these questions.

Warm-Up Questions

1) What is the general formula of an alcohol?
2) Name the first four carboxylic acids in the homologous series.
3) Give the functional group of an ester.
4) What type of polymerisation is involved in the production of a polyester?
5) Name the monomer that proteins are made of.

Exam Questions

1 Which of the following structures, **A**, **B**, **C**, or **D** shows butanol?

A
```
    H   H   H
    |   |   |
H — C — C — C — O — H
    |   |   |
    H   H   H
```

B
```
    H   H   H
    |   |   |    //O
H — C — C — C — C
    |   |   |    \
    H   H   H     OH
```

C
```
    H   H   H   H
    |   |   |   |
H — C — C — C — C — H
    |   |   |   |
    H   H   O   H
            |
            H
```

D
```
    H   H      H
    |   |     /
H — C — C = C
    |   |     \
    H   H      H
```

[1 mark]

2 Carboxylic acids are a widely used family of organic chemicals. Grade 6-7

2.1 Draw bonds and missing atoms to complete the displayed formula of ethanoic acid.

C—C

[2 marks]

2.2 One use of carboxylic acids is in the production of esters.
Name the ester formed when ethanoic acid is reacted with ethanol.

[1 mark]

2.3 Carboxylic acids are weak acids. Explain what this means.

[1 mark]

3 Alcohols are an important group of organic chemicals. The most widely used alcohol is ethanol.

3.1 Name the alcohol in this homologous series that contains three carbon atoms.

[1 mark]

3.2 Ethanol burns in oxygen to give carbon dioxide and water.
Balance the symbol equation for this reaction.

......C_2H_5OH + ...O_2 → ......CO_2 + ...H_2O

[1 mark]

4* Polymers can be made by addition polymerisation and condensation polymerisation reactions.
Discuss the differences between these two types of polymerisation.

[4 marks]

Revision Summary for Topic 7

That wraps up Topic 7 — time to put yourself to the test and find out how much you really know.

- Try these questions and tick off each one when you get it right.
- When you've done all the questions under a heading and are completely happy with it, tick it off.

Hydrocarbons and Crude Oil (p.132-135)

1) What two elements do hydrocarbons contain?
2) What is the general formula for alkanes?
3) Draw the displayed formula of butane.
4) What is the general word equation for the complete combustion of a hydrocarbon?
5) How is crude oil formed?
6) Where are the shortest carbon chains found in the fractional distillation column?
7) Which is the hottest part of a fractional distillation column?
8) Name two methods of cracking?
9) Give a product of cracking that is used for making plastics.

Alkenes and Addition Polymerisation (p.138-142)

10) Give the general formula for an alkene.
11) Give the balanced equation for the incomplete combustion of butane.
12) Name the molecule that an alkene reacts with when it undergoes hydrogenation.
13) Draw the product of the reaction of ethene with chlorine.
14) What is used to test for alkenes?
15) What type of compounds are used as the monomers to make addition polymers?
16) Draw the displayed formula for the repeating unit of poly(ethene).

Alcohols, Carboxylic Acids and More about Polymers (p.144-149)

17) Give the names and displayed formulas of the first four alcohols.
18) Give the balanced equation for the combustion of propanol.
19) Name the gas that is produced when ethanol reacts with sodium.
20) Name the process that converts sugar to ethanol using yeast.
21) Draw the displayed formula of propanoic acid.
22) Give the general word equation for the reaction of an alcohol and a carboxylic acid.
23) How many different products are there in a condensation polymerisation reaction?
24) What are polypeptides?
25) What small molecule is lost when two amino acids react together to form a new bond?

Purity and Formulations

Substances are often not 100% pure — they might have other stuff that you can't see mixed in with them. The purity of a substance might need to be checked, before, say, a drug is made from it.

Purity is Defined Differently in Chemistry to Everyday

1) Usually when you refer to a substance as being pure you mean that nothing has been added to it, so it's in its natural state. For example, pure milk or beeswax.
2) In chemistry, a pure substance is something that only contains one compound or element throughout — not mixed with anything else.

The Boiling or Melting Point Tells You How Pure a Substance Is

1) A chemically pure substance will melt or boil at a specific temperature.
2) You can test the purity of a sample by measuring its melting or boiling point and comparing it with the melting or boiling point of the pure substance (which you can find in a data book).
3) The closer your measured value is to the actual melting or boiling point, the purer your sample is.
4) Impurities in your sample will lower the melting point and increase the melting range of your substance.
5) Impurities in your sample will also increase the boiling point and may result in your sample boiling over a range of temperatures.

Formulations are Mixtures with Exact Amounts of Components

1) Formulations are useful mixtures with a precise purpose that are made by following a 'formula' (a recipe). Each component in a formulation is present in a measured quantity, and contributes to the properties of the formulation so that it meets its required function.

Take a look at p.24 for more on mixtures.

For example, paints are formulations composed of:

Pigment — gives the paint colour (for example titanium oxide is used as a pigment in white paints).

Solvent — used to dissolve the other components and alter the viscosity (runniness).

Binder (resin) — forms a film that holds the pigment in place after it's been painted on.

Additives — added to further change the physical and chemical properties of the paint.

Depending on the purpose of the paint, the chemicals used and their amounts will be changed so the paint produced is right for the job.

2) Formulations are really important in the pharmaceutical industry. For example, by altering the formulation of a pill, chemists can make sure it delivers the drug to the correct part of the body at the right concentration, that it's consumable and has a long enough shelf life.
3) In everyday life, formulations can be found in cleaning products, fuels, cosmetics, fertilisers (see p.190), metal alloys (see p.61) and even food and drink.
4) When you buy a product, you might find that it has information about its composition on the packaging. For example, the ratio or percentage of each component. This tells you the product's a formulation. It also lets you choose a formulation with the right composition for your particular use.

Make sure you can use data to identify pure and impure substances...

Knowing how pure a product is can be vital in industries such as pharmaceuticals and the food industry. Extra stuff in it by mistake could change the properties of the product, and even make it dangerous.

Testing for Gases

PRACTICAL

Ahh... tests, glorious tests. Luckily, these aren't the kind of tests you have to revise for, but you should probably revise these tests for your exam — it's swings and roundabouts really...

There are Tests for Four Common Gases

1) Chlorine

Chlorine bleaches damp litmus paper, turning it white. (If you use blue litmus paper it may turn red for a moment first though — that's because a solution of chlorine is acidic.)

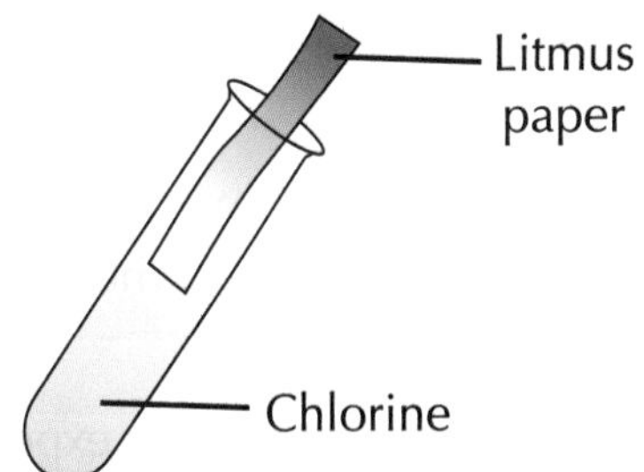

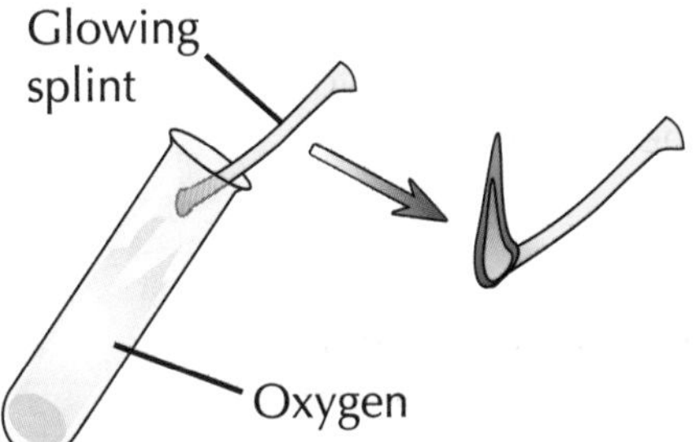

2) Oxygen

If you put a glowing splint inside a test tube containing oxygen, the oxygen will relight the glowing splint.

3) Carbon Dioxide

Bubbling carbon dioxide through (or shaking carbon dioxide with) an aqueous solution of calcium hydroxide (known as limewater) causes the solution to turn cloudy.

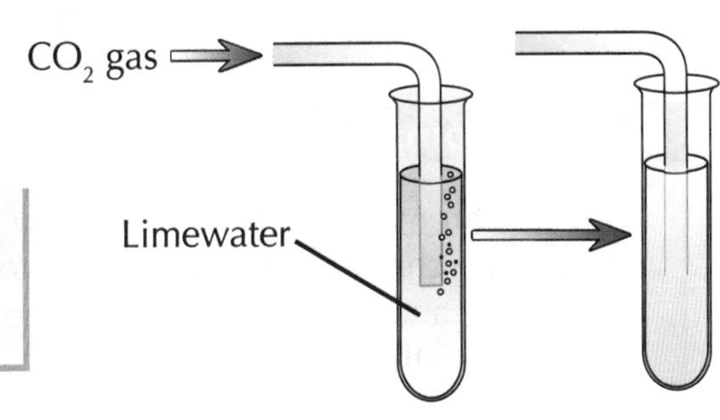

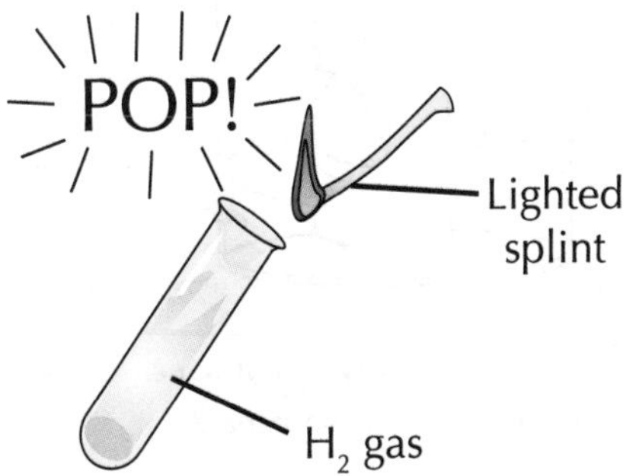

4) Hydrogen

If you hold a burning splint at the open end of a test tube containing hydrogen, you'll get a "squeaky pop". (The noise comes from the hydrogen burning quickly in the oxygen in the air to form water.)

These are all really useful tests to know...

The method you use to collect a gas will depend on whether it's lighter or heavier than air. If it's heavier (like chlorine), you have the test tube the right way up and the gas will sink to the bottom. If it's lighter (like hydrogen), you have the test tube upside-down and the gas will rise to fill it.

Paper Chromatography

You met chromatography on page 25. Now it's time to see how it works.

Chromatography uses Two Phases

Chromatography is an analytical method used to separate the substances in a mixture. You can also use information from a chromatography experiment to help identify the substances that you have separated. There are different types of chromatography, but they all have two 'phases':

> A mobile phase — where the molecules can move. This is always a liquid or a gas.
> A stationary phase — where the molecules can't move. This can be a solid or a really thick liquid.

1) During a chromatography experiment, the substances in the sample constantly move between the mobile and the stationary phases — an equilibrium is formed between the two phases.
2) The mobile phase moves through the stationary phase, and anything dissolved in the mobile phase moves with it.
3) How quickly a chemical moves depends on how it's 'distributed' between the two phases — whether it spends more time in the mobile phase or the stationary phase.
4) The chemicals that spend more time in the mobile phase than the stationary phase will move further through the stationary phase.
5) The components in a mixture will normally separate through the stationary phase, so long as all the components spend different amounts of time in the mobile phase.
6) The separated components form spots. The number of spots formed may change in different solvents as the distribution of the chemical will change depending on the solvent.
7) A pure substance will only ever form one spot in any solvent, since there is only one substance in the sample.

Paper Chromatography

In paper chromatography the stationary phase is the chromatography paper (often filter paper) and the mobile phase is the solvent (e.g. ethanol or water).

The amount of time the molecules spend in each phase depends on two things:

- How soluble they are in the solvent.
- How attracted they are to the paper.

Molecules with a higher solubility in the solvent, and which are less attracted to the paper, will spend more time in the mobile phase — and they'll be carried further up the paper.

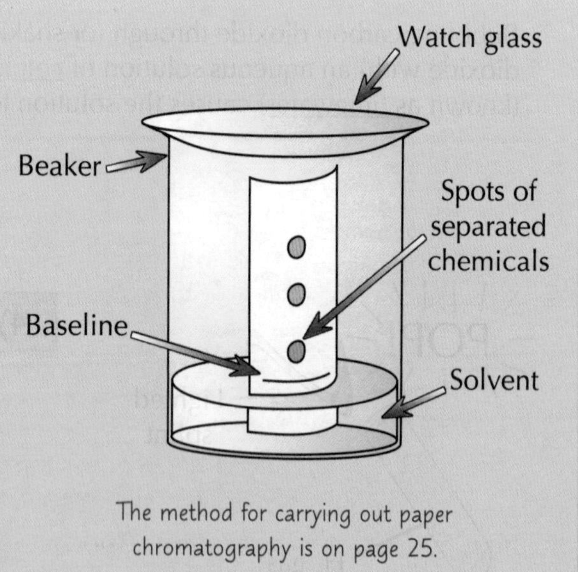

The method for carrying out paper chromatography is on page 25.

Chromatography revision — it's a phase you have to get through...

Chromatography works because each of the chemicals in a mixture spends different amounts of time dissolved in the mobile phase and stuck to the stationary phase. It's great — all you need is some paper and a bit of solvent. There's more about using it to identify chemicals coming up on the next page.

Paper Chromatography

Now that you know a bit of the theory behind how paper chromatography works, here's how you can use a chromatogram to analyse a particular substance and find out what's in it.

You can Calculate the R_f Value for Each Chemical

1) The result of chromatography analysis is called a chromatogram.

Distance moved by solvent (solvent front)

Spot of chemical

Baseline (Origin)

A

B

2) An R_f value is the ratio between the distance travelled by the dissolved substance (the solute) and the distance travelled by the solvent.

3) You can calculate R_f values using this formula:

$$R_f = \frac{\text{distance travelled by substance (B)}}{\text{distance travelled by solvent (A)}}$$

This is the distance from the baseline to the centre of the spot.

4) The further through the stationary phase a substance moves, the larger the R_f value.

5) Chromatography is often carried out to see if a certain substance is present in a mixture. To do this, you run a pure sample of that substance (a reference) alongside the unknown mixture. If the R_f values of the reference and one of the spots in the mixture match, the substance may be present (although you haven't yet proved they're the same).

The R_f Value is Affected by the Solvent

1) The R_f value of a substance is dependent on the solvent you use — if you change the solvent the R_f value for the substance will change.
2) You can test both the mixture and the reference in a number of different solvents.
3) If the R_f value of the reference matches the R_f value of one of the spots in the mixture in all the solvents, then it's likely the reference compound is present in the mixture.
4) If the spot in the mixture and the spot in the reference only have the same R_f value in some of the solvents, then the reference compound isn't present in the mixture.

You need to learn the formula for R_f

Sometimes, when you're doing paper chromatography, you'll end up with a spot left sitting on the baseline, even after your solvent has run all the way up the paper. Any substance that remains on the baseline is insoluble in that solvent — you could rerun the experiment using a different solvent to try and identify it.

Warm-Up & Exam Questions

Look, a chromatography question — those things are fun. Get your investigative hat on and get stuck in...

Warm-Up Questions

1) What is meant by a pure substance in chemistry?
2) What effect will impurities in a substance have on its boiling point?
3) Give an example of a formulation used in everyday life.
4) Describe the test for carbon dioxide.
5) What effect does chromatography have on a mixture?

Exam Questions

PRACTICAL

1 Electrolysis of water gives hydrogen gas and oxygen gas. 

1.1 Describe how you could identify hydrogen gas using a simple laboratory test.

[2 marks]

1.2 Describe how you could identify oxygen gas using a simple laboratory test.

[2 marks]

PRACTICAL

2 A scientist used chromatography to analyse the composition of five food colourings. Four of the colourings were unknown (**A** – **D**). The other was sunrise yellow. The results are shown in **Figure 1**.

Figure 1

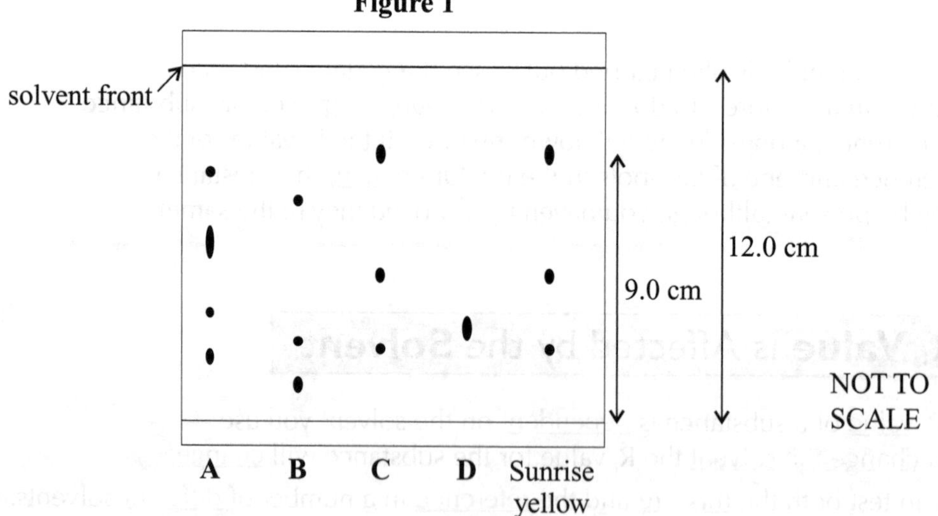

2.1 Which food colouring is most likely to be a pure compound?

[1 mark]

2.2 Which food colouring definitely contains at least four different compounds?

[1 mark]

2.3 Which of the food colourings, **A-D**, could be the same as sunrise yellow?

[1 mark]

2.4 Calculate the R_f value for the spot of chemical in sunrise yellow which is furthest up the chromatogram.

[2 marks]

Tests for Anions

Imagine you have a mystery substance — you don't know what it is, but you want to find out. If it's ionic, it'll have a positive and a negative part. So, first, here's how to test for some common negative ions (anions).

Tests for **Anions** Often **Give Precipitates**

Dilute Acid Can Help Detect **Carbonates**

1) Carbonates are substance that contain CO_3^{2-} ions.
2) You can test for carbonate ions by using a dropping pipette to add a few drops of dilute acid to a test tube containing your mystery substance.
3) Then you can connect this test tube to a test tube containing limewater. If carbonate ions are present, their reaction with the acid will release carbon dioxide, which will turn the limewater cloudy when it bubbles through it (see page 153 for more on this test).
4) For example, here's the equation for the reaction between sodium carbonate and hydrochloric acid:

$$Na_2CO_{3(aq)} + 2HCl_{(aq)} \rightarrow CO_{2(g)} + 2NaCl_{(aq)} + H_2O_{(l)}$$

Test for **Sulfates** with HCl and **Barium Chloride**

1) To identify sulfate ions (SO_4^{2-}), use a dropping pipette to add a couple of drops of dilute hydrochloric acid (HCl) followed by a couple of drops of barium chloride solution ($BaCl_2$) to a test tube containing your mystery solution.
2) If sulfate ions are present, a white precipitate of barium sulfate will form:

$$Ba^{2+}_{(aq)} + SO_4^{2-}{}_{(aq)} \rightarrow BaSO_{4(s)}$$

Hydrochloric acid is added to get rid of any traces of carbonate ions before you do the test. These would also produce a precipitate, so they'd confuse the results.

Test for **Halides (Cl^-, Br^-, I^-)** with **Nitric Acid** and **Silver Nitrate**

To identify a halide ion, add a couple of drops of dilute nitric acid (HNO_3), followed by a couple of drops of silver nitrate solution, $AgNO_3$, to your mystery solution.

$Ag^+_{(aq)} + Cl^-_{(aq)} \longrightarrow AgCl_{(s)}$	A chloride gives a white precipitate of silver chloride.
$Ag^+_{(aq)} + Br^-_{(aq)} \longrightarrow AgBr_{(s)}$	A bromide gives a cream precipitate of silver bromide.
$Ag^+_{(aq)} + I^-_{(aq)} \longrightarrow AgI_{(s)}$	An iodide gives a yellow precipitate of silver iodide.

Hopefully this page won't be too testing for you...

There are three tests to learn here, along with their results. That's quite a bit of information to take in, so don't just stare at the whole page till your eyes are swimming and you never ever want to see the word 'precipitate' again. Just learn the tests one by one and you'll be absolutely fine.

Tests for Cations

I haven't met a person who doesn't like flame tests for positive ions (cations) — everybody loves them...

Flame Tests Identify Metal Ions

1) Compounds of some metals produce a characteristic colour when heated in a flame.
2) So you can test for various metal ions by heating your substance and seeing whether the flame turns a distinctive colour:

   Lithium ions, Li^+, produce a crimson flame.

   Sodium ions, Na^+, produce a yellow flame.

   Potassium ions, K^+, produce a lilac flame.

   Calcium ions, Ca^{2+}, produce a orange-red flame.

   Copper ions, Cu^{2+}, produce a green flame.

Flame emission spectroscopy uses the light emitted by metal ions when they enter a flame to analyse and identify them. There's more on this on the next page.

3) To do the test, you first need to clean a nichrome or platinum wire loop by rubbing with fine emery paper and then holding it in a blue flame from a Bunsen burner. The Bunsen flame might change colour for a bit, but once it's blue again, the loop is clean. Then, dip the loop into the sample you want to test and put it back in the flame. Record the colour of the flame.
4) You can use these colours to detect and identify different ions. However it only works for samples that contain a single metal ion. If the sample tested contains a mixture of metal ions, the flame colours of some ions may be hidden by the colours of others.

Some Metals Form a Coloured Precipitate with NaOH

1) Many metal hydroxides are insoluble and precipitate out of solution when formed. Some of these hydroxides have a characteristic colour.
2) So in this test you add a few drops of sodium hydroxide solution to a solution of your mystery compound — all in the hope of forming an insoluble hydroxide.
3) If you get a coloured insoluble hydroxide you can often tell which metal was in the compound.

Metal Ions	Colour of Precipitate	Ionic Equation for Precipitate Formation
Calcium, Ca^{2+}	White	$Ca^{2+}_{(aq)} + 2OH^-_{(aq)} \rightarrow Ca(OH)_{2\,(s)}$
Copper(II), Cu^{2+}	Blue	$Cu^{2+}_{(aq)} + 2OH^-_{(aq)} \rightarrow Cu(OH)_{2\,(s)}$
Iron(II), Fe^{2+}	Green	$Fe^{2+}_{(aq)} + 2OH^-_{(aq)} \rightarrow Fe(OH)_{2\,(s)}$
Iron(III), Fe^{3+}	Brown	$Fe^{3+}_{(aq)} + 3OH^-_{(aq)} \rightarrow Fe(OH)_{3\,(s)}$
Aluminium, Al^{3+}	White at first. But then redissolves in excess NaOH to form a colourless solution.	$Al^{3+}_{(aq)} + 3OH^-_{(aq)} \rightarrow Al(OH)_{3\,(s)}$
Magnesium, Mg^{2+}	White	$Mg^{2+}_{(aq)} + 2OH^-_{(aq)} \rightarrow Mg(OH)_{2\,(s)}$

You don't need to know the equation for how the aluminium hydroxide precipitate dissolves.

Don't mix up the flame colours and the precipitate colours...

Compounds containing metal cations are used in fireworks. When the fireworks burn, it's these metal cations that give them their pretty colours. November 5th will never seem the same again...

Flame Emission Spectroscopy

As well as producing pretty colours, the science behind flame tests can be used to identify different metal ions in solution accurately and find their concentrations. Colours and science — just my kind of thing...

Every **Metal Ion** Gives a Characteristic **Line Spectrum**

1) During flame emission spectroscopy a sample is placed in a flame. As the ions heat up, their electrons become excited (they move to higher energy levels). When the electrons drop back to their original energy levels, they release energy as light.

2) The light passes through a spectroscope, which can detect different wavelengths of light to produce a line spectrum.

A line spectrum for an ion could look something like this:

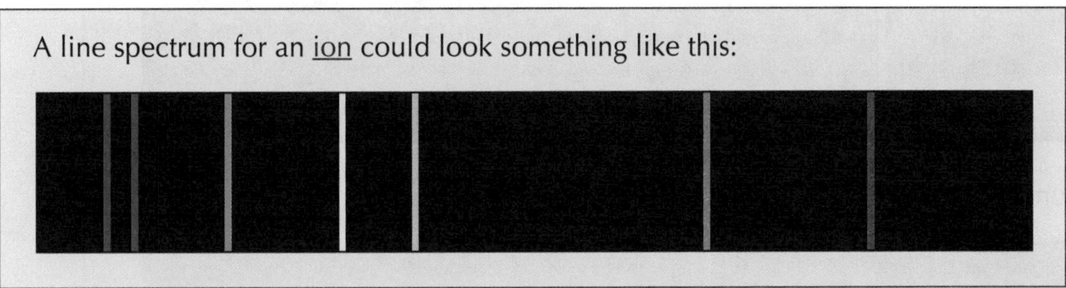

3) The combination of wavelengths emitted by an ion depends on its charge and its electron arrangement (see page 33).

4) Since no two ions have the same charge and the same electron arrangement, different ions emit different wavelengths of light. So each ion produces a different pattern of wavelengths, and has a different line spectrum.

5) The intensity of the spectrum indicates the concentration of that ion in solution.

6) This means that line spectra can be used to identify ions in solution and calculate their concentrations.

It's the same idea as a flame test, but you get more information

When you do a bog standard flame test for metal ions, you get a coloured flame. But determining what colour the flame is can be a bit subjective. Flame emission spectroscopy produces a line spectrum instead. The pattern on the line spectrum is unique for every ion — and it's easy to tell the different patterns of lines apart just by comparing them. So it's a more accurate method of identifying an ion than a flame test.

Flame Emission Spectroscopy

Flame Emission Spectroscopy Works for Mixtures

Flame emission spectroscopy can also be used to identify different ions in mixtures. This makes it more useful than flame tests, which only work for substances that contain a single metal ion.

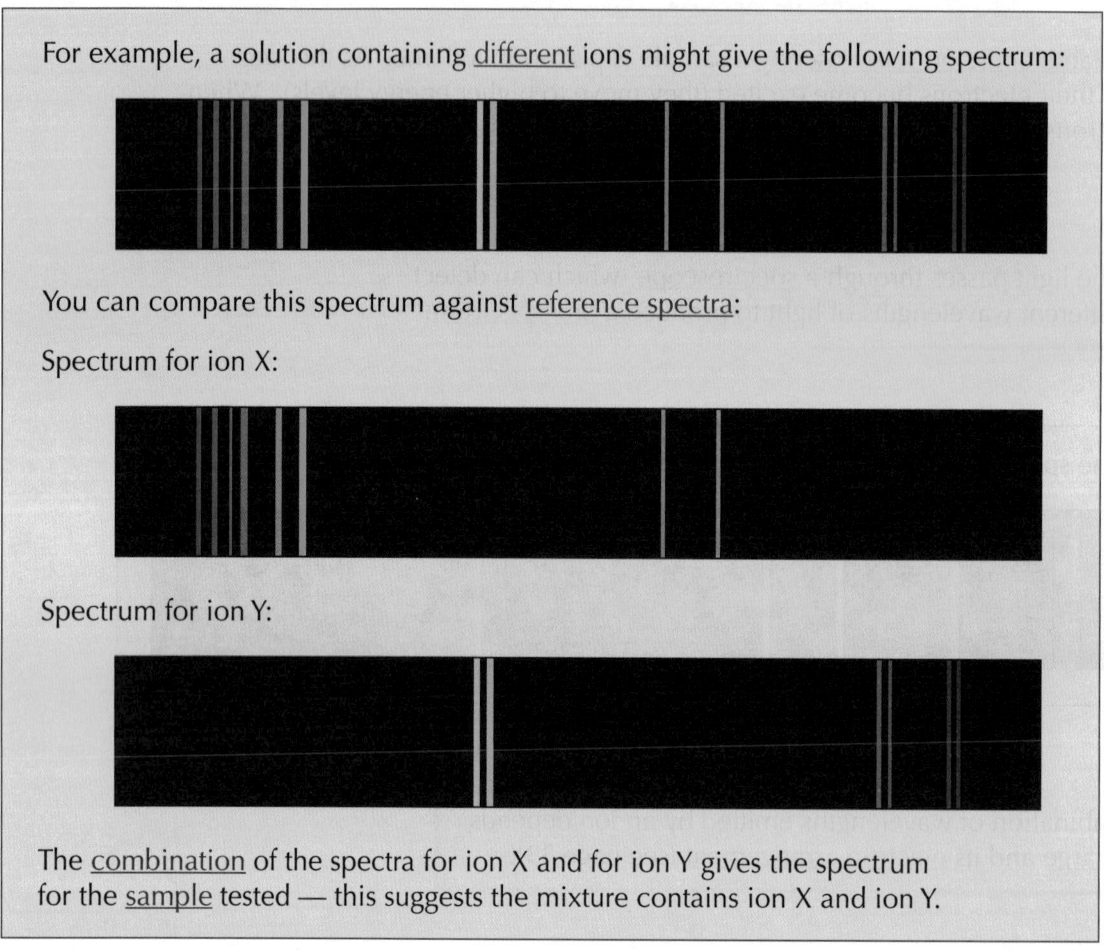

For example, a solution containing different ions might give the following spectrum:
You can compare this spectrum against reference spectra:
Spectrum for ion X:
Spectrum for ion Y:
The combination of the spectra for ion X and for ion Y gives the spectrum for the sample tested — this suggests the mixture contains ion X and ion Y.

Machines can Analyse Unknown Substances

Chemists often use instrumental analysis (i.e. tests that use machines), such as flame emission spectroscopy, instead of conducting manual tests.

Advantages of Using Machines:

- Very sensitive — they can detect even the tiniest amounts of substances.
- Very fast and tests can be automated.
- Very accurate.

Unfortunately, machines can't do the exam for you...

Instrumental analysis is fast and it gives you results that are accurate, even if there's only a tiny amount of a substance present. For example, flame emission spectroscopy is used in environmental monitoring to detect small amounts of any metals that may be present in samples of soil and water.

Warm-Up & Exam Questions

And now, some questions to hone your skills. There are quite a few picky little details to learn for all those ion tests, plus all the flame emission spectroscopy stuff, so make sure you've got it all before you move on.

Warm-Up Questions

1) What metal ions would produce a lilac flame in a flame test?
2) a) What colour precipitate forms when you add sodium hydroxide to a solution containing copper(II) ions?
   b) Write an ionic equation for the reaction between copper ions (Cu^{2+}) and hydroxide ions (OH^-).
3) In flame emission spectroscopy, what does the intensity of the line spectrum tell you?

Exam Questions

PRACTICAL

1 Kelly carried out flame tests on compounds of four different metal ions. Complete **Table 1**, below, which shows her results.

Table 1

Flame colour	Metal ion
green	
	Li^+
yellow	
	Ca^{2+}

[4 marks]

2 The elements caesium and rubidium were both discovered by their line spectra.

2.1 Describe how flame emission spectroscopy can be used to produce a line spectrum for an element.

[3 marks]

2.2 Explain why caesium and rubidium have different line spectra.

[3 marks]

PRACTICAL

3 **Table 2** shows the results of a series of chemical tests conducted on two unknown compounds, **X** and **Y**.

Table 2

Test	Observation	
	Compound X	Compound Y
sodium hydroxide solution	white precipitate	green precipitate
hydrochloric acid and barium chloride solution	white precipitate	no precipitate
nitric acid and silver nitrate solution	no precipitate	yellow precipitate

3.1 What is the chemical name of compound **Y**?

[2 marks]

3.2 A student says "From these results, I can be certain that compound **X** is magnesium sulfate." State whether he is correct. Explain your answer.

[3 marks]

Revision Summary for Topic 8

Here are some questions to work through before you get all geared up for Topic 9. I know you can't wait...

- Try these questions and tick off each one when you get it right.
- When you've done all the questions under a heading and are completely happy with it, tick if off.

Purity, Formulations and Tests for Gases (p.152-153) ☑

1) Describe how you could test the purity of a sample of a chemical. ☑
2) What effect will impurities in a substance have on its melting point? ☑
3) What is a formulation? ☑
4) Describe the test for chlorine gas. ☑

Paper Chromatography (p.154-155) ☑

5) What are the two phases called in chromatography? ☑
6) In paper chromatography, how many spots will a pure substance form on the paper? ☑
7) Give two factors which affect the amount of time a chemical spends in each of the two phases during a paper chromatography experiment. ☑
8) Give the formula for working out the R_f value of a substance. ☑
9) Would you expect the R_f value of a substance to change if you changed the solvent used in a paper chromatography experiment? ☑

Tests for Anions and Cations (p.157-158) ☑

10) Describe a test you could use to test for the presence of carbonate ions. ☑
11) Describe what you would observe if you added dilute hydrochloric acid and barium chloride solution to a substance containing sulfate ions. ☑
12) What colour is the precipitate produced by the reaction between silver nitrate and a solution containing bromide ions? ☑
13) How do you conduct a flame test? ☑
14) What colour flame does potassium burn with? ☑
15) What colour precipitate do iron(III) compounds form with sodium hydroxide solution? ☑
16) a) Describe what would happen if you added a few drops of sodium hydroxide solution to a solution containing Al^{3+} ions.
    b) Describe what would happen next if you added some more sodium hydroxide solution. ☑
17) Give the ionic equation for the formation of a precipitate when a solution containing magnesium ions is reacted with sodium hydroxide. ☑

Flame Emission Spectroscopy (p.159-160) ☑

18) Name two things that affect the combination of wavelengths of light emitted by an ion during flame emission spectroscopy. ☑
19) If a solution contains a mixture of ions, which technique could you use to identify the ions, a flame test or flame emission spectroscopy? Explain your answer. ☑
20) Give three advantages of using machines to conduct chemical analysis. ☑

The Evolution of the Atmosphere

Theories for how the Earth's atmosphere evolved have changed a lot over the years — it's hard to gather evidence from such a long time period and from so long ago (4.6 billion years). Here is one idea we've got:

Phase 1 — **Volcanoes** Gave Out **Gases**

1) The first billion years of Earth's history were pretty explosive — the surface was covered in volcanoes that erupted and released lots of gases.

2) We think this was how the early atmosphere was formed.

3) The early atmosphere was probably mostly made up of carbon dioxide, with virtually no oxygen. This is quite like the atmospheres of Mars and Venus today.

4) Volcanic activity also released nitrogen, which built up in the atmosphere over time, as well as water vapour and small amounts of methane and ammonia.

Phase 2 — **Oceans**, **Algae** and **Green Plants** Absorbed CO_2

1) When the water vapour in the atmosphere condensed, it formed the oceans.

2) Lots of carbon dioxide was removed from the early atmosphere as it dissolved in the oceans. This dissolved carbon dioxide then went through a series of reactions to form carbonate precipitates that formed sediments on the seabed.

3) Later, marine animals evolved. Their shells and skeletons contained some of these carbonates from the oceans.

4) Green plants and algae evolved and absorbed some of the carbon dioxide so that they could carry out photosynthesis (see next page).

Before volcanic activity, the Earth didn't even have an atmosphere

One way scientists can get information about what Earth's atmosphere was like in the past is from Antarctic ice cores. Each year a layer of ice forms with tiny bubbles of air trapped in it. The deeper you go in the ice, the older the air. So analysing bubbles from different layers shows you how the atmosphere has changed.

The Evolution of the Atmosphere

Some **Carbon** Became **Trapped** in **Fossil Fuels** and **Rocks**

Some of the carbon that organisms took in from the atmosphere and oceans became locked up in rocks and fossil fuels after the organisms died.

1) When plants, plankton and marine animals die, they fall to the seabed and get buried by layers of sediment.
2) Over millions of years, they become compressed and form sedimentary rocks, oil and gas — trapping the carbon within them (helping to keep it out of the atmosphere).
3) Things like coal, crude oil and natural gas that are made by this process are called 'fossil fuels'.
4) Crude oil and natural gas are formed from deposits of plankton. These fossil fuels form reservoirs under the seabed when they get trapped in rocks.
5) Coal is a sedimentary rock made from thick plant deposits.
6) Limestone is also a sedimentary rock. It's mostly made of calcium carbonate deposits from the shells and skeletons of marine organisms.

The shells of marine organisms like this one are made of calcium carbonate. They form deposits on the seabed when the organism dies.

Phase 3 — Green Plants and Algae **Produced Oxygen**

1) As well as absorbing the carbon dioxide in the atmosphere, green plants and algae produced oxygen by photosynthesis — this is when plants use light to convert carbon dioxide and water into sugars:

carbon dioxide + water $\xrightarrow{\text{light}}$ glucose + oxygen

$$6CO_2 + 6H_2O \xrightarrow{\text{light}} C_6H_{12}O_6 + 6O_2$$

2) Algae evolved first — about 2.7 billion years ago.
3) Then over the next billion years or so, green plants also evolved.
4) As oxygen levels built up in the atmosphere over time, more complex life (like animals) could evolve.
5) Eventually, about 200 million years ago, the atmosphere reached a composition similar to how it is today:
   - approximately 80% nitrogen,
   - approximately 20% oxygen,
   - small amounts of other gases (each making up less than 1% of the atmosphere), mainly carbon dioxide, noble gases and water vapour.

Not too much CO_2 and enough O_2 — just right for complex life

Another way that scientists can work out what the atmosphere was like billions of years ago is to look at the chemical composition of rocks. They can also look for the appearance of different organisms in the fossil record (including plants and algae) to try to work out what might have caused the changes they see.

Climate Change and Greenhouse Gases

Greenhouse gases are important, but can also cause problems — it's all about keeping a delicate balance.

Carbon Dioxide is a Greenhouse Gas

1) Greenhouse gases like carbon dioxide, methane and water vapour act like an insulating layer in the Earth's atmosphere — this, amongst other factors, allows the Earth to be warm enough to support life.

2) All particles absorb certain frequencies of radiation. Greenhouse gases don't absorb the incoming short wavelength radiation from the Sun — but they do absorb the long wavelength radiation that gets reflected back off the Earth.

3) Then they re-radiate it in all directions — including back towards the Earth.

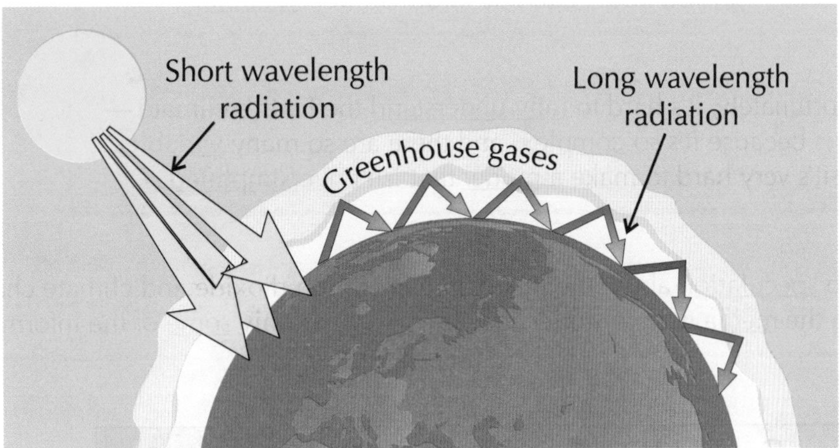

4) The longwave radiation is thermal radiation, so it results in warming of the surface of the Earth. This is the greenhouse effect.

5) Some forms of human activity affect the amount of greenhouse gases in the atmosphere. For example:

- Deforestation — fewer trees means less CO_2 is removed from the atmosphere via photosynthesis.
- Burning fossil fuels — carbon that was 'locked up' in these fuels is released as CO_2.
- Agriculture — more farm animals produce more methane through their digestive processes.
- Creating waste — more landfill sites and more waste from agriculture means more CO_2 and methane released by decomposition of waste.

Greenhouse gases aren't all bad news — we need them to survive

Without greenhouse gases our planet would be incredibly cold — the greenhouse effect warms the Earth enough for it to support living things. Without it we wouldn't be here. But the overall balance of gases in the atmosphere matters — you can have too much of a good thing, as you're about to find out...

Climate Change and Greenhouse Gases

Increasing Carbon Dioxide is Linked to Climate Change

1) The Earth's temperature does vary naturally.

2) But recently, the average temperature of the Earth's surface has been increasing by amounts that are greater than we would expect to see naturally. Most scientists agree that the extra carbon dioxide from human activity is causing this increase, and that increasing global temperature will lead to climate change.

3) Evidence for this has been peer-reviewed (see page 1) — so you know that this information is reliable.

4) Unfortunately, it's hard to fully understand the Earth's climate — this is because it's so complex, and there are so many variables, that it's very hard to make a model that isn't oversimplified.

5) This has led to speculation about the link between carbon dioxide and climate change, particularly in the media (where stories may be biased, or only some of the information given).

See page 2 for more on science in the media.

Climate Change Could Have Dangerous Consequences

The Earth's climate is complex, but it's still important to make predictions about the consequences of climate change so that policy-makers can make decisions now. For example:

1) An increase in global temperature could lead to polar ice caps melting — causing a rise in sea levels, increased flooding in coastal areas and coastal erosion.

2) Changes in rainfall patterns (the amount, timing and distribution) may cause some regions to get too much or too little water. This, along with changes in temperature, may affect the ability of certain regions to produce food.

3) The frequency and severity of storms may also increase.

4) Changes in temperature and the amount of water available in a habitat may affect wild species, leading to differences in their distribution.

People can get quite hot under the collar talking about all this...

That's because climate change could have a massive impact on many people's lives across the world. It's really important that we understand what causes it, as well as how to prevent any damaging consequences.

Carbon Footprints

It's generally accepted that greenhouse gas emissions from human activities are causing climate change. Knowing what causes the biggest emissions of carbon dioxide could help us focus our efforts to reduce them.

Carbon Footprints are Tricky to Measure

1) Carbon footprints are basically a measure of the amount of carbon dioxide and other greenhouse gases released over the full life cycle of something.

2) That something could be a service (e.g. the school bus), an event (e.g. the Olympics), a product (e.g. a toastie maker) — almost anything.

3) Measuring the total carbon footprint of something can be very hard, though — or even impossible.

4) That's because there are so many different factors to consider — for example, you would have to count the emissions released as a result of manufacturing all the parts of your toastie maker and in making it, not to mention the emissions produced when you actually use it and finally dispose of it.

5) Still, a rough calculation can give a good idea of what the worst emitters are, so that people can avoid them in the future.

There are Ways of Reducing Carbon Footprints

You can't always measure a carbon footprint exactly, but there are always ways to reduce it. Anything that reduces the amount of greenhouse gases (e.g. carbon dioxide or methane) given out by a process will also reduce its carbon footprint. Here are some things that can be done:

1) Renewable energy sources or nuclear energy could be used instead of fossil fuels.

2) Using more efficient processes could conserve energy and cut waste. Lots of waste decomposes to release methane, so this will reduce methane emissions.

3) Governments could tax companies or individuals based on the amount of greenhouse gases they emit — e.g. taxing cars based on the amount of carbon dioxide they emit over a set distance could mean that people choose to buy cars that are more fuel-efficient and so less polluting.

4) Governments can also put a cap on emissions of all greenhouse gases that companies make — then sell licences for emissions up to that cap.

5) There's also technology that captures the CO_2 produced by burning fossil fuels before it's released into the atmosphere — it can then be stored deep underground in cracks in the rock, such as old oil wells.

Carbon Footprints

But Making **Reductions** is **Still Difficult**

1) It's easy enough saying that we should cut emissions, but actually doing it — that's a different story.

2) For a start, there's still a lot of work to be done on alternative technologies that result in lower CO_2 emissions. For example:

- Carbon capture and storage (see previous page) is a relatively new idea. At the moment the technology is still at the developmental stage.
- Many renewable energy technologies, e.g. solar panels, are still quite expensive. More development should make them cheaper, so they can be used more widely.

3) A lot of governments are also worried that making changes to reduce carbon dioxide emission could have an impact on the economic growth of their countries — which could be bad for people's well-being. This is particularly important for countries that are still developing.

Forests take in carbon dioxide and burning trees releases carbon dioxide. So preventing deforestation can help to cut a country's carbon footprint.

4) Because not everyone is on board, it's hard to make international agreements to reduce emissions. Most countries don't want to sacrifice their economic development if they think that others won't do the same.

5) It's not just governments, though — individuals (particularly those in developed countries) need to make changes to their lifestyles. For example:

An individual can reduce their personal carbon footprint by:
- choosing to cycle or walk instead of using a car,
- reducing how much they use air travel,
- doing anything that saves energy at home (e.g. turning heating down).

6) But it might be hard to get people to make changes if they don't want to and if there isn't enough education provided about why the changes are necessary and how to make them.

Lifestyle changes can be tough, but they might be essential in future

You need to be able to describe how carbon footprints could be reduced — don't forget that governments, as well as individuals, can take actions to reduce carbon emissions. Make sure that you can also give reasons why it can be hard to get countries and people to take actions to reduce their carbon footprints.

Air Pollution

Increasing carbon dioxide is causing climate change. But CO_2 isn't the only gas released when fossil fuels burn — you also get other nasty gases, like oxides of nitrogen, sulfur dioxide and carbon monoxide.

Combustion of Fossil Fuels Releases Gases and Particles

1) Fossil fuels, such as crude oil and coal, contain hydrocarbons. During combustion, the carbon and hydrogen in these compounds are oxidised, so that carbon dioxide and water vapour are released back into the atmosphere.

Hydrocarbons are compounds that only contain hydrogen and carbon (see p.132).

2) When there's plenty of oxygen, all the fuel burns — this is called complete combustion.
3) If there's not enough oxygen, some of the fuel doesn't burn — this is called incomplete combustion. Under these conditions, solid particles (called particulates), made up of soot (carbon) and unburned hydrocarbons, are released and carbon monoxide can be produced as well as carbon dioxide.
4) Particulates in the air can cause all sorts of problems:

- If particulates are inhaled, they can get stuck in the lungs and cause damage. This can then lead to respiratory problems.
- They're bad for the environment too. Particulates (and the clouds they help to produce), reflect sunlight back into space. This means that less light reaches the Earth, causing global dimming.

5) It's not just particulates from incomplete combustion that cause problems. Carbon monoxide is pretty nasty too.

- Carbon monoxide (CO) is really dangerous because it can stop your blood from doing its job of carrying oxygen around the body.
- It does this by binding to the haemoglobin in your blood that normally carries O_2 — so less oxygen is able to be transported round your body.
- A lack of oxygen in the blood can lead to fainting, a coma or even death.
- Carbon monoxide doesn't have any colour or smell, so it's very hard to detect. This makes it even more dangerous.

There's more about complete combustion on p.133 and incomplete combustion on p.169.

Sulfur Dioxide and Oxides of Nitrogen Can be Released

1) Sulfur dioxide (SO_2) is released during the combustion of fossil fuels, such as coal, that contain sulfur impurities — the sulfur in the fuel becomes oxidised.
2) Nitrogen oxides are created from a reaction between the nitrogen and oxygen in the air, caused by the heat of burning. (This can happen in the internal combustion engines of cars.)
3) When these gases mix with water in clouds they form dilute sulfuric acid or dilute nitric acid. This then falls as acid rain.
4) Acid rain kills plants and damages buildings and statues. It also makes metal corrode.
5) Not only that, but sulfur dioxide and nitrogen oxides can also be bad for human health — they cause respiratory problems if they're breathed in.

You can test for sulfur impurities in a fuel by bubbling the gases from combustion through universal indicator solution. If the fuel contains sulfur, the gases will contain SO_2, which will form sulfuric acid and turn the universal indicator red.

Fossil fuels are bad news — but we depend on them for many things...

...so a big reduction in their use is likely to be hard to achieve. Make sure you know the different pollutants that are given out when fuels burn, as well as the differences between complete and incomplete combustion.

Warm-Up & Exam Questions

There's lots of important information in this section, from the Earth's atmosphere to climate change and pollution. Answer these questions to see what you can remember and what you need to go over again.

Warm-Up Questions

1) Where do scientists think the gases that made up Earth's early atmosphere came from?
2) Give an example of a sedimentary rock.
3) Give an example of a fossil fuel.
4) Suggest one way that an individual could reduce their personal carbon footprint.
5) Name two potential pollutants that could be released as a result of incomplete combustion of hydrocarbons, that wouldn't be released as a result of complete combustion.

Exam Questions

1 Use words from the box to complete the sentences below. (Grade 4-6)

oxygen	carbon	carbon dioxide	photosynthesis	respiration

Once green plants had evolved, they thrived in Earth's early atmosphere which was rich in
.... These plants also produced .... by the process of ....
Some of the .... from dead plants eventually became 'locked up' in fossil fuels.

[4 marks]

2 Burning fossil fuels can produce pollutants like carbon dioxide, sulfur dioxide and particulate matter. (Grade 6-7)

2.1 Describe how sulfur dioxide can cause acid rain.

[2 marks]

2.2 Give **one** environmental problem caused by acid rain.

[1 mark]

2.3 Explain how burning fossil fuels can lead to global dimming.

[3 marks]

2.4 Carbon dioxide is a greenhouse gas. Describe how the greenhouse effect is responsible for warming the surface of the Earth.

[4 marks]

3 Many people believe that it is necessary for individuals to reduce their carbon footprints in order to prevent climate change from happening in the future. (Grade 7-9)

3.1 Define the term 'carbon footprint'.

[1 mark]

3.2* Describe some of the effects that an increase in global temperature could have on the environment. Explain the impact these effects may have on humans.

[6 marks]

Revision Summary for Topic 9

That's all for Topic 9, but before you breathe a sigh of relief, there are some questions to get through.

- Try these questions and tick off each one when you get it right.
- When you've done all the questions under a heading and are completely happy with it, tick it off.

The Evolution of the Atmosphere (p.163-164) ☐

1) Scientists believe that the early atmosphere was mostly carbon dioxide. Describe how the level of carbon dioxide in the atmosphere were reduced. ☐
2) Name four gases, other than carbon dioxide, that scientists think were present in the early atmosphere. ☐
3) Describe how limestone was formed from organisms millions of years ago. ☐
4) Write the balanced chemical equation for photosynthesis. ☐
5) State the approximate composition of the atmosphere today. ☐

Climate Change and Greenhouse Gases (p.165-166) ☐

6) Name three greenhouse gases. ☐
7) Explain how the greenhouse effect keeps the Earth warm. ☐
8) State three ways in which human activity is leading to an increase in carbon dioxide in the atmosphere. ☐
9) Describe the recent trend that scientists have linked to increasing levels of carbon dioxide in the Earth's atmosphere. ☐

Carbon Footprints (p.167-168) ☐

10) Give two ways in which a government could act to try to reduce the carbon footprints of companies in their country. ☐
11) Explain why it can be difficult to persuade countries to reduce their carbon dioxide emissions. ☐

Air Pollution (p.169) ☐

12) Describe how the following air pollutants are produced:
    a) particulates,
    b) carbon monoxide,
    c) sulfur dioxide,
    d) nitrogen oxides. ☐
13) Why is carbon monoxide dangerous? ☐
14) Describe a test that you could use to detect sulfur impurities in a fuel. ☐
15) Apart from acid rain, give one other problem which may be caused by sulfur dioxide and nitrogen oxides. ☐

Materials and their Properties

Over time, humans have developed and adopted materials which we use to make our lives easier.

Ceramics Come in Many Different Forms

Ceramics are non-metal solids with high melting points that aren't made from carbon-based compounds.

1) Some ceramics can be made from clay.
2) Clay is a soft material when it's dug up out of the ground, so can be moulded into different shapes.
3) When it's fired at high temperatures, it hardens to form a clay ceramic.
4) Its ability to be moulded when wet and then hardened makes clay ideal for making pottery and bricks.
5) Another example of a ceramic is glass. Glass is generally transparent, can be moulded when hot and can be brittle when thin.
6) Most glass made is soda-lime glass, which is made by heating a mixture of limestone, sand and sodium carbonate (soda) until it melts. When the mixture cools it comes out as glass.
7) Borosilicate glass has a higher melting point than soda-lime glass. It's made in the same way as soda-lime glass, using a mixture of sand and boron trioxide.

Composites are Generally Made of Two Different Materials

1) Composites are made of one material embedded in another. Fibres or fragments of a material (known as the reinforcement) are surrounded by a matrix acting as a binder.
2) The properties of a composite depend on the properties of the materials it is made from. For example:

Fibreglass consists of fibres of glass embedded in a matrix made of polymer (plastic). It has a low density (like plastic) but is very strong (like glass). It's used for things like skis, surfboards and boats.

Carbon fibre composites also have a polymer matrix. The reinforcement is either made from long chains of carbon atoms bonded together (carbon fibres) or from carbon nanotubes (see p.60). These composites are very strong and light so are used in aerospace and sports car manufacturing.

Concrete is made from aggregate (any material made from fragments — usually sand and gravel are used in concrete) embedded in cement. It's very strong. This makes it ideal for use as a building material, e.g. in skate parks.

Wood is a natural composite of cellulose fibres held together by an organic polymer matrix.

Materials and their Properties

Polymers Can Have Very Different Properties

Two important things can influence the properties of a polymer — how it's made and what it's made from.

1) For example, the properties of poly(ethene) depend on the catalyst that was used and the reaction conditions (the temperature and pressure) that it was made under.

- Low density (LD) poly(ethene) is made from ethene at a moderate temperature under a high pressure. It's flexible and is used for bags and bottles.
- High density (HD) poly(ethene) is also made from ethene but at a lower temperature and pressure with a catalyst. It's more rigid and is used for water tanks and drainpipes.

2) The monomers that a polymer is made from determine the type of bonds that form between the polymer chains. These weak bonds between the different molecule chains determine the properties of the polymer:

- Thermosetting polymers contain monomers that can form cross-links between the polymer chains, holding the chains together in a solid structure. Unlike thermosoftening polymers, these polymers don't soften when they're heated. Thermosetting polymers are strong, hard and rigid.
- Thermosoftening polymers contain individual polymer chains entwined together with weak forces between the chains. You can melt these polymers and remould them.

Different Materials are Suited to Different Jobs

What materials are used for depends on their properties. You need to be able to interpret information about the properties of materials and decide how suitable these materials would be for different purposes.

Ceramics include glass and clay ceramics such as porcelain and bricks. They're insulators of heat and electricity, brittle (they aren't very flexible and break easily) and stiff.

Polymers are insulators of heat and electricity, they can be flexible (they can be bent without breaking) and can be easily moulded. Polymers have many applications including in clothing and insulators in electrical items.

The properties of composites depend on the matrix/binder and the reinforcement used to make them, so they have many different uses.

Metals are generally malleable, good conductors of heat and electricity, ductile (they can be drawn into wires), shiny and stiff (see p.37-38). Metals have many uses, including in electrical wires, car bodywork, and cutlery.

EXAMPLE: **Suggest which material would be best for making the hull of a large ship.**

Material	Brittleness	Water resistance
Clay ceramic	brittle	porous (absorbs water)
Rigid PVC (a polymer)	brittle	non-porous, corrosion resistant
Low carbon steel	tough	non-porous, corrodes in water, but cheap to protect against corrosion.

1) A ship's hull needs to be non-porous so it doesn't absorb water, and shouldn't corrode.
This rules out clay ceramic — it absorbs water so the ship could sink.
Low carbon steel corrodes when unprotected, but it's cheap to protect so could still be suitable.

2) If the hull's brittle, it could suddenly break.
Rigid PVC is brittle, so not suitable.

Low carbon steel is non-porous, strong and can be easily protected against corrosion. ← Low carbon steel looks like the best choice.

Alloys

Pure Metals Don't Always Have the **Properties Needed**

1) As you saw on page 61, the regular structure of pure metals makes them <u>soft</u> — often too soft for use in everyday life.
2) <u>Alloys</u> are made by adding another element to the metal. This disrupts the structure of the metal, making alloys <u>harder</u> than pure metals.
3) For example, alloys of iron called <u>steels</u> are often used instead of pure iron. Steels are made by adding <u>small</u> amounts of <u>carbon</u> and sometimes <u>other metals</u> to the pure iron.

For more on the structure of metals have a look at p.61.

Type of Steel	Properties	Uses
Low carbon steel (0.1-0.3% carbon)	easily shaped	car bodies
High carbon steel (0.22-2.5% carbon)	very hard, inflexible	blades for cutting tools, bridges
Stainless steel (chromium added, and sometimes nickel)	corrosion-resistant	cutlery, containers for corrosive substances

4) Many other <u>alloys</u> are used in everyday life.

Bronze = Copper + Tin

Bronze is <u>harder</u> than copper.
It's used to make medals, decorative ornaments and statues.

Brass = Copper + Zinc

Brass is more <u>malleable</u> than bronze and is used in situations where <u>lower friction</u> is required, such as in water taps and door fittings.

Gold alloys are used to make jewellery

Pure gold is <u>very soft</u>. Metals such as zinc, copper, and silver are used to <u>harden</u> the gold. Pure gold is described as <u>24 carat</u>, so 18 carats means that 18 out of 24 parts of the alloy are pure gold. In other words, 18 carat gold is 75% gold.

Aluminium alloys are used to make aircraft

Aluminium has a <u>low density</u> which is an important property in aircraft manufacture. But pure aluminium is <u>too soft</u> for making aeroplanes so it's <u>alloyed</u> with small amounts of other metals to make it <u>stronger</u>.

Alloys are really important in industry

If the properties of a metal aren't quite right to a job, an alloy is often used <u>instead</u>. By mixing a metal with another element, the properties are <u>changed</u>. For example, the finished alloy can be a lot <u>harder</u> or less <u>brittle</u> — the properties can be varied so that an alloy can be made to suit a particular job really well.

Corrosion

Some metals react with substances in their environment, and this can make them flake away.

Iron and Steel **Corrode Much More** than Aluminium

Corrosion is where metals react with substances in their environment and are gradually destroyed.

1) Iron corrodes easily. In other words, it rusts.
2) In order to rust, iron needs to be in contact with both oxygen and water, which are present in air.

The word "rust" is only used for the corrosion of iron and its alloys (e.g. steel), not other metals.

3) The stuff we call rust is actually the compound hydrated iron(III) oxide. Here's the equation for it's formation:

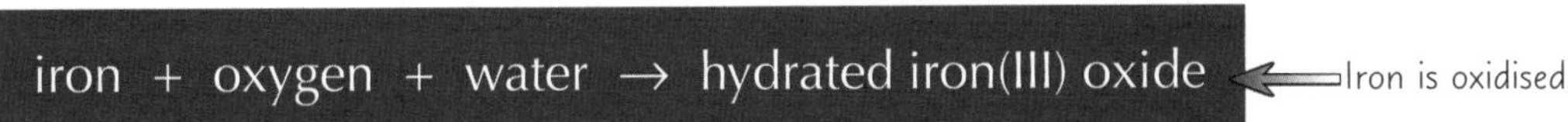
iron + oxygen + water → hydrated iron(III) oxide
Iron is oxidised.

4) Corrosion only happens on the surface of a material, where it's exposed to the air.
5) Unfortunately, rust is a soft crumbly solid that soon flakes off to leave more iron available to rust again. This means that, eventually, all the iron in an object corrodes away even if it wasn't initially at the surface.

Iron exposed to the surface as rust flakes away.
Iron(III) oxide flakes
Iron

6) Aluminium also corrodes when exposed to air. Unlike iron objects, things made from aluminium aren't completely destroyed by corrosion. This is because the aluminium oxide that forms when aluminium corrodes doesn't flake away. In fact, it forms a nice protective layer that sticks firmly to the aluminium below and stops any further reaction taking place.

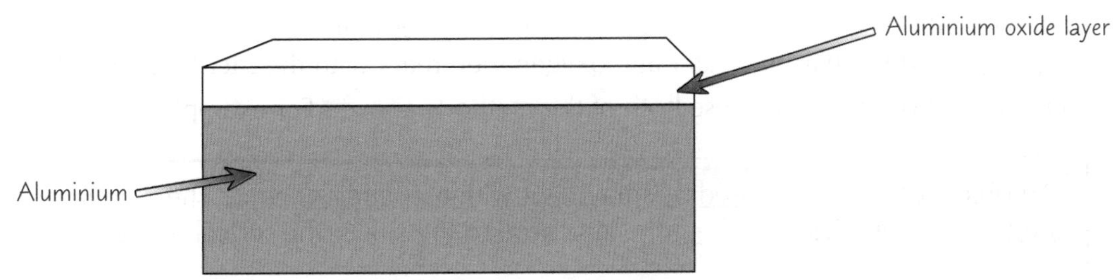
Aluminium oxide layer
Aluminium

REVISION TIP

When aluminium corrodes a layer of aluminium oxide forms

Make sure you learn the stuff on this page including the differences in the corrosion of iron and aluminium — it might be useful to memorise the diagrams to get to grips with how they corrode.

Corrosion

Corrosion can be very destructive, but there are a number of ways that we can protect metals from its effects.

Both **Air** and **Water** are Needed for **Iron** to **Rust**

Experiments can show that both oxygen and water are needed for iron to rust.

1) To show that water alone is not enough, you can put an iron nail in a boiling tube with just water and the nail won't rust. (The water is boiled to remove oxygen and oil is used to stop air getting in.)
2) To show that oxygen alone is not enough, you can put an iron nail in a boiling tube with just air and the nail won't rust. (Calcium chloride can be used to absorb any water from the air.)
3) However, if you put an iron nail in a boiling tube with air and water, it will rust.

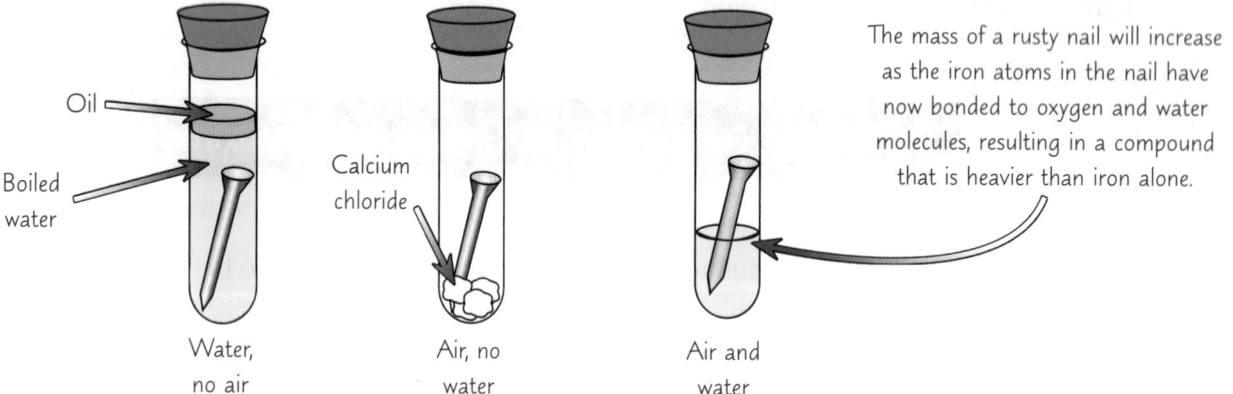
Oil
Boiled water
Calcium chloride
The mass of a rusty nail will increase as the iron atoms in the nail have now bonded to oxygen and water molecules, resulting in a compound that is heavier than iron alone.
Water, no air
Air, no water
Air and water

There are **Two** Main Ways to **Prevent Rusting**

1) The obvious way to prevent rusting is to coat the iron with a barrier to keep out the water and oxygen. This can be done by:

> Painting/Coating with plastic — ideal for big and small structures alike. It can be decorative too.

> Electroplating — this uses electrolysis to reduce metal ions onto an iron electrode. It can be used to coat the iron with a layer of a different metal that won't be corroded away.

> Oiling/Greasing — this has to be used when moving parts are involved, like on bike chains.

2) Another method is the sacrificial method. This involves placing a more reactive metal, such as zinc or magnesium, with the iron. Water and oxygen then react with the sacrificial metal instead of with the iron.
3) Some protection techniques use both of the methods above. For example:

> An object can be galvanised by spraying it with a coating of zinc. The zinc layer is firstly protective, but if it's scratched, the zinc around the site of the scratch works as a sacrificial metal.

Coatings and the sacrificial method can be used to prevent rust

Make sure you can describe experiments that investigate rusting. Remember, for rusting to occur, both oxygen and water need to be in contact with iron or steel, but there are ways to stop it happening.

Warm-Up & Exam Questions

There's tonnes of important information on these pages. If you don't think you've got it covered, go back through and check any areas that you might be a bit unsure of, then have a go at the questions on this page.

Warm-Up Questions

1) What are ceramics?
2) Why don't thermosetting polymers melt?
3) What two metals is brass made from?
4) Write the word equation for the rusting of iron.
5) How does electroplating iron prevent it from rusting?

Exam Questions

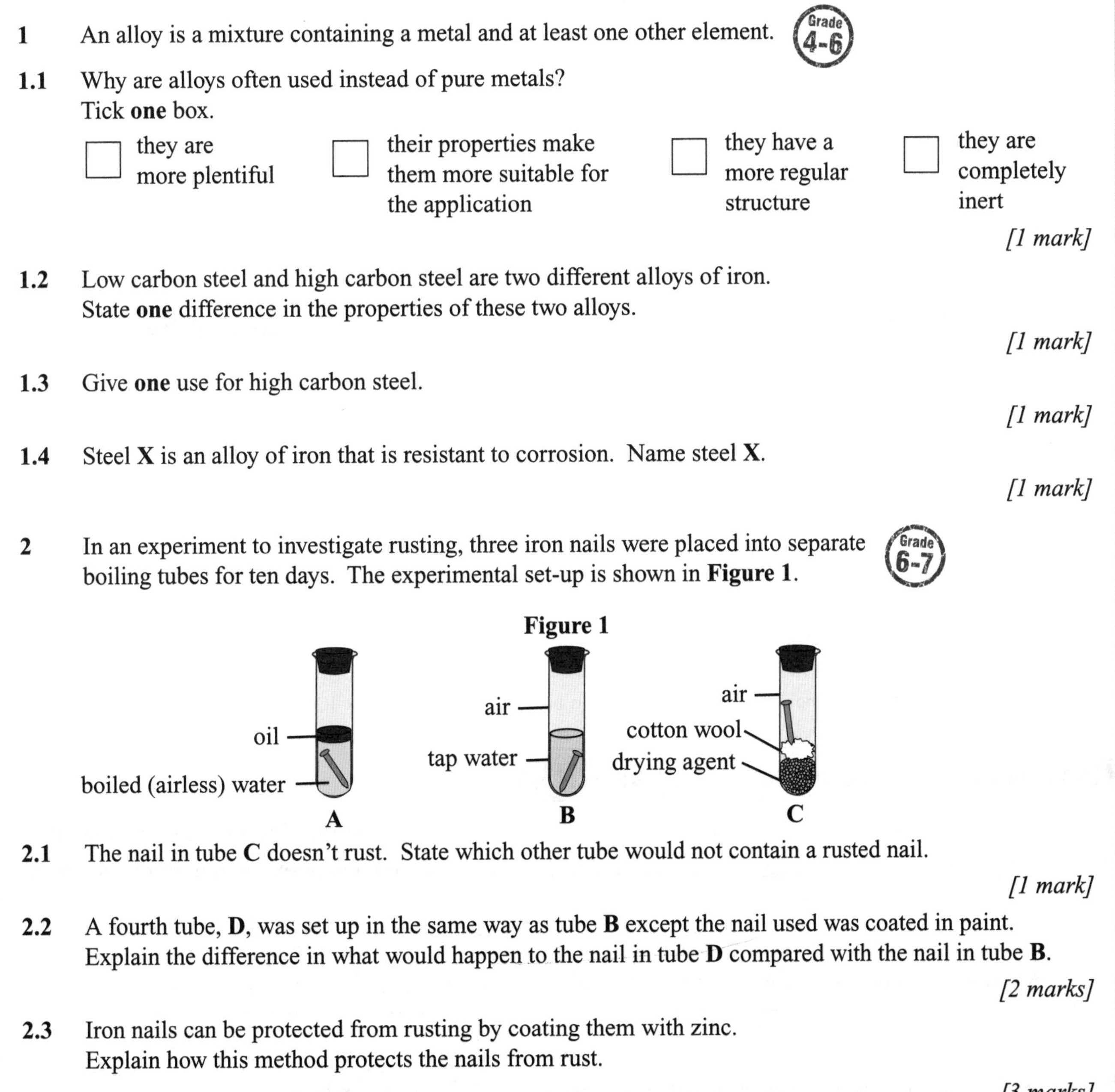

1 An alloy is a mixture containing a metal and at least one other element. (Grade 4-6)

1.1 Why are alloys often used instead of pure metals?
Tick **one** box.

- ☐ they are more plentiful
- ☐ their properties make them more suitable for the application
- ☐ they have a more regular structure
- ☐ they are completely inert

[1 mark]

1.2 Low carbon steel and high carbon steel are two different alloys of iron.
State **one** difference in the properties of these two alloys.

[1 mark]

1.3 Give **one** use for high carbon steel.

[1 mark]

1.4 Steel **X** is an alloy of iron that is resistant to corrosion. Name steel **X**.

[1 mark]

2 In an experiment to investigate rusting, three iron nails were placed into separate boiling tubes for ten days. The experimental set-up is shown in **Figure 1**. (Grade 6-7)

Figure 1

2.1 The nail in tube **C** doesn't rust. State which other tube would not contain a rusted nail.

[1 mark]

2.2 A fourth tube, **D**, was set up in the same way as tube **B** except the nail used was coated in paint.
Explain the difference in what would happen to the nail in tube **D** compared with the nail in tube **B**.

[2 marks]

2.3 Iron nails can be protected from rusting by coating them with zinc.
Explain how this method protects the nails from rust.

[3 marks]

Finite and Renewable Resources

There are lots of different resources that humans use to provide energy for things like heating or travelling, as well as for building materials and food. Unfortunately, some of these resources will run out one day.

Natural Resources Come From the Earth, Sea and Air

1) Natural resources form without human input. They include anything that comes from the earth, sea or air. For example, cotton for clothing or oil for fuel.
2) Some of these natural products can be replaced by synthetic products or improved upon by man-made processes. For example, rubber is a natural product that can be extracted from the sap of a tree, however man-made polymers have now been made which can replace rubber in uses such as tyres.
3) Agriculture provides conditions where natural resources can be enhanced for our needs. E.g. the development of fertilisers have meant we can produce a high yield of crops.

Some Natural Resources will Run Out

1) Renewable resources reform at a similar rate to, or faster than, we use them.
2) For example, timber is a renewable resource as trees can be planted following a harvest and only take a few years to regrow. Other examples of renewable resources include fresh water and food.
3) Finite (non-renewable) resources, aren't formed quickly enough to be considered replaceable.
4) Finite resources include fossil fuels and nuclear fuels such as uranium and plutonium. Minerals and metals found in ores in the earth are also non-renewable materials.
5) After they've been extracted, many finite resources undergo man-made processes to provide fuels and materials necessary for modern life. E.g. crude oil can undergo fractional distillation (see p.134) to produce products such as petrol, and metal ores can be reduced to produce pure metals (see p.96).

Tables, Charts and Graphs Give You an Insight Into Different Resources

You may be asked to interpret information about resources in the exam.

EXAMPLE:

The table below shows information for two resources, coal and timber. Identify which resource is which.

	Energy Density (MJ/m^3)	Time it takes to form
Resource 1	7600-11 400	10 years
Resource 2	23 000-26 000	10^6 years

The time it takes for Resource 1 to reform is 10^5 times shorter than Resource 2 suggesting it is a renewable resource. Resource 1 also has a much lower energy density than Resource 2, so is more likely to be timber than coal.

Resource 1 is timber and Resource 2 is coal.

10^6 is a shorthand way of showing 1 000 000. This is because $10^6 = 10 \times 10 \times 10 \times 10 \times 10 \times 10 = 1\,000\,000$.

Extracting Finite Resources has Risks

1) Many modern materials are made from raw, finite resources, e.g. most plastics and building materials.
2) People have to balance the social, economic and environmental effects of extracting finite resources.

For example, mining metal ores is good because useful products can be made. It also provides local people with jobs and brings money into the area. However, mining ores is bad for the environment as it uses loads of energy, scars the landscape, produces lots of waste and destroys habitats.

Natural resources can be either renewable or finite

Before you move on, make sure you learn a few examples of finite and renewable resources.

Sustainability

Many materials used in the modern world are limited. Scientists are constantly trying to find new and improved ways to extract and use natural resources more sustainably.

Chemistry is **Improving Sustainability**

1) Sustainable development is an approach to development that takes account of the needs of present society while not damaging the lives of future generations.
2) As you saw on the previous page, not all resources are renewable so it's unsustainable to keep using them.
3) As well as using resources, extracting resources can be unsustainable due to the amount of energy used and waste produced. Processing the resources into useful materials, such as glass or bricks, can be unsustainable too as the processes often use energy that's made from finite resources.
4) If people reduce how much they use of a finite resource, that resource is more likely to last longer. Reducing usage of these resources will also reduce the use of anything needed to produce them.
5) We can't stop using finite resources altogether, but chemists can develop and adapt processes that use lower amounts of finite resources and reduce damage to the environment. For example, chemists have developed catalysts that reduce the amount of energy required for certain industrial processes.

Copper-Rich Ores are in **Short Supply**

1) Copper is a finite resource — the supply of copper-rich ores is limited.
2) One way to improve its sustainability is by extracting it from low-grade ores (ores without much copper in). Scientists are looking into new ways of doing this:

Bioleaching

Bacteria are used to convert copper compounds in the ore into soluble copper compounds, separating out the copper from the ore in the process. The leachate (the solution produced by the process) contains copper ions, which can be extracted, e.g. by electrolysis see p.100) or displacement (see p.98) with a more reactive metal, e.g. scrap iron.

These methods can be used to extract other metals too.

Phytomining

Phytomining involves growing plants in soil that contains copper. The plants can't use or get rid of the copper so it gradually builds up in the leaves. The plants can be harvested, dried and burned in a furnace. The ash contains soluble copper compounds from which copper can be extracted by electrolysis or displacement using scrap iron.

3) Traditional methods of copper mining are pretty damaging to the environment (see next page). These new methods of extraction have a much smaller impact, but the disadvantage is that they're slow.

The extraction of resources can be unsustainable

If we keep using finite resources at the rate that we currently are, one day they will run out. We need to find alternatives for finite resources, but in the meantime we can use methods to make them last as long as possible. For example, bioleaching and phytomining are a way of making reserves of metal ores last longer.

Recycling

Once an object made from a finite resource is worn out, it's usually more sustainable to recycle it than to use new raw materials to replace it. Read on to find out why...

Recycling Metals is Important

1) Mining and extracting metals takes lots of energy, most of which comes from burning fossil fuels.

2) Recycling metals often uses much less energy than is needed to mine and extract new metal, conserves the finite amount of each metal in the earth and cuts down on the amount of waste getting sent to landfill.

Recycling is a way to reduce our need for copper-rich ores.

3) Metals are usually recycled by melting them and then casting them into the shape of the new product.

4) Depending on what the metal will be used for after recycling, the amount of separation required for recyclable metals can change. For example:

Waste steel and iron can be kept together as they can be both added to iron in a blast furnace to reduce the amount of iron ore required.

A blast furnace is used to extract iron from its ore at a high temperature using carbon.

Glass can Also be Recycled

Glass recycling can help sustainability by reducing the amount of energy needed to make new glass products, and also the amount of waste created when used glass is thrown away.

1) Glass bottles can often be reused without reshaping.

2) Other forms of glass can't be reused so they're recycled instead. Usually the glass is separated by colour and chemical composition before being recycled.

3) The glass is crushed and then melted to be reshaped for use in glass products such as bottles or jars. It might also be used for a different purpose such as insulating glass wool for wall insulation in homes.

Recycling is key to sustainability — it's useful in lots of ways...

Remember that recycling doesn't just reduce the use of raw materials, it reduces the amount of energy used, the amount of damage to the environment and the amount of waste produced.

Life Cycle Assessments

If a company wants to manufacture a new product, they carry out a life cycle assessment (LCA).

Life Cycle Assessments Show Total Environmental Costs

1) A life cycle assessment (LCA) looks at every stage of a product's life to assess the impact it would have on the environment.
2) Here are the different stages that need to be considered:

1) Getting the Raw Materials

1) Extracting raw materials needed for a product can damage the local environment, for example, mining metals. Extraction can also result in pollution due to the amount of energy needed.
2) Raw materials often need to be processed to extract the desired materials and this often needs large amounts of energy. For example, extracting metals from ores or fractional distillation of crude oil.

2) Manufacturing and Packaging

1) Manufacturing products and their packaging can use a lot of energy resources and can also cause a lot of pollution. For example, harmful fumes such as carbon monoxide or hydrogen chloride.
2) You also need to think about any waste products and how to dispose of them. The chemical reactions used to make compounds from their raw materials can produce waste products. Some waste can be turned into other useful chemicals, reducing the amount that ends up polluting the environment.

3) Using the Product

1) The use of a product can damage the environment. For example, burning fuels releases greenhouse gases and other harmful substances. Fertilisers can leach into streams and rivers causing damage to ecosystems.
2) How long a product is used for or how many uses it gets is also a factor — products that need lots of energy to produce but are used for ages may mean less waste in the long run.

4) Product Disposal

1) Products are often disposed of in landfill sites. This takes up space and pollutes land and water. E.g. paint may wash off a product in landfill and pollute a river.
2) Energy is used to transport waste to landfill, which causes pollutants to be released into the atmosphere.
3) Products might be incinerated (burnt), which causes air pollution.

LCAs look at the environmental impact of a product's entire life

Doing a life cycle assessment for a product can be a very time consuming and expensive process, as there is so much to take account of. Some things can only be predicted, for example, how a consumer will use a product. Get to grips with each stage and what factors have to be taken into account at every step.

Life Cycle Assessments

You Can Compare Life Cycle Assessments for Plastic and Paper Bags

The table below shows how you might carry out a LCA of plastic and paper bags.

Life Cycle Assessment Stage	Plastic Bag	Paper Bag
Raw Materials	Crude oil	Timber
Manufacturing and Packaging	The compounds needed to make the plastic are extracted from crude oil by fractional distillation, followed by cracking and then polymerisation. Waste is reduced as the other fractions of crude oil have other uses.	Pulped timber is processed using lots of energy. Lots of waste is made.
Using the Product	Can be reused. Can be used for other things as well as shopping, for example bin liners.	Usually only used once.
Product Disposal	Recyclable but not biodegradable and will take up space in landfill and pollute land.	Biodegradable, non-toxic and can be recycled.

Life cycle assessments have shown that even though plastic bags aren't biodegradable, they take less energy to make and have a longer lifespan than paper bags, so may be less harmful to the environment.

There are Problems with Life Cycle Assessments

1) The use of energy, some natural resources and the amount of certain types of waste produced by a product over it's lifetime can be easily quantified. But the effect of some pollutants is harder to give a numerical value to. E.g. it's difficult to apply a value to the negative visual effects of plastic bags in the environment compared to paper ones.

2) So, producing an LCA is not an objective method as it takes into account the values of the person carrying out the assessment. This means LCAs can be biased.

3) Selective LCAs, which only show some of the impacts of a product on the environment, can also be biased as they can be written to deliberately support the claims of a company in order to give them positive advertising.

LCAs aren't all they are cracked up to be...

In the exam, you may have to analyse LCAs comparing different products and come to a conclusion about which product is the most environmentally friendly. But don't forget that not all environmental impacts can be measured in an LCA. Also, the opinions of the assessors have an effect on the result of the assessment and the results can be misused to suit an organisations aims.

Warm-Up & Exam Questions

That's some more revision done and dusted and now it's time to test yourself on how much you've taken in. Have a go at the questions on this page to see if you need to revisit some topics.

Warm-Up Questions

1) Give an example of a natural resource which has been replaced by a man-made alternative.
2) What is sustainable development?
3) Give three positive effects of recycling metals.
4) What are the four stages that need to be considered to conduct a life cycle assessment?

Exam Questions

1 Natural resources are formed without human input and are used for construction, fuel and food. Grade 4-6

1.1 What is a finite natural resource?
Tick **one** box.

A natural resource that can never be remade. ☐

A natural resource that will never run out. ☐

A natural resource that doesn't renew itself quickly enough to be considered replaceable. ☐

A natural resource that can only be used in certain conditions. ☐

[1 mark]

1.2 Aluminium is used to make soft drink cans. Extracting aluminium is a very energy intensive process.
Suggest **two** ways that the use of soft drink cans can be made more sustainable.

[2 marks]

2 Copper needs to be extracted from its ore before it can be used. Grade 6-7

2.1 Explain why scientists have developed ways of extracting copper from low-grade ores.

[1 mark]

2.2 Copper can be extracted from low-grade ores by a process called bioleaching.
Explain how the process works.

[2 marks]

2.3 Give **one** advantage of using processes such as bioleaching rather than traditional mining.

[1 marks]

2.4 Give **one** disadvantage of using processes such as bioleaching rather than traditional mining.

[1 mark]

3 A life cycle assessment looks at the environmental impact of a product over its lifetime. Grade 6-7

3.1 Describe why it can be difficult to give a complete assessment of a product over its entire life cycle.

[1 mark]

3.2 A company carried out a LCA on one of their products but didn't take into account its disposal.
However, an independent LCA found the disposal of the product to have the greatest environmental impact of any part of the product's life cycle.
Suggest why the LCA performed by the company didn't include details of the disposal of the product.

[2 marks]

Potable Water and Water Treatment

We all need safe drinking water. The way that water's made safe depends on local conditions.

Potable Water is Water You Can Drink

1) Potable water is water that's been treated or is naturally safe for humans to drink — it's essential for life.
2) Chemists wouldn't call it pure, though. Pure water only contains H_2O molecules whereas potable water can contain lots of other dissolved substances.
3) The important thing is that the levels of dissolved salts aren't too high, that it has a pH between 6.5 and 8.5 and also that there aren't any nasties (like bacteria or other microbes) swimming around in it.

How Potable Water is Produced Depends on Where You Are

1) Rainwater is a type of fresh water. Fresh water is water that doesn't have much dissolved in it.
2) When it rains, water can either collect as surface water (in lakes, rivers and reservoirs) or as groundwater (in rocks called aquifers that trap water underground).
3) In the UK, the source of fresh water used depends on location. Surface water tends to dry up first, so in warm areas, e.g. the south-east, most of the domestic water supply comes from groundwater.
4) Even though it only has low levels of dissolved substances, water from these fresh water sources still needs to be treated to make it safe before it can be used. This process includes:

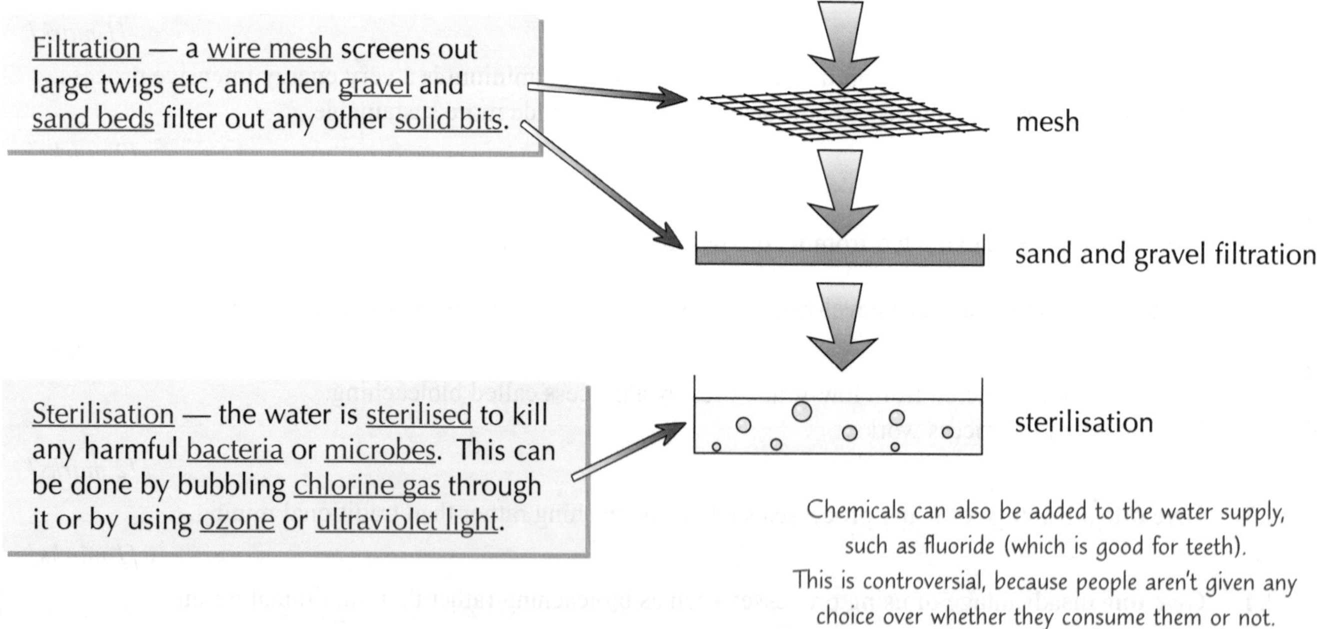

Filtration — a wire mesh screens out large twigs etc, and then gravel and sand beds filter out any other solid bits.
mesh
sand and gravel filtration
Sterilisation — the water is sterilised to kill any harmful bacteria or microbes. This can be done by bubbling chlorine gas through it or by using ozone or ultraviolet light.
sterilisation
Chemicals can also be added to the water supply, such as fluoride (which is good for teeth). This is controversial, because people aren't given any choice over whether they consume them or not.

5) In some very dry countries, e.g. Kuwait, there's not enough surface or groundwater and instead sea water must be treated by desalination to provide potable water. In these countries distillation (see next page) can be used to desalinate sea water.
6) Sea water can also be treated by processes that use membranes — like reverse osmosis. The salty water is passed through a membrane that only allows water molecules to pass through. Ions and larger molecules are trapped by the membrane so separated from the water.
7) Both distillation and reverse osmosis need loads of energy, so they're really expensive and not practical for producing large quantities of fresh water.

Potable Water and Water Treatment

You can **Test** and **Distil Water** in the **Lab**

PRACTICAL

1) First, test the pH of the water using a pH meter. If the pH is too high or too low, you'll need to neutralise it by adding acid (if the pH is high) or alkali (if the pH is low). Use a pH meter to tell you when the solution is between 6.5 and 8.5.

Alternatively you could use a titration to neutralise the water, but this will involve adding an indicator to the water so will contaminate it.

2) You should also test the water for the presence of sodium chloride (the main salt in seawater):
   - To test for sodium ions, do a flame test on a small sample (see p.158). If sodium ions are present the flame will turn yellow.
   - To test for chloride ions, take another sample of your water and add a few drops of dilute nitric acid, followed by a few drops of silver nitrate solution. If chloride ions are present, a white precipitate will form.
3) To distil the water, pour the salty water into a distillation apparatus, like the one on p.28. Heat the flask from below. The water will boil and form steam, leaving any dissolved salts in the flask. The steam will condense back to liquid water in the condenser and can be collected as it runs out.
4) Then, retest the distilled water for sodium chloride to check that it has been removed. Also retest the pH of the water with a pH meter to check that it's between pH 6.5 and 8.5.

Waste Water Comes from Lots of **Different Sources**

1) We use water for lots of things at home — like having a bath, going to the toilet, doing the washing-up, etc. When you flush this water down the drain, it goes into the sewers and towards sewage treatment plants.
2) Agricultural systems also produce a lot of waste water including nutrient run-off from fields and slurry from animal farms.
3) Sewage from domestic or agricultural sources has to be treated to remove any organic matter and harmful microbes before it can be put back into fresh water sources like rivers or lakes. Otherwise it would make them very polluted and would pose health risks.
4) Industrial processes, such as the Haber Process (see p.188), also produce a lot of waste water that has to be collected and treated.
5) As well as organic matter, industrial waste water can also contain harmful chemicals — so it has to undergo additional stages of treatment before it is safe to release it into the environment.

The water that comes out of our taps has been treated

Location is a really important factor in determining how water is treated. For example, in the UK, there is a lot of fresh water available so this is filtered and then sterilised. However, in very dry countries a more expensive process may have to be used, such as the distillation or reverse osmosis of sea water.

Potable Water and Water Treatment

It may not be glamourous but dealing with waste water is vital to stop us polluting our environment.

Sewage **Treatment** Happens in **Several Stages**

Some of the processes involved in treating waste water at sewage treatment plants include:

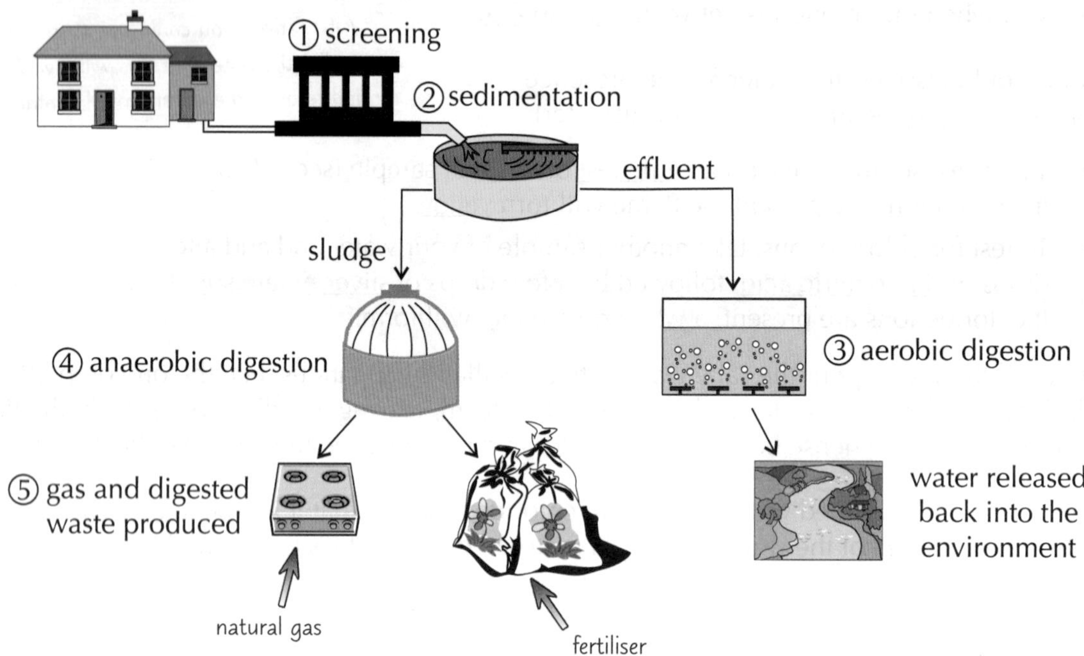

1) Before being treated, the sewage is screened — this involves removing any large bits of material (like twigs or plastic bags) as well as any grit.
2) Then it's allowed to stand in a settlement tank and undergoes sedimentation — the heavier suspended solids sink to the bottom to produce sludge while the lighter effluent floats on the top.
3) The effluent in the settlement tank is removed and treated by biological aerobic digestion. This is when air is pumped through the water to encourage aerobic bacteria to break down any organic matter — including other microbes in the water.
4) The sludge from the bottom of the settlement tank is also removed and transferred into large tanks. Here it gets broken down by bacteria in a process called anaerobic digestion.
5) Anaerobic digestion breaks down the organic matter in the sludge, releasing methane gas in the process. The methane gas can be used as an energy source and the remaining digested waste can be used as a fertiliser.
6) For waste water containing toxic substances, additional stages of treatment may involve adding chemicals (e.g. to precipitate metals), UV radiation or using membranes.

Aerobic just means with oxygen, whereas anaerobic means without oxygen.

Sewage treatment requires more processes than treating fresh water but uses less energy than the desalination of salt water, so could be used as an alternative in areas where there's not much fresh water. For example, Singapore is treating waste water and recycling it back into drinking supplies. However, some people don't like the idea of drinking water that used to be sewage.

Waste water can be recycled to produce potable water

To learn the stages of water treatment, cover the diagram, write out each step and then check it.

Warm-Up & Exam Questions

Now you know all there is to possibly know about water it's time to test yourself with some questions.

Warm-Up Questions

1) Describe two features of potable water that make it safe for drinking.
2) What is the first step in treating fresh water?
3) In addition to organic matter, what else may have to be removed from industrial waste water?
4) What is produced following the sedimentation stage of water treatment?

Exam Questions

1 **Figure 1** shows how water in some parts of the UK is treated during the production of potable water.

Figure 1

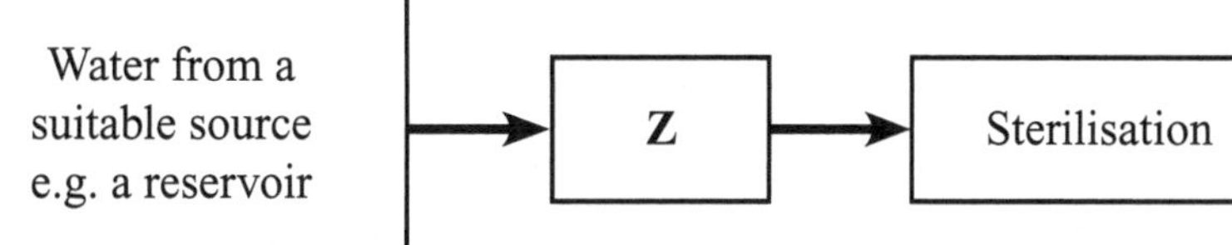

1.1 Describe the difference between potable water and pure water.

[1 mark]

1.2 Describe what happens during stage **Z** in **Figure 1**.

[2 marks]

1.3 Suggest **three** agents that could be used during the sterilisation stage of the process.

[3 marks]

2 Waste water from human activities must undergo treatment before being released into the environment. The stages of sewage treatment include screening, sedimentation, anaerobic digestion and aerobic digestion.

2.1 State the stage at which large pieces of material, such as twigs, are removed.

[1 mark]

2.2 During sedimentation, effluent is produced.
Describe what happens during the next step in the treatment of effluent following sedimentation.

[2 marks]

2.3 In areas where fresh water is limited, treated waste water and sea water can be used as drinking water.
Describe two disadvantages of using these processes compared to the treatment of fresh water.

[2 marks]

PRACTICAL

3 In some dry countries potable water is produced by the distillation of salt water.
A student carries out distillation of salt water in the lab.

3.1 Describe **two** tests which can be used to determine whether or not the distilled water contains any sodium or chloride ions.

[4 marks]

3.2 Suggest a further test that would need to be carried out before the water could be considered potable.

[1 mark]

The Haber Process

This is an important industrial process. It produces ammonia (NH_3), which is used to make fertilisers.

Nitrogen and Hydrogen are Needed to Make Ammonia

The Haber process is used to make ammonia from hydrogen and nitrogen using the following reaction:

$$\text{nitrogen} + \text{hydrogen} \rightleftharpoons \text{ammonia} \quad (+\text{ heat})$$
$$N_{2(g)} + 3H_{2(g)} \rightleftharpoons 2NH_{3(g)}$$

See pages 127-129 for more on reversible reactions.

This reaction is well suited for an industrial scale as the reactants aren't too difficult or expensive to obtain.

1) The nitrogen is obtained easily from the air, which is 78% nitrogen.
2) The hydrogen mainly comes from reacting methane (from natural gas) with steam to form hydrogen and carbon dioxide.
3) The reactant gases are passed over an iron catalyst. A high temperature (450 °C) and a high pressure (200 atmospheres) are used.

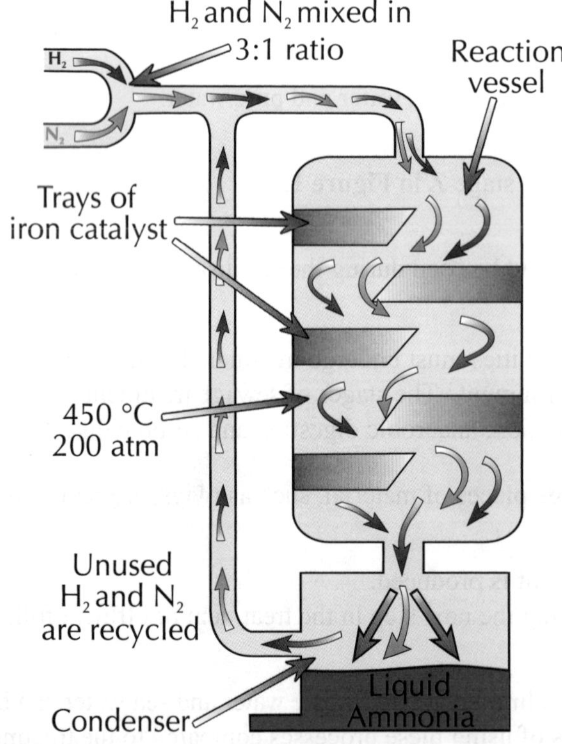

Industrial conditions: pressure = 200 atmospheres; temperature = 450 °C; catalyst: iron.

4) Because the reaction is reversible (it occurs in both directions), some of the ammonia produced converts back into hydrogen and nitrogen again. It eventually reaches a dynamic equilibrium.
5) The ammonia is formed as a gas, but as it cools in the condenser it liquefies and is removed. The unused hydrogen (H_2), and nitrogen, (N_2), are recycled, so nothing is wasted.
6) The ammonia produced can then be used to make ammonium nitrate — a very nitrogen-rich fertiliser (see page 190).

The Haber Process

As the Haber process is reversible, the position of equilibrium is vital in generating a large enough yield. Choosing the best reaction conditions to suit the aims of the reaction in industry can be tricky...

The Reaction is **Reversible**, So There's a **Compromise** to be Made

1) You might remember from page 118 that certain conditions, such as the temperature and pressure, can affect the rate of a reaction.
2) These factors can also affect the position of equilibrium for a reversible reaction — and sometimes there is a trade-off between increasing the rate and maximising the yield (see page 83) of a reaction. E.g.:

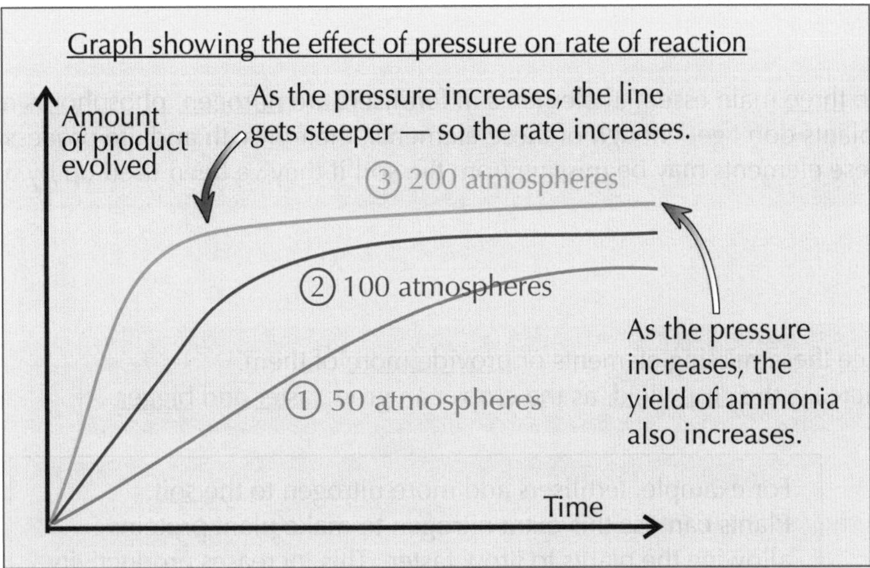

1) The forward reaction in the Haber process is exothermic. That means that increasing the temperature will move the equilibrium the wrong way — away from ammonia and towards nitrogen and hydrogen. So the yield of ammonia would be greater at lower temperatures.
2) The trouble is, lower temperatures mean a slower rate of reaction (and so equilibrium is reached more slowly).

For more on factors that affect the position of equilibrium, see page 128.

3) The 450 °C is a compromise between maximum yield and speed of reaction. It's better to wait just 20 seconds for a 10% yield than to have to wait 60 seconds for a 20% yield.
4) Higher pressures move the position of equilibrium towards the products since there are four molecules of gas on the left-hand side for every two molecules on the right. So increasing pressure maximises the percentage yield. It also increases the rate of reaction.
5) So the pressure is set as high as possible, without making the process too expensive to or dangerous to build and maintain. Hence the 200 atmospheres operating pressure.
6) And finally, the iron catalyst makes the reaction go faster, but doesn't affect the yield.

The Haber process is a really important reaction

Without the Haber process, we wouldn't have the fertiliser which has allowed the human population to grow as quickly as it has over the last 100 years. Make sure you learn the word equation, the sources of hydrogen and nitrogen for the process and the conditions for the reaction. Also understand how the conditions are a compromise between yield and rate.

Fertilisers

Fertilisers help us to produce more food by allowing us to keep growing crops on the same land every year.

NPK Fertilisers Provide Plants with Elements Needed for Growth

1) Farmers used to use manure to fertilise fields. Formulated fertilisers are better as they're more widely available, easier to use, don't smell and have just enough of each nutrient so more crops can be grown.

2) The three main essential elements in fertilisers are nitrogen, phosphorus and potassium. If plants don't get enough of these elements, their growth and life processes are affected. These elements may be missing from the soil if they've been used up by a previous crop.

3) Fertilisers replace these missing elements or provide more of them. This helps to increase the crop yield, as the crops can grow faster and bigger.

For example, fertilisers add more nitrogen to the soil. Plants can use this extra nitrogen to make plant proteins — allowing the plants to grow faster. This increases productivity.

4) NPK fertilisers are formulations containing salts of nitrogen (N), phosphorus (P) and potassium (K) (hence, "NPK") in the right percentages of the elements.

There's more on formulations on p.152.

Ammonia is Used to Produce Nitrogen-Containing Compounds

1) Ammonia can be reacted with oxygen and water in a series of reactions to make nitric acid.
2) You can also react ammonia with acids, including nitric acid, to get ammonium salts, which can be used as fertilisers.
3) Ammonia and nitric acid react together to produce ammonium nitrate — this is an especially good compound to use in a fertiliser because it has nitrogen from two sources.

$$NH_{3\,(aq)} + HNO_{3\,(aq)} \rightarrow NH_4NO_{3\,(aq)}$$

Ammonia + Nitric acid → Ammonium nitrate

Fertilisers

Ammonium Nitrate can be Produced in Different Ways

Fertilisers are often made differently in industry than in the lab. In your exam, you may be asked to compare differences between the methods of production carried out in the lab and in industry, for example, ammonium nitrate:

In Industry

1) The reaction is carried out in giant vats, at high concentrations resulting in a very exothermic reaction.
2) The heat released is used to evaporate water from the mixture to make a very concentrated ammonium nitrate product.

If you get asked to compare reactions in the exams, you'll be given all the details you need.

In the Lab

1) The reaction is carried out on a much smaller scale by titration and crystallisation.
2) The reactants are at a much lower concentration than in industry, so less heat it produced by the reaction and it's safer for a person to carry it out.
3) After the titration, the mixture then needs to be crystallised to give pure ammonium nitrate crystals.
4) Crystallisation isn't used in industry because it's very slow.

For how to carry out a titration or crystallisation have a look at p.88 and p.26.

Phosphate and Potassium are Sourced from Mined Compounds

1) Potassium chloride and potassium sulphate can be mined and used as a source of potassium.
2) Phosphate rock is also mined. However, because the phosphate salts in the rock are insoluble, plants can't directly absorb them and use them as nutrients.
3) Reacting phosphate rock with a number of different types of acids produces soluble phosphates:

Reaction with nitric acid produces phosphoric acid and calcium nitrate.

Reaction with sulfuric acid produces calcium sulfate and calcium phosphate (this mixture is known as single superphosphate).

Reaction with phosphoric acid only produces calcium phosphate (the product of this reaction can be called triple superphosphate).

NPK = nitrogen + phosphorous + potassium

NPK fertilisers are made of compounds which contain all three elements, getting the right percentages of each element in a formulation is the technical part. Getting it right is important for delivering the correct nutrients to crops that farmers are growing. With the right mix of nutrients the growth of crops will be maximised, yields will be high and farmers will have lots to sell. Good news for food production.

Warm-Up & Exam Questions

These pages have all the knowledge you need to know about the Haber process and fertilisers. But don't assume you're the bee's knees — test yourself with this lovely selection of questions.

Warm-Up Questions

1) What two reactants that are need to manufacture ammonia?
2) True or false? The forward reaction in the Haber process is exothermic.
3) Name the catalyst used in the Haber process.
4) What are NPK fertilisers?
5) Name a compound commonly used as a source of potassium in NPK fertilisers.

Exam Questions

1 Farmers often use fertilisers. 

1.1 Fertilisers usually contain nitrogen. Name the **two** other elements they usually contain.

[2 marks]

1.2 Explain why farmers apply fertilisers to their crops.

[2 marks]

2 Fertilisers contain compounds derived from ammonia and phosphate rock. 

2.1 Ammonium nitrate is a fertiliser made by neutralising nitric acid and ammonia.
Suggest why ammonium nitrate is a particularly useful ingredient in a fertiliser.

[1 mark]

2.2 Phosphate rock can't be used as a fertiliser directly. Explain why.

[1 mark]

2.3 Name the compound commonly found in fertilisers made by reacting sulfuric acid with phosphate rock.

[1 mark]

3 Ammonia is made by combining nitrogen and hydrogen at a pressure of 200 atm, a temperature of 450 °C and in the presence of a catalyst. 

A flow diagram is shown for the reaction in **Figure 1**.

Figure 1

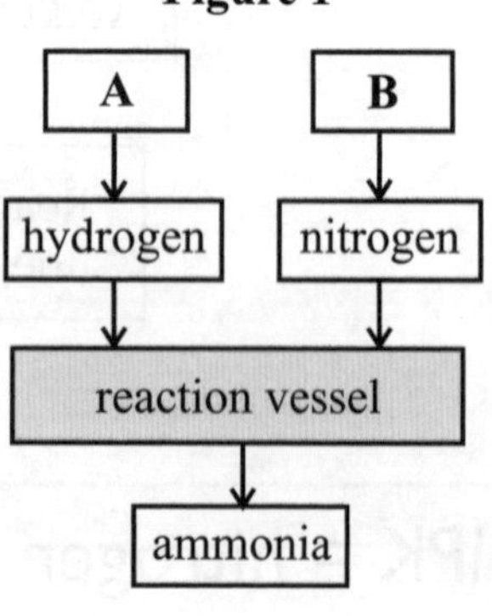

3.1 **A** and **B** represent the sources of nitrogen and hydrogen.
State appropriate sources for nitrogen and hydrogen used in the reaction.

[2 marks]

3.2 Balance the equation for the reaction shown below.

$$..... H_{2(g)} + ... N_{2(g)} \rightleftharpoons ..... NH_{3(g)}$$

[1 mark]

3.3 The reaction is exothermic. What would happen to the yield of ammonia if the temperature was lowered? Explain your answer.

[2 marks]

3.4 Increasing the pressure increases the yield of the reaction.
Suggest why a pressure greater than 200 atm isn't used.

[1 mark]

Revision Summary for Topic 10

That wraps up Topic 10 — time to try these questions to see if you've got this topic in the bag.

- Try these questions and tick off each one when you get it right.
- When you've done all the questions under a heading and are completely happy with it, tick it off.

Materials and Corrosion (p.172-176)

1) How is soda-lime glass made?
2) Give three examples of composite materials.
3) Describe the differences in the properties of low density poly(ethene) and high density poly(ethene).
4) Name two elements that can be added to iron to make stainless steel.
5) What two metals is bronze made of?
6) Give one use of brass.
7) What two substances does iron react with when it rusts?
8) Explain why aluminium is not completed destroyed by corrosion.
9) Suggest three ways of coating iron to form a barrier that protects it from rusting.
10) Suggest one metal that can be used as a sacrificial metal to prevent rusting.

Sustainability and Recycling (p.178-182)

11) Give two examples of finite resources
12) Give two examples of renewable resources.
13) State two methods that can be used to extract copper from low grade ores.
14) Other than recycling, how can glass bottles be used in a sustainable way?
15) Describe what life cycle assessments (LCAs) do?

Potable Water and Waste Water Treatment (p.184-186)

16) What is potable water?
17) Suggest a source of fresh water.
18) Name a process that can be used to make seawater safe to drink.
19) Name three different sources of waste water.
20) Why is it important to treat waste water before releasing it into the environment?
21) What two useful products can be obtained by the anaerobic digestion of sewage sludge?

The Haber Process and Fertilisers (p.188-191)

22) State the word equation for the Haber process.
23) What happens to the unused hydrogen and nitrogen in the Haber process?
24) What three elements are present in NPK fertilisers?
25) What substance can react with ammonia to produce ammonium nitrate?
26) What substance can react with sulfuric acid to produce calcium phosphate and calcium sulphate?

Taking Measurements

You've got a few practicals that you need to know about for your exams. This section is full of lots of information and advice for how to carry out practicals really well. First up — measuring things...

Three Ways to Measure Liquids

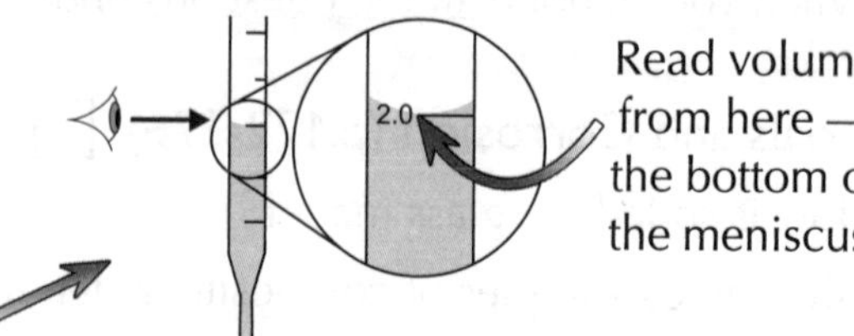

There are a few methods you might use to measure the volume of a liquid. Whichever method you use, always read the volume from the bottom of the meniscus (the curved upper surface of the liquid) when it's at eye level.

pipette filler

Pipettes are long, narrow tubes that are used to suck up an accurate volume of liquid and transfer it to another container. A pipette filler attached to the end of the pipette is used so that you can safely control the amount of liquid you're drawing up. Pipettes are often calibrated to allow for the fact that the last drop of liquid stays in the pipette when the liquid is ejected. This reduces transfer errors.

Burettes measure from top to bottom (so when they're filled to the top of the scale, the scale reads zero). They have a tap at the bottom which you can use to release the liquid into another container (you can even release it drop by drop). To use a burette, take an initial reading, and once you've released as much liquid as you want, take a final reading. The difference between the readings tells you how much liquid you used.

Burettes are used a lot for titrations. There's loads more about titrations on page 88.

Measuring cylinders are the most common way to measure out a liquid. They come in all different sizes. Make sure you choose one that's the right size for the measurement you want to make. It's no good using a huge 1000 cm^3 cylinder to measure out 2 cm^3 of a liquid — the graduations will be too big, and you'll end up with massive errors. It'd be much better to use one that measures up to 10 cm^3.

If you only want a couple of drops of liquid, and don't need it to be accurately measured, you can use a dropping pipette to transfer it. For example, this is how you'd add a couple of drops of indicator into a mixture.

Gas Syringes Measure Gas Volumes

1) Gases can be measured with a gas syringe.
2) They should be measured at room temperature and pressure as the volume of a gas changes with temperature and pressure.
3) You should also use a gas syringe that's the right size for the measurement you're making.
4) Before you use the syringe, you should make sure it's completely sealed and that the plunger moves smoothly.

Remember — burettes release liquid and pipettes suck it up

Pipettes, burettes and measuring cylinders may look quite similar from simple diagrams, but the differences between them are much more obvious when you use them in real life. The best kind of revision for these practical techniques is real hands-on experience.

Taking Measurements

Solids Should Be Measured Using a Balance

1) To weigh a solid, start by putting the container you're weighing your substance into on the balance.
2) Set the balance to exactly zero and then start weighing out your substance.
3) It's no good carefully weighing out your solid if it's not all transferred to your reaction vessel — the amount in the reaction vessel won't be the same as your measurement. Here are a couple of methods you can use to make sure that none gets left in your weighing container...

- If you're dissolving the solid in a solvent to make a solution, you could wash any remaining solid into the new container using the solvent. This way you know that all the solid you weighed has been transferred.
- You could set the balance to zero before you put your weighing container on the balance. Then reweigh the weighing container after you've transferred the solid. Use the difference in mass to work out exactly how much solid you added to your experiment.

You May Have to Measure the Time Taken for a Change

1) You should use a stopwatch to time experiments. These measure to the nearest 0.1 s so are sensitive.
2) Always make sure you start and stop the stopwatch at exactly the right time. For example, if you're investigating the rate of an experiment, you should start timing at the exact moment you mix the reagents and start the reaction. If you're measuring the time taken for a precipitate to form, you should watch the reaction like a hawk so you can stop timing the moment it goes cloudy.

Measure pH to Find Out How Acidic or Alkaline a Solution Is

You need to be able to decide the best method for measuring pH, depending on what your experiment is.

1) Indicators are dyes that change colour depending on whether they're in an acid or an alkali. You use them by adding a couple of drops of the indicator to the solution you're interested in. They're useful for titration reactions, when you want to find the point at which a solution is neutralised.
2) Universal indicator is a mixture of indicators that changes colour gradually as pH changes. It doesn't show a sudden colour change. It's useful for estimating the pH of a solution based on its colour.
3) Indicators can be soaked into paper and strips of this paper can be used for testing pH. If you use a dropping pipette to spot a small amount of a solution onto some indicator paper, it will change colour depending on the pH of the solution.

- Litmus paper turns red in acidic conditions and blue in basic conditions.
- Universal indicator paper can be used to estimate the pH based on its colour.

There's loads more about pH on page 87.

4) Indicator paper is useful when you don't want to change the colour of all of the substance, or if the substance is already coloured so might obscure the colour of the indicator. You can also hold a piece of damp indicator paper in a gas sample to test its pH.
5) pH probes are attached to pH meters which have a digital display that gives a numerical value for the pH of a solution. They're used to give an accurate value of pH.

Have your stopwatch ready before you try to time something

To make sure your results can be trusted by other scientists, you have to make sure all your measurements are accurate. So make sure you learn all the tips on these two pages for improving accuracy — e.g. using the right size of gas syringe for the measurement you're taking.

Safety and Heating Substances

Safety is really important when carrying out experiments. This is because experiments in chemistry can often involve using hazardous chemicals or heating substances. Read on to find out how to be safe in the lab.

Be Careful When You **Handle** or **Mix** Substances

1) There are lots of hazards in chemistry experiments, so before you start any experiment, you should read any safety precautions to do with your method or the chemicals you're using.
2) The substances used in chemical reactions are often hazardous. For example, they might catch fire easily (they're flammable), or they might irritate or burn your skin if you come into contact with them.
3) Whenever you're doing an experiment, you should wear a lab coat, safety goggles and gloves.
4) Always be careful that the chemicals you're using aren't flammable before you go lighting any Bunsen burners, and make sure you're working in an area that's well ventilated.
5) If you're doing an experiment that might produce nasty gases (such as chlorine), you should carry out the experiment in a fume hood so that the gas can't escape out into the room you're working in.
6) Never directly touch any chemicals (even if you're wearing gloves). Use a spatula to transfer solids between containers. Carefully pour liquids between different containers, using a funnel to avoid spills.
7) Be careful when you're mixing chemicals, as a reaction might occur. If you're diluting a liquid, add the concentrated substance to the water (not the other way around) or the mixture could get very hot.

The **Temperature** of **Water Baths & Electric Heaters** Can Be **Set**

1) A water bath is a container filled with water that can be heated to a specific temperature.

- Set the temperature on the water bath, and allow the water to heat up.
- Place the vessel containing your substance in the water bath using a pair of tongs.
- The level of the water outside the vessel should be just above the level of the substance inside the vessel.
- The substance will then be warmed to the same temperature as the water.
- As the substance in the vessel is surrounded by water, the heating is very even. Water boils at 100 °C though, so you can't use a water bath to heat something to a higher temperature than this — the water won't get hot enough.

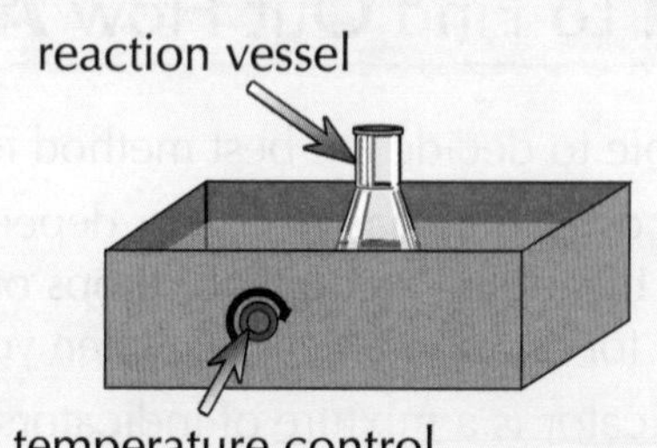

Handle any glassware you've heated with tongs until you're sure it's cooled down.

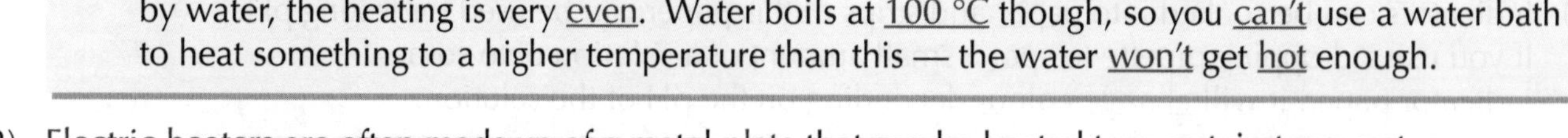

2) Electric heaters are often made up of a metal plate that can be heated to a certain temperature. The vessel containing the substance you want to heat is placed on top of the hot plate. You can heat substances to higher temperatures than you can in a water bath but, as the vessel is only heated from below, you'll usually have to stir the substance inside to make sure it's heated evenly.

Measure **Temperature** Accurately

You can use a thermometer to measure the temperature of a substance:

- Make sure the bulb of your thermometer is completely submerged in any mixture you're measuring.
- If you're taking an initial reading, you should wait for the temperature to stabilise first.
- Read your measurement off the scale on a thermometer at eye level to make sure it's correct.

Heating Substances

Sometimes water baths and electric heaters aren't appropriate for what you want to do — e.g. if you want to heat a sample of a compound to see what colour flame it produces, you should use a Bunsen burner instead.

Bunsen Burners Have a Naked **Flame**

1) Bunsen burners are good for heating things quickly. You can easily adjust how strongly they're heating.
2) But you need to be careful not to use them if you're heating flammable compounds as the flame means the substance would be at risk of catching fire.
3) Here's how to use a Bunsen burner:

   1) Connect the Bunsen burner to a gas tap, and check that the hole is closed. Place it on a heat-proof mat.
   2) Light a splint and hold it over the Bunsen burner. Now, turn on the gas. The Bunsen burner should light with a yellow flame.
   3) The more open the hole is, the more strongly the Bunsen burner will heat your substance. Open the hole to the amount you want. As you open the hole more, the flame should turn more blue.

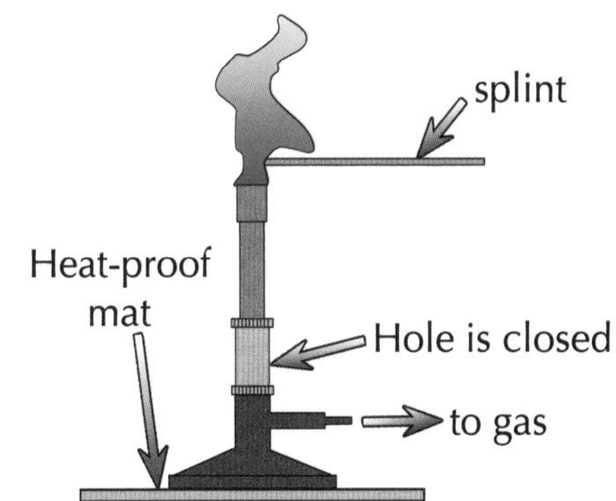

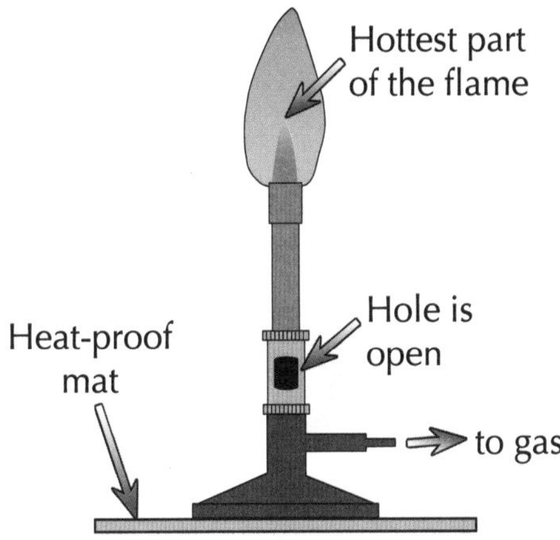

   4) The hottest part of the flame is just above the blue cone, so you should heat things here.
   5) If your Bunsen burner is alight but not heating anything, make sure you close the hole so that the flame becomes yellow and clearly visible.
   6) If you're heating something so that the container (e.g. a test tube) is in the flame, you should hold the vessel at the top, furthest away from the substance (and so the flame) using a pair of tongs.
   7) If you're heating something over the flame (e.g. an evaporating dish), you should put a tripod and gauze over the Bunsen burner before you light it, and place the vessel on this.

4) You'd use a Bunsen burner to carry out flame tests to identify metal ions in a compound (see page 158).

   - A sample of the compound is placed on a metal wire that you hold just above the cone of a Bunsen burner with a blue flame.
   - The flame should then change colour depending on what metal ion is in the sample.

You might not expect the blue flame to be hotter, but it is

As you can see, it's not just when you're heating substances that you have to be safe — you should also be careful, for example, when you're using hazardous chemicals or if chemical reactions are likely to give off a lot of heat. Always read through any safety precautions you're given carefully before you start, otherwise you might do something dangerous.

Setting Up Equipment

Setting up the equipment for an experiment correctly is important if you want to take accurate measurements.

To Collect **Gases**, the System Needs to be **Sealed**

1) There are times when you might want to collect the gas produced by a reaction. For example, to investigate the rate of reaction.
2) The most accurate way to measure the volume of a gas that's been produced is to collect it in a gas syringe (see page 121).

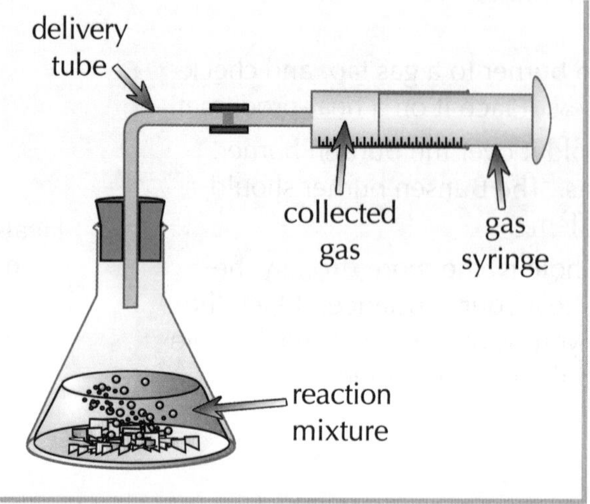

3) You could also collect it by displacing water from a measuring cylinder. Here's how you do it...

    1) Fill a measuring cylinder with water, and carefully place it upside down in a container of water.
    2) Record the initial level of the water in the measuring cylinder.
    3) Position a delivery tube coming from the reaction vessel so that it's inside the measuring cylinder, pointing upwards. Any gas that's produced will pass through the delivery tube and into the measuring cylinder. As the gas enters the measuring cylinder, the water is pushed out.
    4) Record the level of water in the measuring cylinder and use this value, along with your initial value, to calculate the volume of gas produced.

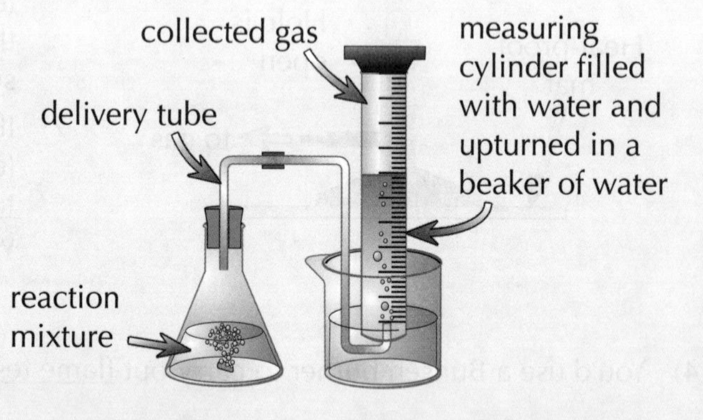

4) This method is less accurate than using a gas syringe to measure the volume of gas produced. This is because some gases can dissolve in water, so less gas ends up in the measuring cylinder than is actually produced.
5) If just want to collect a sample to test (and don't need to measure a volume), you can collect it over water as above using a test tube. Once the test tube is full of gas, you can stopper it and store the gas for later.

If the delivery tube is underneath the measuring cylinder rather than inside it then some of the gas might escape out into the air.

You can test the gas you've collected to find out what it is

Having a good knowledge of practical techniques won't just make your investigations more reliable, it might come in handy for questions in your exams. It's possible that you could be asked to comment on how equipment has been set up, e.g. suggest improvements.

Setting Up Equipment

Some experiments involve slightly more complicated equipment that might be less familiar to you. For example, electrolysis experiments use electrodes — this page shows you how to set them up.

You May Have to **Identify** the **Products** of Electrolysis

There's more about electrolysis on p.100-103.

1) When you electrolyse an aqueous solution, the products of electrolysis will depend on how reactive the ions in the solution are compared to the H^+ and OH^- ions that come from water.
2) At the cathode you'll either get a pure metal coating the electrode or bubbles of hydrogen gas.
3) At the anode, you'll get bubbles of oxygen gas unless a halide ion is present, when you'll get the halogen.
4) You may have to predict and identify what's been made in an electrolysis experiment. To do this, you need to be able to set up the equipment correctly so that you can collect any gas that's produced. The easiest way to collect the gas is in a test tube.
5) Here's how to set up the equipment...

The tests for gases are described on page 153.

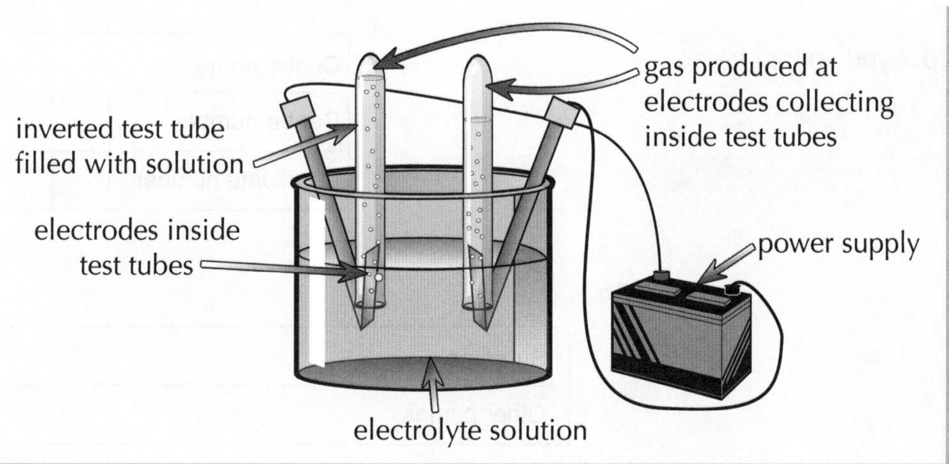

Make Sure You Can **Draw Diagrams** of Your Equipment

1) When you're writing out a method for your experiment, it's always a good idea to draw a labelled diagram showing how your apparatus will be set up.
2) The easiest way to do this is to use a scientific drawing, where each piece of apparatus is drawn as if you're looking at its cross-section. For example:

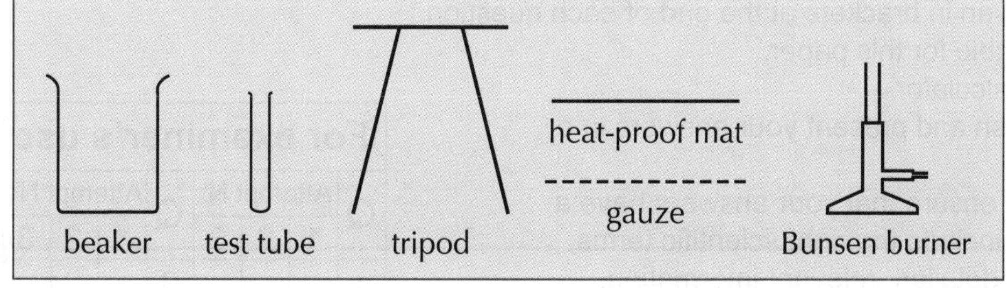

The pieces of glassware are drawn without tops so they aren't sealed. If you want to draw a closed system, remember to draw a bung in the top.

These simple diagrams are clear and easy to draw

Have a go at drawing out some of the practical diagrams on these pages using a simplified cross-section drawing of each piece of equipment. It'll give you some practice at doing them and you can revise how to set up the experiments as well. Win-win.

Practice Exams

Once you've been through all the questions in this book, you should feel pretty confident about the exams. As final preparation, here is a set of **practice exams** to really get you set for the real thing. The time allowed for each paper is 1 hour 45 minutes. These papers are designed to give you the best possible preparation for your exams.

GCSE Chemistry

Paper 1

Higher Tier

In addition to this paper you should have:
- A ruler.
- A calculator.
- A periodic table

Centre name				
Centre number				
Candidate number				

Time allowed:
- 1 hour 45 minutes

Surname
Other names
Candidate signature

Instructions to candidates
- Write your name and other details in the spaces provided above.
- Answer **all** questions in the spaces provided.
- Do all rough work on the paper.
- Cross out any work you do not want to be marked.

Information for candidates
- The marks available are given in brackets at the end of each question.
- There are 100 marks available for this paper.
- You are allowed to use a calculator.
- You should use good English and present your answers in a clear and organised way.
- For Questions 3.5 and 9.1, ensure that your answers have a clear and logical structure, include the right scientific terms, spelt correctly, and include detailed, relevant information.

Advice to candidates
- In calculations show clearly how you worked out your answers.

For examiner's use							
Q	Attempt Nº			Q	Attempt Nº		
	1	2	3		1	2	3
1				6			
2				7			
3				8			
4				9			
5							
				Total			

1 The elements in Group 1 of the periodic table are known as the alkali metals.

1.1 Group 1 metals can react with non-metals to form ionic compounds.
What is the charge on a Group 1 ion in an ionic compound?
Tick **one** box.

☐ +2

☐ +3

☐ –1

☐ +1

[1 mark]

1.2 A white solid is formed when sodium metal reacts with fluorine gas.
What is the chemical formula of this compound?
Tick **one** box.

☐ Na_2F

☐ NaF

☐ NaF_2

☐ NaF_3

[1 mark]

A student watched his teacher carefully place small pieces of lithium, sodium and potassium into cold water. His observations are recorded in **Table 1**.

Table 1

Metal	Observations
lithium	Fizzes, moves across surface.
sodium	Fizzes strongly, moves quickly across surface, melts.
potassium	Fizzes violently, moves very quickly across surface, melts and a flame is seen.

He decides that the order of reactivity of the three metals is:

- potassium (most reactive)
- sodium
- lithium (least reactive)

Question 1 continues on the next page

1.3 Give **two** pieces of evidence from **Table 1** that support the student's conclusion.

....

....

....

....

[2 marks]

1.4 Explain the pattern of reactivity that the student has noticed in terms of the outer electrons of the atoms.

....

....

....

....

[3 marks]

1.5 Complete and balance the chemical equation for the reaction between lithium and water.

$2Li + 2H_2O \rightarrow$ .... + ...

[2 marks]

1.6 Choose the statement that explains why the solution produced when lithium reacts with water is alkaline.
Tick **one** box.

- ☐ It contains lithium ions.
- ☐ It contains water.
- ☐ It contains an ionic compound.
- ☐ It contains hydroxide ions.

[1 mark]

2 Atoms contain protons, neutrons and electrons.

2.1 Whose experimental work led to the discovery of neutrons within the nucleus?
Tick **one** box

- [ ] Niels Bohr
- [ ] Ernest Rutherford
- [ ] James Chadwick
- [ ] Ernest Marsden

[1 mark]

2.2 Complete **Table 2** to show the relative charges and relative masses of protons, neutrons and electrons.

Table 2

Particle	Relative mass	Relative charge
Proton	1	
Neutron		
Electron	Very small	

[2 marks]

2.3 **Table 3** shows the numbers of protons, neutrons and electrons present in six different atoms.

Table 3

Atom	Number of protons	Number of neutrons	Number of electrons
A	5	6	5
B	7	7	7
C	6	8	6
D	6	6	6
E	10	10	10
F	4	5	4

Which **two** atoms are isotopes of the same element? Explain your answer.

....

....

....

[2 marks]

Question 2 continues on the next page

2.4 The nuclear symbol for an atom of zinc is shown below.

$$^{64}_{30}\text{Zn}$$

How many protons, neutrons and electrons are there in this atom of zinc?

Protons = ...

Neutrons = ...

Electrons = ...

[3 marks]

Atoms can bond with each other to form compounds.

2.5 What is a compound?

....

....

[2 marks]

2.6 Iron(II) sulfate is a compound with the formula $FeSO_4$.
Calculate the relative formula mass of iron(II) sulfate.

....

....

relative formula mass = ...

[2 marks]

3 The structure and bonding of elements and compounds affects their properties.

3.1 Sodium and chlorine can react together to form an ionic compound (sodium chloride). **Figure 1**, below, is a dot and cross diagram illustrating this reaction.

Complete the right-hand side of **Figure 1** by adding the charges of both ions and adding the electrons to the outer shell of the chloride ion.

Figure 1

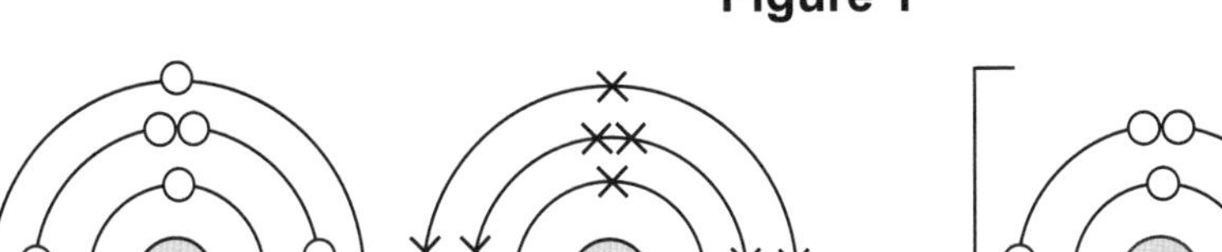

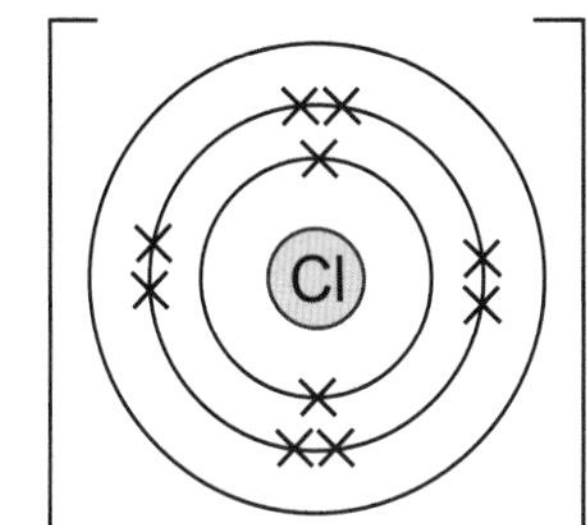

[2 marks]

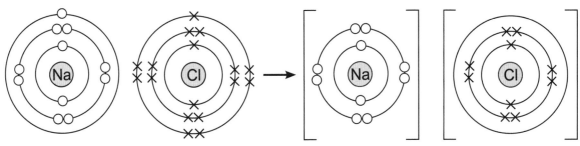

3.2 Describe the structure of sodium chloride.
Include a description of the forces that hold the structure together.

....

....

....

[2 marks]

Table 4 shows some properties of five substances, **A-E**.

Table 4

Substance	Melting point in °C	Boiling point in °C	Does it conduct electricity when solid?	Does it conduct electricity when dissolved or molten?
A	–210	–196	No	No
B	–219	–183	No	No
C	801	1413	No	Yes
D	115	445	No	No
E	1083	2567	Yes	Yes

3.3 Substance **A** consists of small, covalently bonded molecules.
Explain why it has a relatively low melting point.

....

....

[2 marks]

Question 3 continues on the next page

Turn over ▶

3.4 Look at **Table 4**. One of the substances, **A-E**, is an ionic compound.
Use the information in the table to suggest which of the substances is ionic.
Explain your answer.

....

....

....

[2 marks]

3.5 **Table 5** contains information about some of the properties of diamond and graphite.

Table 5

	Hardness	Melting point	Conducts electricity?
Diamond	Hard	High	No
Graphite	Soft	High	Yes

Explain these properties of diamond and graphite
in terms of their structure and bonding.

....

....

....

....

....

....

....

....

....

....

....

[6 marks]

4 **Figure 2** shows the apparatus used by a student to measure the temperature change that occurred when she added a piece of magnesium ribbon to dilute hydrochloric acid.

Figure 2

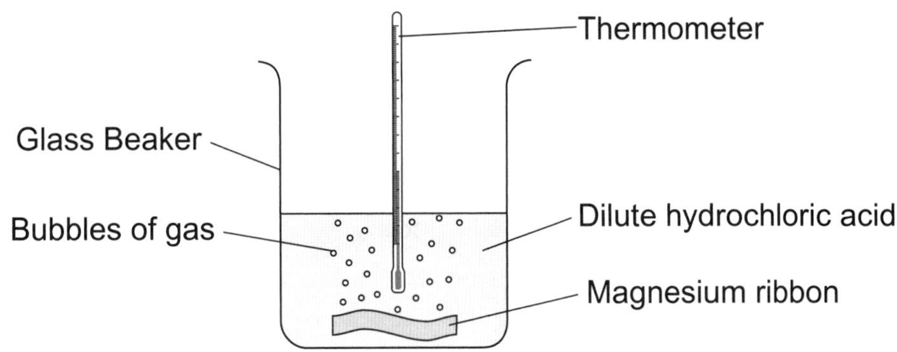

4.1 Suggest **two** changes that the student could make to the apparatus in order to reduce heat loss from her experiment.

....

....

....

....

[2 marks]

4.2 A close up of the thermometer used during the experiment is shown in **Figure 3.**

Figure 3

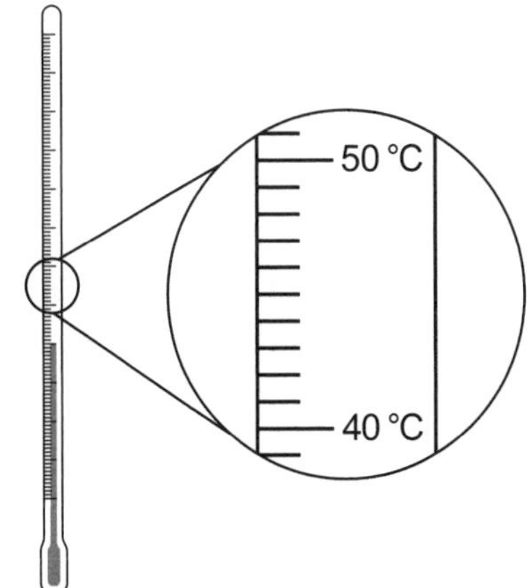

What value does each **small division** on the scale of the thermometer represent?
Tick **one** box.

☐ 0.1 °C

☐ 10 °C

☐ 1 °C

☐ 2 °C

[1 mark]

Question 4 continues on the next page

Turn over ▶

The student recorded the initial temperature of the dilute hydrochloric acid. She added the magnesium ribbon to the acid. She then measured the temperature of the reaction mixture every 10 seconds. The student's results are shown on the graph in **Figure 4**.

Figure 4

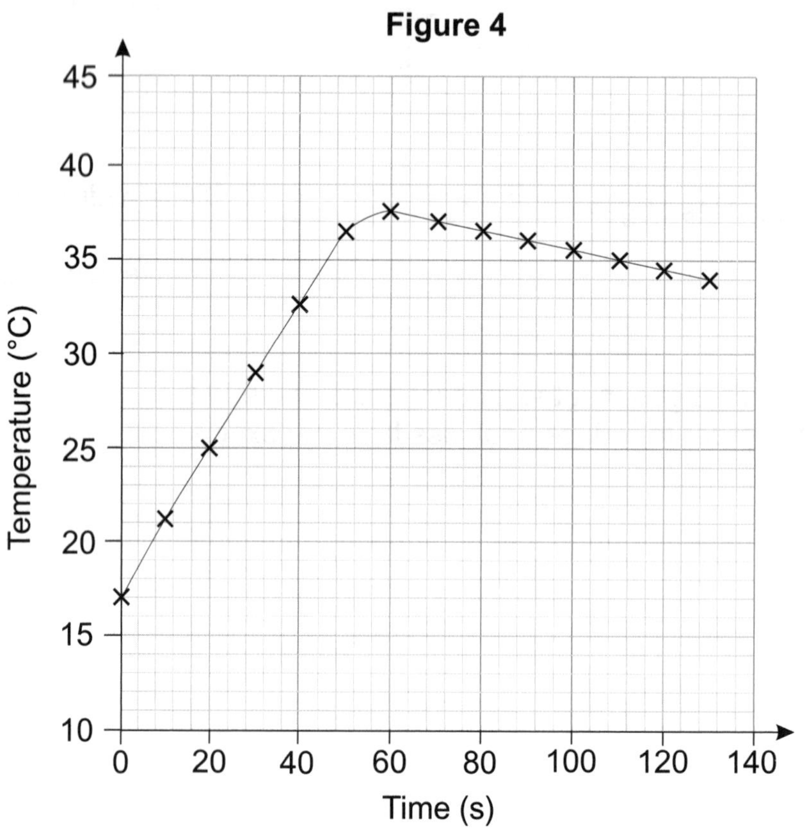

4.3 Using the graph in **Figure 4**, give the highest temperature of the mixture that the student recorded.

Highest temperature = ... °C

[1 mark]

4.4 The initial temperature of the acid was 17 °C. Use this information and your answer from **4.3** to estimate the total change in temperature of the reaction mixture.

....

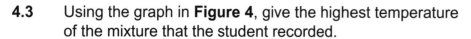

Temperature change = ... °C

[2 marks]

4.5 State whether this reaction was exothermic or endothermic. Explain your answer.

....

....

[1 mark]

5 A student reacts four different metals with dilute sulfuric acid.
She controls all of the relevant variables to make sure that the test is fair.
She collects the gas given off by each reaction in a gas syringe.
Figure 5 shows all four reactions after 30 seconds

Figure 5

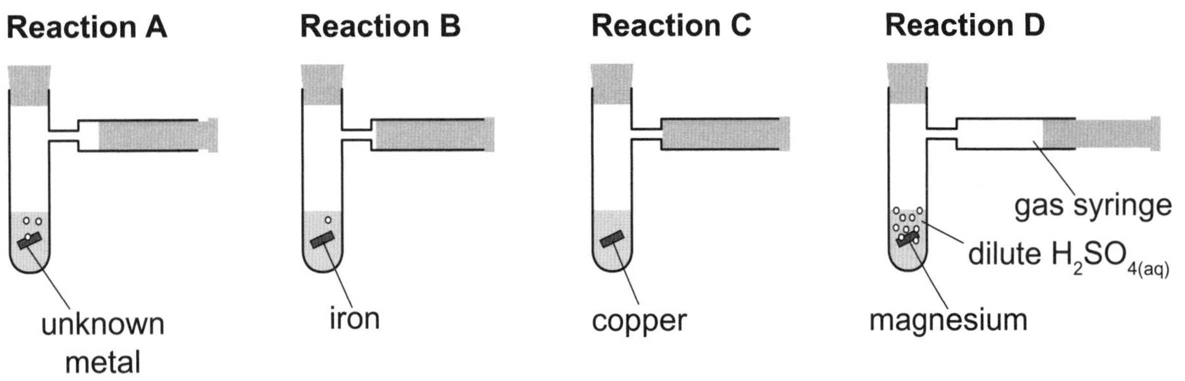

5.1 Name the gas that is being collected in the gas syringes.

....

[1 mark]

5.2 Which reaction, **A**, **B**, **C**, or **D**, contains the **most reactive** metal?
Explain how you can tell.

....

....

[3 marks]

5.3 Use your knowledge of the reactivity series to suggest a possible identity for the unknown metal used in reaction **A**.

....

[1 mark]

5.4 The student wanted to include sodium in the experiment, but her teacher said this would not be safe.
Suggest why it would not be safe to include sodium in the experiment.

....

....

[1 mark]

Question 5 continues on the next page

5.5 In another experiment, the student placed pieces different metals in metal salt solutions. She left them for 10 minutes. The student then recorded whether any reaction had occurred. The results of this experiment are shown in **Table 6**.

Table 6

	Did any reaction occur with:		
	iron sulfate	**magnesium sulfate**	**copper sulfate**
iron	No	No	....
magnesium	....	No	Yes
copper	No	No	No

Complete **Table 6** by filling in the gaps.

[2 marks]

5.6 The equation for the reaction between magnesium and copper sulfate solution is:

$$Mg_{(s)} + CuSO_{4(aq)} \rightarrow MgSO_{4(aq)} + Cu_{(s)}$$

Write the ionic equation for this reaction.

....

[2 marks]

5.7 In the reaction described in **5.6**, which species was oxidised?

....

[1 mark]

6 A student added an excess of zinc oxide (ZnO) to dilute hydrochloric acid (HCl). They reacted to produced zinc chloride, a soluble salt.

6.1 **Figure 6** shows the student's experiment just after the zinc oxide was added.

Figure 6

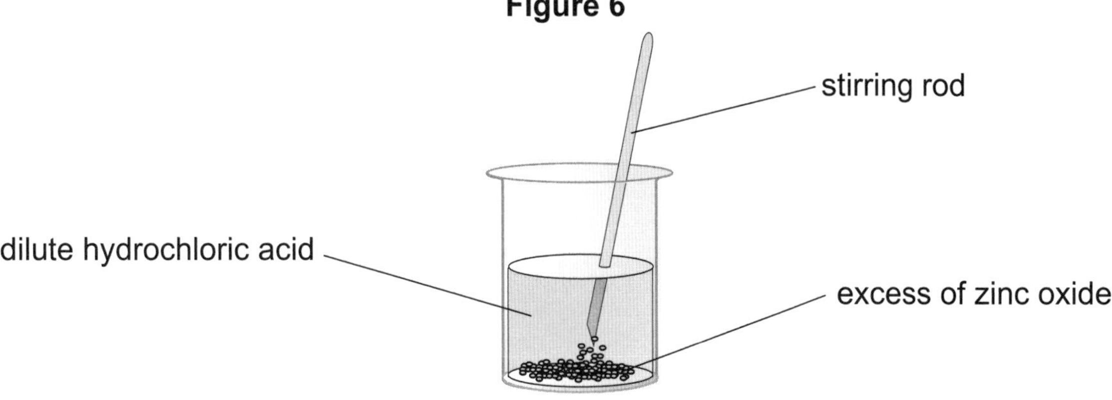

How could the student tell when all of the acid had been neutralised?

....

....

[1 mark]

6.2 In addition to the zinc chloride, one other product is formed by this reaction. Name the other product of the reaction.

....

[1 mark]

6.3 Give the chemical formula of zinc chloride.

....

[1 mark]

6.4 Describe how you could produce pure, dry crystals of zinc chloride from the reaction mixture after the reaction had finished.

....

....

....

....

....

....

....

[4 marks]

Question 6 continues on the next page

Turn over ▶

6.5 Another student added solid sodium carbonate (Na_2CO_3) to dilute hydrochloric acid.
Name the salt formed by this reaction.

....

[1 mark]

6.6 Which gas was given off by the reaction described in **6.5**?
Tick **one** box.

- ☐ Hydrogen
- ☐ Oxygen
- ☐ Chlorine
- ☐ Carbon Dioxide

[1 mark]

6.7 The solution formed at the end of the reaction described in **6.5** is neutral.
Suggest a method you could use to test the pH of the solution.
State what you would expect to observe.

....

....

....

[2 marks]

7 Iron can be extracted from iron(III) oxide, Fe_2O_3, by reduction, using carbon. Here is the equation for this reaction:

$$2Fe_2O_3 + 3C \rightarrow 4Fe + 3CO_2$$

7.1 Give **one** reason why this reaction is described as a reduction reaction.

....

[1 mark]

7.2 Calculate the mass of iron that could be extracted from 40 g of iron(III) oxide. Relative atomic masses, A_r: Fe = 56, O = 16.

....

....

....

....

....

Mass of iron = ... g

[4 marks]

7.3 Explain why reduction with carbon isn't used to extract aluminium from its ore.

....

....

[2 marks]

7.4 Small amounts of carbon can be added to iron to make an alloy called steel. Explain why steel is harder than pure iron.

....

....

....

....

....

....

....

[4 marks]

Question 7 continues on the next page

7.5 A student used the apparatus shown in **Figure 7** to compare how well steel, carbon and iron conduct thermal energy.

Figure 7

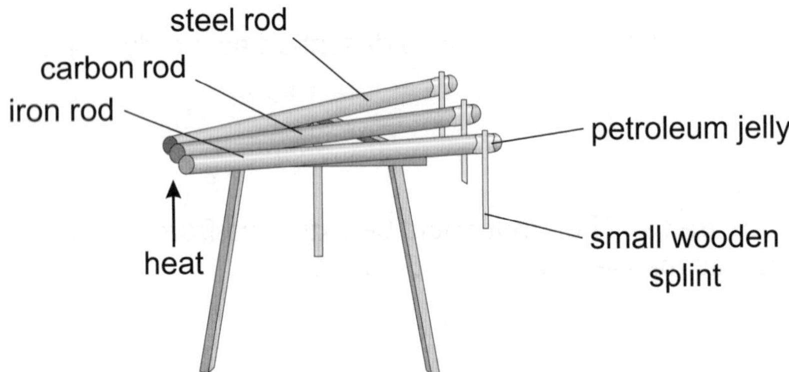

The student placed one rod of each material on a tripod, with a small wooden splint stuck to one end with petroleum jelly. The student then heated the rods at the other end and timed how long it took for each wooden splint to fall off as the petroleum jelly melted.

Describe **three** things the student should keep the same for each material to make the experiment a fair test.

....

....

....

....

....

....

[3 marks]

8 A student carried out an experiment to investigate the electrolysis of copper sulfate solution, $CuSO_{4\,(aq)}$, using inert carbon electrodes. The diagram in **Figure 8** shows how his experiment was set up.

Figure 8

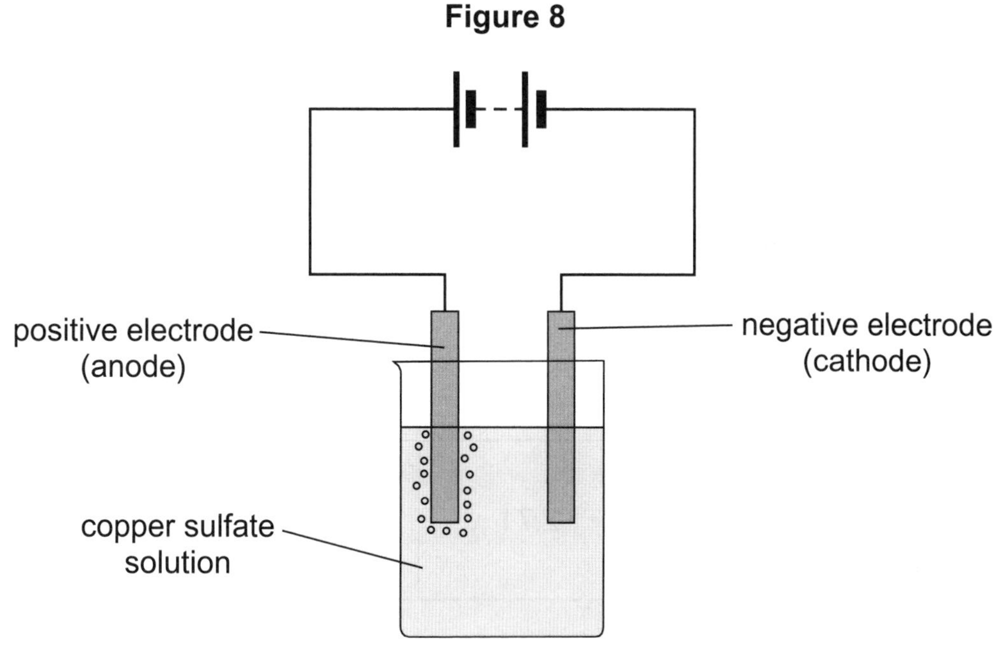

8.1 Explain why a solution of copper sulfate can conduct electricity, but solid copper sulfate cannot.

....

....

[2 marks]

8.2 List all **four** of the ions that are present in aqueous copper sulfate solution.

....

....

[1 mark]

8.3 Complete the half equation shown below for the reaction taking place at the positive electrode (anode).

.... $OH^{-} \rightarrow O_2$ + ... H_2O + 4 ....

[1 mark]

Question 8 continues on the next page

The student repeated the experiment three times, running the electrolysis for exactly 30 minutes each time. He weighed the dry electrodes before starting each run. At the end of each run, he dried both electrodes and weighed them again.

Table 7 shows the change in mass of both electrodes for each of the three runs.

Table 7

	Change in mass of electrode in g			
	Run 1	**Run 2**	**Run 3**	**Mean**
Positive electrode (anode)	0.00	0.00	0.00	0.00
Negative electrode (cathode)	2.36	2.71	2.55	....

8.4 Complete **Table 7** by finding the mean change in mass for the negative electrode.

[1 mark]

8.5 Explain why the negative electrode increased in mass in this experiment. Your answer should include the half equation for the reaction taking place at the negative electrode.

....

....

....

....

....

[3 marks]

8.6 The student decided to run the experiment for 60 minutes instead of 30 minutes. Predict the effect that this would have on the change in mass of the negative electrode.

....

[1 mark]

9 Atom economy and percentage yield are both important factors that are considered when chemical firms decide how to produce a chemical industrially.

9.1 Discuss why it is better to use reactions with high atom economies than reactions with low atom economies in industry.
In your answer you should talk about:

- the environmental benefits of using high atom economy reactions,
- the economic benefits of using high atom economy reactions.

....

....

....

....

....

....

....

....

....

....

[6 marks]

9.2 Lithium chloride can be made by reacting lithium carbonate with hydrochloric acid.
The equation for this reaction is:.

$$Li_2CO_3 + 2HCl \rightarrow 2LiCl + H_2O + CO_2$$

Calculate the atom economy for producing lithium chloride using this reaction.
Give your answer to 2 significant figures.
Relative formula masses, M_r: Li_2CO_3 = 74, HCl = 36.5, LiCl = 42.5, H_2O = 18, CO_2 = 44.

....

....

....

....

.... %

[4 marks]

Question 9 continues on the next page

Turn over ▶

9.3 Lithium chloride can also be formed following the reaction of lithium metal with chlorine gas. The equation for this reaction is shown below.

$$2Li + Cl_2 \rightarrow 2LiCl$$

What is the atom economy for the above reaction? Tick **one** box.

☐ 100%

☐ 50%

☐ 200%

☐ 10%

[1 mark]

9.4 A student carried out the reaction shown in **9.2** using 185 g of lithium carbonate and hydrochloric acid in excess. The theoretical yield of this reaction was 212.5 g.
The student actually produced 170 g of lithium chloride.
Calculate the percentage yield of the reaction.

....

....

.... %

[1 mark]

END OF QUESTIONS

GCSE Chemistry

Paper 2

Higher Tier

In addition to this paper you should have:
- A ruler.
- A calculator.
- A periodic table.

Centre name					
Centre number					
Candidate number					

Time allowed:
- 1 hour 45 minutes

Surname	
Other names	
Candidate signature	

Instructions to candidates
- Write your name and other details in the spaces provided above.
- Answer **all** questions in the spaces provided.
- Do all rough work on the paper.
- Cross out any work you do not want to be marked.

Information for candidates
- The marks available are given in brackets at the end of each question.
- There are 100 marks available for this paper.
- You are allowed to use a calculator.
- You should use good English and present your answers in a clear and organised way.
- For Questions 3.3 and 5.5, ensure that your answers have a clear and logical structure, include the right scientific terms, spelt correctly, and include detailed, relevant information.

Advice to candidates
- In calculations show clearly how you worked out your answers.

For examiner's use

Q	Attempt Nº 1	2	3	Q	Attempt Nº 1	2	3
1				6			
2				7			
3				8			
4				9			
5							
				Total			

1 Alkanes are hydrocarbon compounds found in crude oil. **Table 1** shows how the boiling points of some alkanes change as the molecules get bigger.

Table 1

Alkane	Molecular formula	Boiling point (°C)
Propane	C_3H_8	–42
Butane	C_4H_{10}	–0.5
Pentane	C_5H_{12}	
Hexane	C_6H_{14}	69
Heptane	C_7H_{16}	98

1.1 Using the data in **Table 1**, plot a graph of the number of carbon atoms in an alkane molecule against boiling point on the axes below. Draw a smooth curve through the points you have plotted.

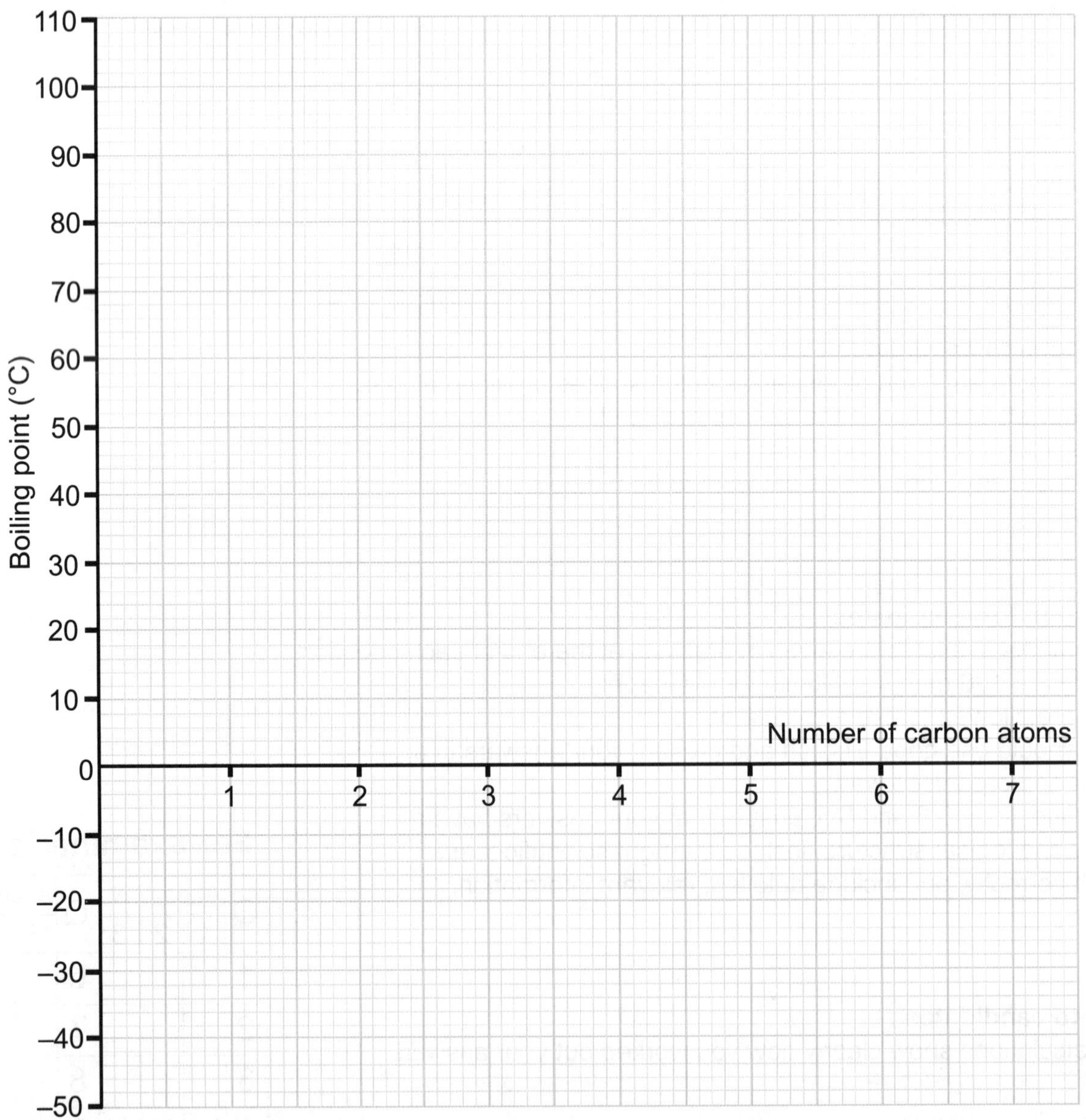

[2 marks]

1.2 Use your graph to estimate the boiling point of pentane.

.... °C

[1 mark]

1.3 Describe what the graph shows about the size of alkane molecules and their boiling points.

....

....

[1 mark]

1.4 What is the general formula of the alkanes? Tick **one** box.

- ☐ C_nH_{2n}
- ☐ C_nH_{2n+1}
- ☐ C_nH_{2n+2}
- ☐ C_nH_{2n-1}

[1 mark]

1.5 Propane is an alkane with three carbon atoms.
Draw the displayed formula of propane.

[1 mark]

1.6 Propene is an alkene with three carbon atoms.
Describe a test you could use to distinguish between propene and propane.
Say what you would observe in each case.

....

....

....

[3 marks]

1.7 Propane burns in the presence of oxygen.
Complete this equation for the complete combustion of propane.

$$C_3H_8 + 5O_2 \rightarrow 3CO_2 + \text{..........................}$$

[2 marks]

1.8 Decane, $C_{10}H_{22}$, is an alkane with ten carbon atoms.
Which of the following statements about decane is correct? Tick **one** box.

- ☐ Decane is more flammable than methane.
- ☐ The complete combustion of decane results in the formation of carbon monoxide.
- ☐ Decane is less viscous than propane.
- ☐ Decane has a higher boiling point than pentane.

[1 mark]

Turn over for the next question

2 Nitrogen dioxide is an atmospheric pollutant that irritates the respiratory system. It is thought that there is a link between exposure to nitrogen dioxide and the severity of asthma attacks in people with asthma.

Figure 1 shows the results of a study carried out by a group of scientists that compared atmospheric nitrogen dioxide levels with the severity of asthma attacks suffered by men under the age of 40 working in a city centre.

Figure 1

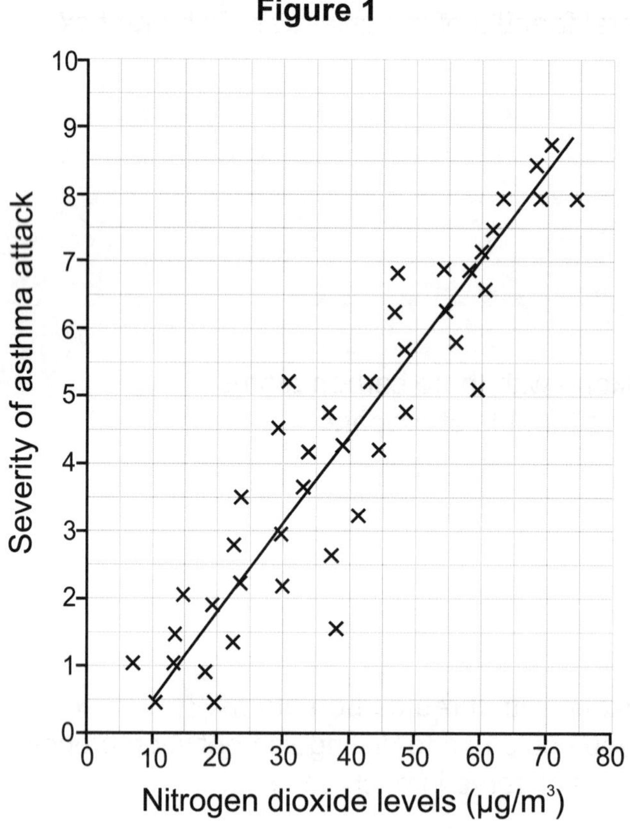

2.1 Describe the relationship between the level of nitrogen dioxide in the air and the severity of asthma attacks shown in **Figure 1**.

....

....

[1 mark]

2.2 The scientists decided they could not draw a general conclusion about the link between nitrogen dioxide levels and the severity of asthma attacks that would apply to everyone.
Which of these reasons **does not** describe why this might be the case? Tick **one** box.

☐ The scientists would need to collect data from women too.

☐ The severity of asthma attacks might be affected by another pollutant which happens to be abundant in the same areas as nitrogen dioxide.

☐ The sample of people that the scientists collected data from was too big.

☐ The scientists would need to collect data from other age groups too.

[1 mark]

2.3 Where do the oxides of nitrogen in the air come from?
Suggest why levels of nitrogen oxides can be particularly high in cities.

....

....

....

....

[3 marks]

2.4 Nitrogen dioxide dissolves in water to form acid rain.
What is the substance formed when nitrogen dioxide dissolves in water?

....

[1 mark]

2.5 Name **one** other pollutant gas that can cause acid rain.

....

[1 mark]

2.6 The scientists who collected the data shown on the graph in **Figure 1** also found higher than normal levels of carbon monoxide in the samples taken in a city centre.
Explain why carbon monoxide is a toxic gas.

....

....

[2 marks]

Turn over for the next question

3 In the UK, the majority of our drinking water is produced from treating groundwater or surface water. Drinking water can also be made by treating sea water or waste water.

3.1 Producing water that is safe to drink from sea water is expensive.
Suggest why some countries produce drinking water by this method.

....

[1 mark]

3.2 What is the name of the process used to remove salt from sea water?
Tick **one** box

☐ Filtration

☐ Desalination

☐ Sterilisation

☐ Cracking

[1 mark]

3.3 A teacher gives a student a sample of sea water,
and asks her to produce a sample of pure water from it.
Outline a method that the student could use to remove the salt from the sea water.

....

....

....

....

....

....

....

....

....

....

....

....

[6 marks]

3.4 Sewage treatment plants process sewage and release clean, treated water back into the environment. The first step in the treatment of sewage is screening. What happens in the screening step?
Tick **one** box.

- [ ] Chlorine is added to the sewage to kill microorganisms.
- [ ] The sewage is placed into large storage tanks and allowed to settle.
- [ ] Anaerobic digestion is used to break down the sewage.
- [ ] Large bits of material (such as grit) are removed from the sewage.

[1 mark]

3.5 Following screening, sedimentation is used to separate the effluent from the sludge. Describe what happens to the effluent before it can be returned to the environment.

....

....

....

[2 marks]

Turn over for the next question

4 A student is investigating the properties of vinegar. Ethanoic acid is found in vinegar.

4.1 Give the functional group of the homologous series that ethanoic acid belongs to.

....

[1 mark]

The student reacted some of the vinegar with solid sodium carbonate.
The apparatus that he used is shown in **Figure 2**.

Figure 2

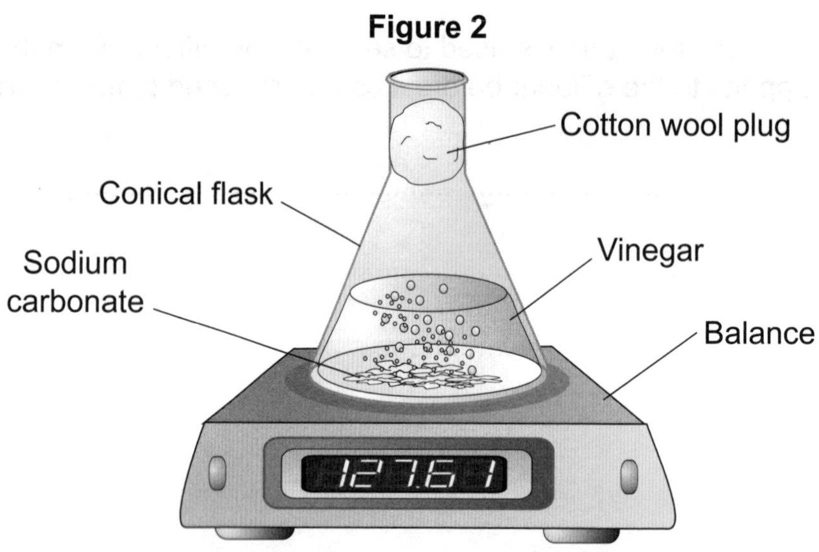

The student recorded the initial mass of the flask and then recorded it again after 60 s.
He repeated the experiment three times. His results are shown in **Table 2**.

Table 2

	Initial Mass in g	Final Mass in g	Loss of Mass in g
Run 1	128.00	127.61	0.39
Run 2	128.50	127.95	0.55
Run 3	128.35	127.90	0.45

4.2 Calculate the mean loss of mass after 60 s.

....

mean mass lost = ... g

[2 marks]

4.3 Why would it be inappropriate to use a mass balance that has a resolution of 1 g for this experiment?

....

....

[1 mark]

4.4 The student repeats the experiment a fourth time, but this time he collects the gas released by the reaction in a gas syringe. Name the gas that the student collects. Describe a test that he could use to confirm the identity of the gas.

....

....

....

[3 marks]

4.5 The teacher asks the student to test the solid sodium carbonate that he was given to confirm that it contains sodium ions. Suggest what kind of test the student could perform and state what observation you would expect him to make.

....

....

....

[2 marks]

Turn over for the next question

5 When a company is choosing what materials to use to make a particular product, they will consider the properties, availability and cost of the materials that they could use.

5.1 **Table 3** shows some properties of four different materials.

Table 3

Material	Density (g/cm³)	Electrical conductivity	Flexibility	Corrosion resistance
Steel	7.8	High	Low	Low
Aluminium alloy	3	Very High	Low	High
PVC	1.4	Very Low	High	High
Glass	2.5	Very Low	Low	High

Use data from the table to explain why PVC is particularly suitable for covering electrical wiring.

....

....

....

[2 marks]

5.2 What kind of material is glass?
Tick **one** box.

☐ Polymer

☐ Ceramic

☐ Composite

☐ Alloy

[1 mark]

5.3 Steel and aluminium alloy are both very strong. Many aeroplane parts need to withstand high forces, so they must be made from strong materials.
Using the information in **Table 3**, explain why aluminium alloy is often used to make aeroplane parts instead of steel.

....

....

....

[2 marks]

5.4 Copper is used to make electrical wiring due to its high electrical conductivity.
Phytomining can be used to extract copper from low-grade ores.
Describe how copper is extracted from low-grade ores using phytomining.

....

....

....

....

....

[4 marks]

5.5 A sports company is deciding on the best material to use for making the shaft of a professional golf club.
Look at **Table 4**, below. Discuss which of the four materials shown in the table would be the most suitable choice for making the golf club shaft.
Use the information in the table to justify your choice.

Table 4

Material	Density in g/cm^3	Strength in MPa	Corrosion resistance	Cost
Carbon fibre	1.5	4100	Good	High
Iron	7.8	200	Poor	Low
Stainless Steel	7.8	780	Good	Low
Lead	11.3	12	Good	Low

....

....

....

....

....

....

....

....

....

[6 marks]

Turn over for the next question

6 This question is about the evolution of the atmosphere.

The pie charts in **Figure 3** show the composition of Earth's atmosphere as it may have been 4 billion years ago and as it is at the present time.

Figure 3

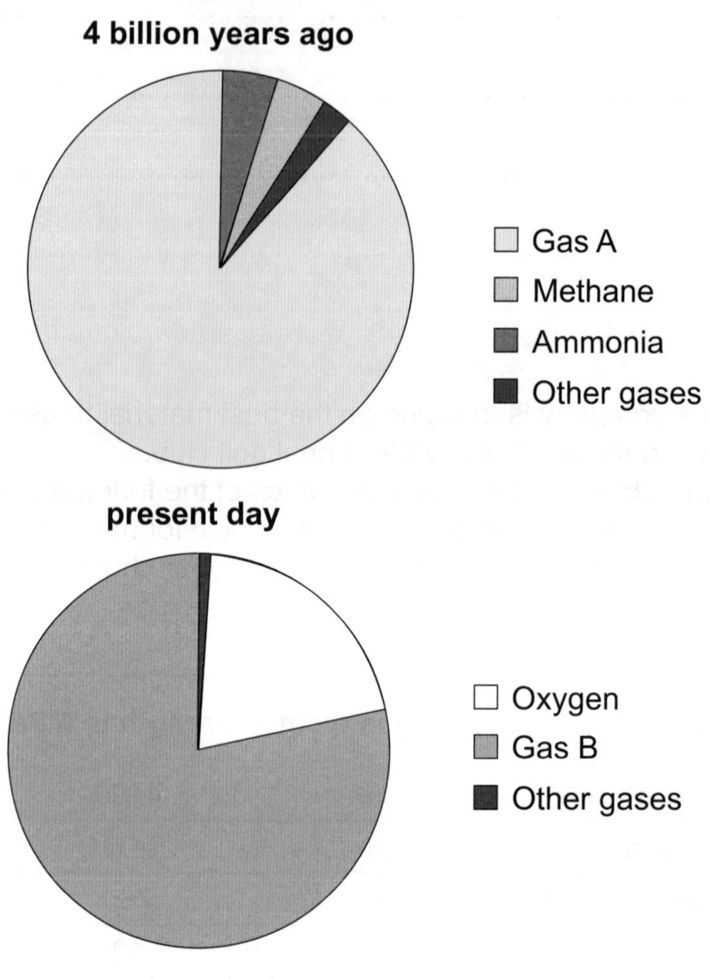

6.1 Name Gas A.

....

[1 mark]

6.2 Name Gas B.

....

[1 mark]

6.3 The evolution of algae and plants affected the composition of gases in the atmosphere. Explain how algae and plants changed the atmosphere.

....

....

....

....

[3 marks]

6.4 Human activity is affecting the composition of the atmosphere.
Suggest **one** reason why each of the following factors affects the level of carbon dioxide in the atmosphere.

Deforestation ....

....

Increased energy consumption ....

....

Increasing population ....

....

[3 marks]

6.5 A student says:

"I think that if the amount of carbon dioxide in the atmosphere continues to rise, this may lead to a drop in global food production in the future."

Explain why the student may be correct.

....

....

....

....

....

....

[3 marks]

Turn over for the next question

7 A student is investigating how the rate of the reaction between calcium carbonate and hydrochloric acid is affected by the concentration of the acid.

The student uses the following method:

- Weigh out 0.7 g of calcium carbonate.
- Add the calcium carbonate to an excess of 0.2 mol/dm^3 hydrochloric acid in a conical flask.
- Use a gas syringe to collect the gas given off by the reaction.
- Measure and record the volume of gas produced every 10 s.
- Repeat the experiment using hydrochloric acid with a concentration of 0.4 mol/dm^3.

7.1 Suggest **two** variables other than the mass of calcium carbonate that the student would have to keep the same for each run to make it a fair test.

....

....

[2 marks]

Figure 4 shows the student's results for the first experiment, using 0.2 mol/dm^3 hydrochloric acid.

Figure 4

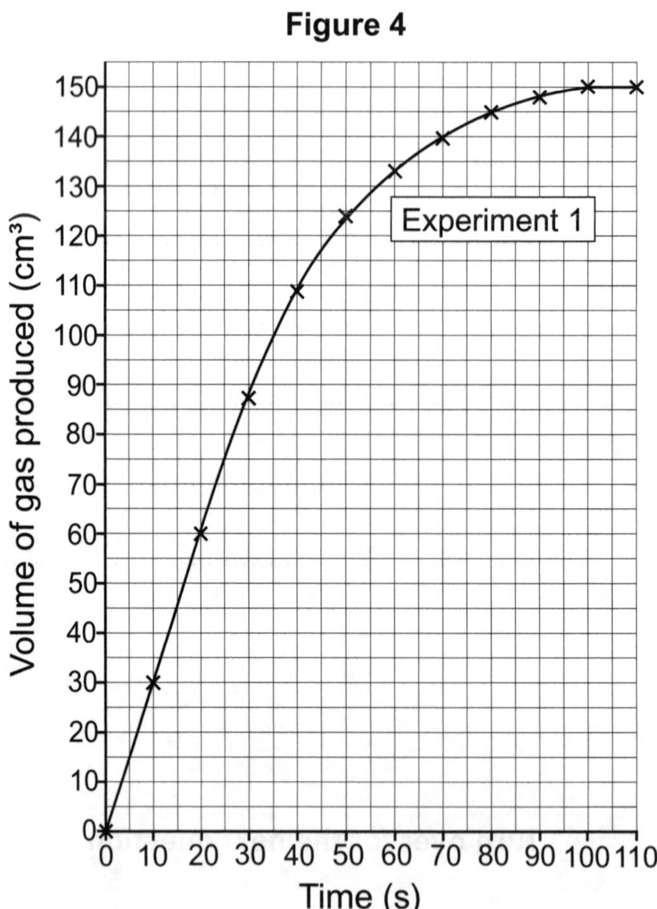

7.2 What does the graph in **Figure 4** show about how the rate of reaction changes as the reaction proceeds?

....

....

[1 mark]

7.3 Draw a tangent to the graph in **Figure 4** at 50 seconds.
Using the gradient of the tangent, find the rate of the reaction at this time.
Include the units of the rate in your answer.

....

....

....

....

....

Rate of reaction = ... Unit = ...

[4 marks]

7.4 Another student repeated the experiment using 0.35 g of calcium carbonate and 0.2 mol/dm^3 hydrochloric acid.
Predict the total amount of gas produced in this experiment.
You can assume that the acid was in excess in both experiments.

....

[1 mark]

Turn over for the next question

The first student added the results of his second experiment, using 0.4 mol/dm^3 hydrochloric acid, to his graph.
This graph is shown in **Figure 5**, below.

Figure 5

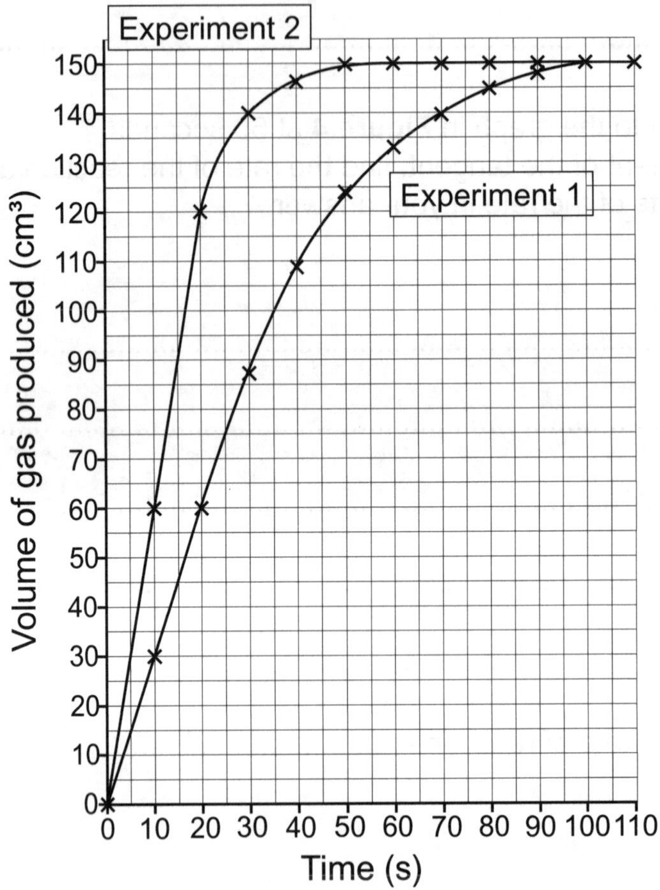

7.5 Explain how the student could tell from his graph that increasing the concentration of the acid had increased the reaction rate.

....

....

[1 mark]

7.6 In terms of collision theory, explain why concentration affects the rate of a reaction.

....

....

....

....

[2 marks]

8 This question is about synthetic and natural polymers.

8.1 **Figure 6** shows the displayed formula of the polymer poly(ethene).

Figure 6

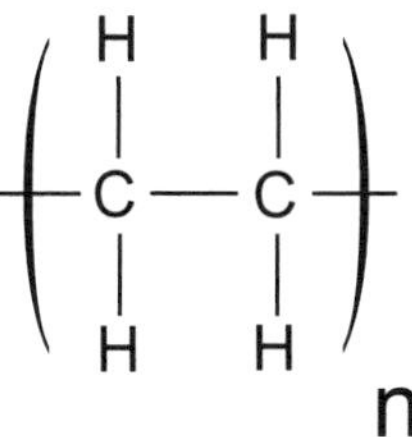

Draw the displayed formula of the monomer used to make this polymer.

[1 mark]

8.2 Which homologous series does the monomer molecule belong to?

....

[1 mark]

8.3 Some synthetic polymers are thermosetting, while others are thermosoftening.
Explain the difference in structure between thermosetting and thermosoftening polymers.
Describe how this affects the way they respond to being heated.

....

....

....

....

....

....

[4 marks]

Question 8 continues on the next page

Some polymers are naturally occurring. **Figure 7** shows the polymerisation of the molecule glycine to form a polypeptide.

Figure 7

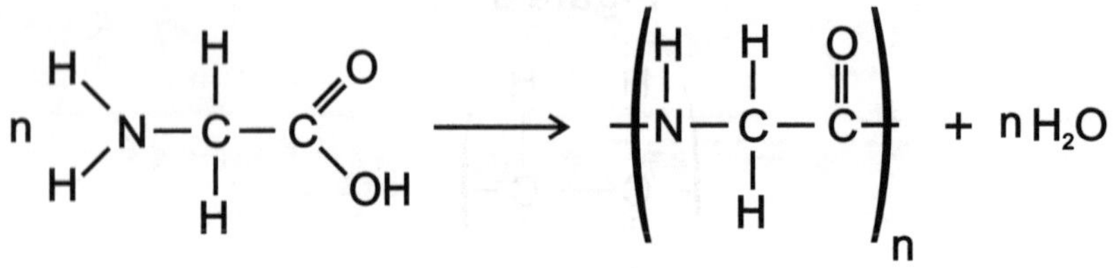

$$n\ H_2N-CH_2-COOH \longrightarrow \left(-NH-CH_2-CO- \right)_n + nH_2O$$

8.4 Name the type of polymerisation shown in **Figure 7.**
Explain how you can tell that the type of polymerisation you have named is occurring.

....

....
[2 marks]

8.5 What type of molecule is glycine?

....
[1 mark]

8.6 Name **one** other naturally occurring polymer.

....
[1 mark]

9 A company uses a process called the contact process to make sulfuric acid. The first step of the process is the exothermic reaction of sulfur dioxide and oxygen to form sulfur trioxide. The equation for the reaction is shown below.

$$2SO_{2(g)} + O_{2(g)} \rightleftharpoons 2SO_{3(g)}$$

9.1 This reaction can reach equilibrium. When a reaction is at equilibrium, what does this tell you about the forward and reverse reactions?

....

....

[1 mark]

9.2 You can predict how changing the conditions will affect this reaction using Le Chatelier's principle. What is Le Chatelier's principle?

....

....

[1 mark]

9.3 State what will happen to the yield of sulfur trioxide in this reaction if the pressure inside the reaction vessel is increased. Explain your answer.

....

....

....

....

[3 marks]

Question 9 continues on the next page

9.4 In order to increase the rate of the reaction, the company decides to increase the temperature of the reaction vessel.
State what effect this will have on the yield of sulfur trioxide. Explain your answer.

....

....

....

....

[3 marks]

9.5 A measured mass of vanadium oxide was added to the reaction vessel at the start of the process. The same mass of vanadium oxide is present at the end of the reaction.
Suggest what role vanadium oxide is playing in this reaction.

....

[1 mark]

9.6 The company used to produce 7500 kg of sulfur trioxide each day.
After modifying the plant to allow the sulfur trioxide to be removed from the reaction vessel as soon as it is made, the yield of the plant increased by 15%.
How much sulfur trioxide can the plant make in five days now?

....

....

....

mass made in five days = ... kg

[3 marks]

END OF QUESTIONS

Topic 1 — Atomic Structure and the Periodic Table

Pages 22-23

Warm-Up Questions

1) The total number of protons and neutrons.
2) An element is a substance that consists of only one type of atom.
3) A compound is a substance made from two or more elements in fixed proportions throughout the compound and they're held together by chemical bonds.
4) ions
5) $C_6H_{12}O_6 + 6O_2 \rightarrow 6CO_2 + 6H_2O$

Exam Questions

1.1

Name of particle	Relative charge
Proton	**+1**
Neutron	**0**

[2 marks — 1 mark for each correct charge]

1.2 nucleus ***[1 mark]***
1.3 8 ***[1 mark]***
1.4 1 ***[1 mark]***
2.1 7 ***[1 mark]***
2.2 The number of protons is the same as the atomic number ***[1 mark]***
2.3 N ***[1 mark]***
3.1 methane and oxygen ***[1 mark]***
3.2 carbon dioxide and water ***[1 mark]***
3.3 oxygen/O_2 ***[1 mark]***
3.4 $CH_4 + \mathbf{2O_2} \rightarrow CO_2 + \mathbf{2}H_2O$
[1 mark for correct missing reactant, 1 mark for correctly balancing the equation]
4.1 $\mathbf{H_2SO_4} + \mathbf{2NH_3} \rightarrow (NH_4)_2SO_4$ ***[1 mark for each correct reactant, 1 mark for correctly balancing the equation]***
4.2 15 ***[1 mark]***
There are eight atoms of hydrogen, one atom of sulfur, four atoms of oxygen, and two atoms of nitrogen.
4.3 8
There are 4 hydrogen atoms in NH_4 and there are two of these.
5.1 An isotope is a different atomic form of the same element ***[1 mark]***, which has the same number of protons ***[1 mark]*** but a different number of neutrons ***[1 mark]***.
5.2 Protons = 6 ***[1 mark]***, neutrons = 7 ***[1 mark]***, electrons = 6 ***[1 mark]***
5.3 It has a different atomic number from carbon ***[1 mark]***.
5.4 $(79 \times 13) + (21 \times 14) = 1321$
Relative atomic mass = $\frac{1321}{100} = 13.21 = \mathbf{13.2}$
[4 marks for correct final answer, otherwise give 1 mark for each correct stage of working]

Page 30

Warm-Up Questions

1) Two or more elements or compounds mixed together.
2) evaporation / crystallisation
3) fractional distillation

Exam Questions

1.1 E.g. put the paper in a beaker of solvent, e.g. water ***[1 mark]***.
1.2 Pencil marks are insoluble (so won't dissolve into the solvent) ***[1 mark]***.
2.1 E.g. mix the lawn sand with water to dissolve the ammonium sulfate ***[1 mark]***. Filter the mixture using filter paper to remove the sharp sand ***[1 mark]***. Pour the remaining solution into an evaporating dish and slowly heat it to evaporate the water until you have a dry product ***[1 mark]***.
2.2 E.g. the products were not completely dry ***[1 mark]***.
3.1 The boiling points of water and methanoic acid are too close together to allow them to be separated by simple distillation ***[1 mark]***.
3.2

Temperature on thermometer	Contents of the flask	Contents of the beaker
30 °C	both liquids	no liquid
65 °C	water	propanone
110 °C	no liquid	both liquids

[3 marks for whole table correct otherwise 1 mark for each correct row]

Page 36

Warm-Up Questions

1) The 'plum pudding model' — a ball of positive charge with electrons stuck in it.
2) a) 2
b) 8
3) The discovery of isotopes.

Exam Questions

1.1

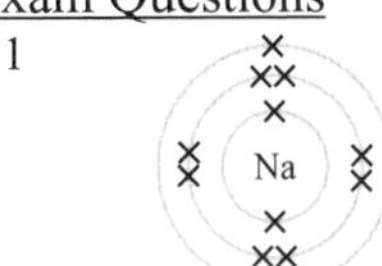

[1 mark]

1.2 Any one of: lithium / potassium / rubidium / caesium / francium ***[1 mark]***
1.3 1 ***[1 mark]***
2.1 By atomic mass ***[1 mark]***
2.2 By atomic number ***[1 mark]***
2.3 Similar chemical properties due to having the same number of electrons in their outer shell ***[1 mark]***.
3 How to grade your answer:
Level 0: No description is given. ***[No marks]***
Level 1: Brief description of how the theory of atomic structure has changed. ***[1 to 2 marks]***
Level 2: Some detail given of how the theory of atomic structure has changed. ***[3 to 4 marks]***
Level 3: A clear and detailed description of how the theory of atomic structure has changed. ***[5 to 6 marks]***

Here are some points your answer may include:
At the start of the 19th century John Dalton described atoms as solid spheres that make up the different elements.
In 1897 JJ Thomson concluded that atoms weren't solid spheres and that an atom must contain smaller, negatively charged particles (electrons). He called this the 'plum pudding model'.
In 1909 Ernest Rutherford conducted a gold foil experiment, firing positively charged particles at an extremely thin sheet of gold. Most of the particles went straight through, so he concluded that there was a positively charged nucleus at the centre, surrounded by a 'cloud' of negative electrons.
Niels Bohr proposed a new model of the atom where all the electrons were contained in shells. He suggested that electrons can only exist in fixed orbits, or shells.
Further experimentation by Rutherford and others showed that the nucleus could be divided into smaller particles, each with the same charge as a hydrogen atom, these became known as protons.
James Chadwick carried out an experiment which provided evidence of the existence of neutral particles within the nucleus, which became known as neutrons.

Pages 44-45

Warm-Up Questions

1) positive ions
2) In the centre block / between Group 2 and Group 3.
3) Any three of: e.g. strong / can be bent or hammered into different shapes / conduct heat / conduct electricity / high melting point / high boiling point.
4) Any three of: e.g. have more than one ion (with different charges) / form coloured compounds / make useful catalysts / good conductors of heat/electricity / dense / strong / shiny.
5) lithium oxide / Li_2O
6) negative
7) Group 0

Exam Questions

1.1 Highest (X): fluorine ***[1 mark]***. Lowest (Y): iodine ***[1 mark]***.

1.2 iodine ***[1 mark]***

2.1 transition metals ***[1 mark]***

2.2 To the right of the line ***[1 mark]***. Since it does not conduct electricity, it must be a non-metal ***[1 mark]***.

3.1 ionic bonds ***[1 mark]***

3.2 E.g. displacement reaction ***[1 mark]***

3.3 iodine/I_2 ***[1 mark]*** as it is displaced from potassium iodide ***[1 mark]***.

3.4 The products of the reaction will be aqueous potassium chloride/$KCl_{(aq)}$ ***[1 mark]*** and bromine gas/$Br_{2(g)}$ ***[1 mark]***. Chlorine is more reactive than bromine as its outer shell is closer to the nucleus ***[1 mark]***. This results in chlorine displacing bromine from potassium bromide ***[1 mark]***.

3.5 Group 0 elements are generally inert ***[1 mark]***. This is because they are energetically stable and don't need to lose or gain electrons to have a full outer shell ***[1 mark]***.

4.1 They have a single outer electron which is easily lost so they are very reactive ***[1 mark]***.

4.2 As you go down Group 1, the outer electron is further from the nucleus ***[1 mark]***. So the attraction between the nucleus and the electron decreases ***[1 mark]***. This means the outer electron is more easily lost and the metal is more reactive ***[1 mark]***.

4.3 hydrogen ***[1 mark]*** and potassium hydroxide ***[1 mark]***

4.4 alkaline ***[1 mark]***

Topic 2 — Bonding, Structure and Properties of Matter

Page 52

Warm-Up Questions

1) A charged atom or group of atoms.
2) +2
3) –1
4) high
5) When ionic compounds dissolve in water, the ions separate and are all free to move in the solution, so they'll carry an electric charge.
6) $Al(OH)_3$

Exam Questions

1.1 Sodium will lose the electron from its outer shell to form a positive ion ***[1 mark]***. Fluorine will gain an electron to form a negative ion ***[1 mark]***.

1.2 Ionic compounds have a giant ionic lattice structure ***[1 mark]***. The ions form a closely packed regular lattice arrangement ***[1 mark]***, held together by strong electrostatic forces of attraction between oppositely charged ions ***[1 mark]***.

2.1 lithium oxide ***[1 mark]***

2.2

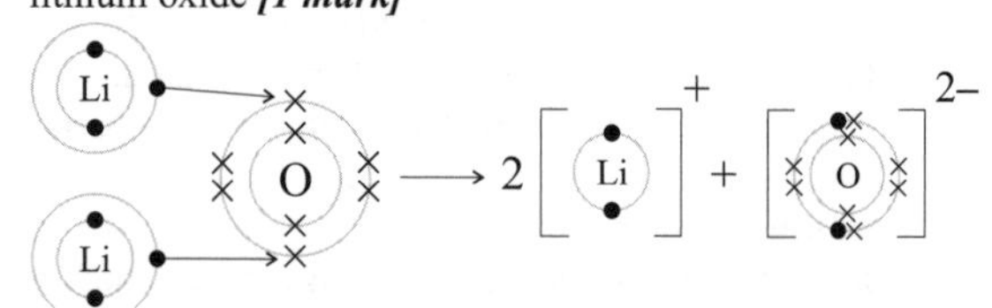

[1 mark for arrows shown correctly, 1 mark for correct electron arrangement and charge on lithium ion, 1 mark for correct electron arrangement and charge on oxygen ion]

2.3 When molten, the Li^+ and O^{2-} are able to move ***[1 mark]***. These ions are able to carry electric charge ***[1 mark]***.

2.4 The formula of the compound is **LiCl** ***[1 mark]***. Lithium is in Group 1, so forms +1 ions. Chlorine is in Group 7, so forms –1 ions. Therefore there needs to be one lithium ion for every chlorine ion for the compound to be neutral ***[1 mark]***.

Page 56

Warm-Up Questions

1) A chemical bond made by the sharing of a pair of electrons between two atoms.
2) 3
3) gas or liquid
4) covalent bonds
5) intermolecular forces

Exam Questions

1 ***[1 mark for showing a pairs of shared electrons between C and H. 1 mark for no other electrons shown]***

2.1

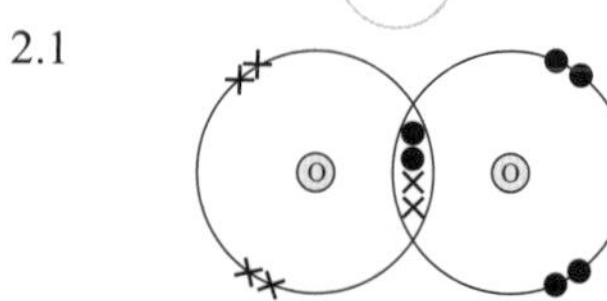

[1 mark for showing 2 pairs of shared electrons between O atoms. 1 mark for showing two non-bonding pairs of electrons on each oxygen]

2.2 3 ***[1 mark]***

3.1 Hydrogen chloride doesn't contain any ions or delocalised electrons to carry a charge ***[1 mark]***.

3.2 E.g. a pair of electrons (one from the hydrogen atom and one from the chlorine atom) is shared between the two atoms ***[1 mark]***. The atoms are held together by the strong attraction between this shared pair of negatively charged electrons and the positively charged nuclei of the atoms ***[1 mark]***.

3.3 Chlorine is a larger molecule than hydrogen chloride ***[1 mark]*** and therefore there are greater intermolecular forces between molecules ***[1 mark]***. As the intermolecular forces are greater between chlorine molecules, its boiling point is higher than hydrogen chloride ***[1 mark]***.

Chlorine has three shells of electrons, whereas hydrogen only has one. So Cl_2 is a bigger molecule than HCl.

Page 62

Warm-Up Questions

1) By covalent bonding
2) solid
3) E.g. diamond is much harder than graphite. Diamond doesn't conduct electricity whereas graphite does.
4) Buckminsterfullerene
5) The layers of atom in metals are able to slide over each other.

Exam Questions

1.1 In a giant covalent structure, all of the atoms are bonded to each other with strong covalent bonds ***[1 mark]***. It takes lots of energy to break these bonds and melt the solid ***[1 mark]***.

1.2 E.g. diamond / graphite / silicon dioxide ***[1 mark]***

2.1 Graphite is formed of layers of carbon atoms that are held together weakly ***[1 mark]***. The layers can easily slide over each other, making graphite soft and slippery ***[1 mark]***.

2.2 In diamond, each carbon atom forms four covalent bonds in a very rigid structure ***[1 mark]***. This makes diamond very hard, so it would be good at cutting other substances ***[1 mark]***.

2.3 Any one from: e.g. delivering drugs / catalysts / lubricants / strengthening materials ***[1 mark]***

2.4 E.g. each carbon atom has a delocalised electron that is able to carry the charge ***[1 mark]***.

3.1 E.g. ***[1 mark for showing two different sizes of atoms, 1 mark for showing irregular arrangement]***

3.2 The regular arrangement of atoms in iron means that they can slide over each other, resulting in iron being soft ***[1 mark]***. Steel contains different sized atoms which distort the layers of iron atoms ***[1 mark]*** making it more difficult for them to slide over each other ***[1 mark]***.

Page 68

Warm-Up Questions

1) The material, the temperature and the pressure.
2) solid
3) The particles move faster.
4) $Na^+_{(aq)}$
5) Between 2500 nm (2.5×10^{-6}m) and 10 000 nm (1×10^{-5}m).
6) The surface ratio to volume ratio increases.

Exam Questions

1.1 In solids, there are **strong** ***[1 mark]*** forces of attraction between particles, which hold them in fixed positions in a **regular** ***[1 mark]*** arrangement. The particles don't **move** ***[1 mark]*** from their positions, so solids keep their shape. The **hotter** ***[1 mark]*** the solid becomes, the more the particles in the solid vibrate.

1.2 The particles gain energy, so they move faster ***[1 mark]***. The intermolecular bonds are weakened and, when they have enough energy, they break ***[1 mark]***. This means the particles are free to move far apart from each other and the liquid becomes a gas ***[1 mark]***.

2 E.g. some people are concerned about the use of nanoparticles as not enough is known about their effects on the body following long-term use ***[1 mark]***. Testing over a long time period could be carried out before products containing nanoparticles are used by humans to help minimise the risk ***[1 mark]***.

3.1 The particles are free to move about / have virtually no forces of attraction between them ***[1 mark]***, so they move randomly, spreading out to fill the container ***[1 mark]***.

3.2 gas ***[1 mark]***

Topic 3 — Quantitative Chemistry

Page 74

Warm-Up Questions

1) relative formula mass
2) Avogadro constant
3) A mole is an amount of a substance that contains as many particles as the Avogadro constant (6.02×10^{23}).
4) 56
5) One mole of O_2 weighs $16 \times 2 = 32$ g

Exam Questions

1.1 M_r of $BF_3 = 11 + (19 \times 3) = 68$ ***[1 mark]***

1.2 M_r of $B(OH)_3 = 11 + (17 \times 3) = 62$ ***[1 mark]***

2 number of moles $= \frac{\text{mass in grams}}{M_r}$ ***[1 mark]***

3 M_r of KOH $= (39 + 16 + 1) = 56$ ***[1 mark]***
Mass of 4 moles of KOH $= 4 \times 56 = 224$ g ***[1 mark]***
Extra mass needed $= 224 - 140 = 84$ g ***[1 mark]***

4.1 Oxygen in the air, which previously couldn't be weighed, has now reacted with metal **X** to form a solid oxide ***[1 mark]***.

4.2 $100 - 40 = 60\%$ ***[1 mark]***

4.3 Mass of metal **X** in 8.0 g of mixture:
$= 8.0 \times (24 \div 100) = 1.92$ g ***[1 mark]***
The metal oxide contains 60% metal **X** by mass, so the mass of metal oxide used to make the mixture $= 1.92 \div (60 \div 100) =$ **3.2 g** ***[1 mark]***.

Page 78

Warm-Up Questions

1) The number of moles of each substance taking part in or made by the reaction.
2) To make sure that all of the other reactant reacts.
3) It halves.
4) Number of moles of gas = volume of gas (in dm^3) ÷ 24
$= 2.4 \div 24 =$ **0.1 moles**

Exam Questions

1.1 Li: $3.5 \div 7 =$ **0.5 mol**
O_2: $4.0 \div 32 =$ **0.125 mol**
Li_2O: $7.5 \div 30 =$ **0.25 mol**
[2 marks for all three correct answers, otherwise 1 mark for two correct answers]

1.2 Divide by the smallest number of moles (0.125):
Li: $\frac{0.5}{0.125} = 4$ O_2: $\frac{0.125}{0.125} = 1$ Li_2O: $\frac{0.25}{0.125} = 2$ ***[1 mark]***
The balanced symbol equation is:
$4Li + O_2 \rightarrow 2Li_2O$ ***[1 mark]***

2.1 M_r of $NaHCO_3 = 23 + 1 + 12 + (3 \times 16) = 84$ ***[1 mark]***
no. of moles of $NaHCO_3 = 6.0 \div 84 = 0.0714...$ mol ***[1 mark]***
1 mole of Na_2SO_4 is made for every 2 moles of $NaHCO_3$ that reacts, so no. of moles of $Na_2SO_4 = 0.0714... \div 2$
$= 0.0357...$ mol ***[1 mark]***
M_r of $Na_2SO_4 = (2 \times 23) + 32 + (4 \times 16) = 142$ ***[1 mark]***
Mass of $Na_2SO_4 = 142 \times 0.0357... = 5.071... =$ **5.1g** (2 s.f.) ***[1 mark]***

2.2 The limiting reactant is the reactant that gets used up and therefore limits the amount of product formed ***[1 mark]***.

3 $M_r(CaCO_3) = 40 + 12 + (3 \times 16) = 100$ ***[1 mark]***
no. of moles of $CaCO_3 = 25 \div 100 = 0.25$ mol ***[1 mark]***
1 mole of CO_2 is formed for every mole of $CaCO_3$ that reacts.
So 0.25 moles of $CaCO_3$ reacts to form 0.25 moles of CO_2.
Volume of $CO_2 = 0.25 \times 24$ dm^3 ***[1 mark]*** $= 6.0$ dm^3 ***[1 mark]***

Page 81

Warm-Up Questions

1) $160 \div 1000 = 0.160$ dm^3
concentration = mass ÷ volume $= 20.0 \div 0.160 =$ **125 g/dm³**
2) number of moles = concentration × volume
$= 0.1 \times (25 \div 1000) =$ **0.0025 mol**
3) $M_r(Na_2CO_3) = (2 \times 23) + 12 + (3 \times 16) = 106$
Concentration in grams per $dm^3 = 0.025 \times 106 =$ **2.65 g/dm³**

Exam Questions

1.1 number of moles = concentration × volume
number of moles $= 1.00 \times (30.4 \div 1000)$
= **0.0304 mol** ***[1 mark]***

1.2 H_2SO_4 and NaOH react in a 1:2 ratio, so number of moles of H_2SO_4
$= 0.0304 \div 2 =$ **0.0152 mol** ***[1 mark]***

1.3 concentration = number of moles ÷ volume ***[1 mark]***
volume $= 25 \div 1000 = 0.025$
concentration $= 0.0152 \div 0.025 =$ **0.608 mol/dm³** ***[1 mark]***

1.4 M_r of $H_2SO_4 = (2 \times 1) + 32 + (4 \times 16) = 98$ ***[1 mark]***
Concentration in g/dm³ = moles × M_r
Concentration in g/dm³ $= 0.608 \times 98$
= **59.6 g/dm³** (3 s.f.) ***[1 mark]***

2 Moles of NaOH = concentration × volume
$= 0.10 \times (9.0 \div 1000) = 0.00090$ mol ***[1 mark]***
HA and NaOH react in a 1:1 ratio, so moles of HA = 0.00090 moles ***[1 mark]***
Concentration of HA = number of moles ÷ volume ***[1 mark]***
volume $= 25 \div 1000 = 0.025$
Concentration of HA $= 0.00090 \div 0.025$
= **0.036 mol/dm³** ***[1 mark]***

Page 85

Warm-Up Questions

1) 100%
All reactions with one product will have 100% atom economy.
2) Because a low atom economy means resources are used up quickly, so any non-renewable resources are likely to run out. It also means lots of waste is likely to be produced, and this has to be stored somewhere.
3) percentage yield = (actual yield ÷ theoretical yield) × 100
4) percentage yield $= (4 \div 5) \times 100 =$ **80%**

Exam Questions

1 To calculate the atom economy of a reaction, you need to know the mass of products you expected to form, and the mass of products that actually formed ***[1 mark]***.

2 M_r of ethanol $= (12 \times 2) + 6 + 16 = 46$,
M_r of ethene $= (12 \times 2) + 4 = 28$ ***[1 mark]***
Atom economy $= (28 \div 46) \times 100$ ***[1 mark]***
= **61 %** (2 s.f.) ***[1 mark]***

3 How to grade your answer:
Level 0: No reasons why yields are always less than 100% are given. ***[No marks]***
Level 1: Brief description of one reason why yields are less than 100% is given. ***[1 to 2 marks]***
Level 2: Two reasons why yields are always less than 100% are described clearly. ***[3 to 4 marks]***
Level 3: Three reasons why yields are always less than 100% are described clearly. ***[5 to 6 marks]***

Here are some points your answer may include:
Some reactions are reversible. Some of the reactants will never be completely converted to products because the reaction goes both ways.
If a liquid is filtered to remove solid particles, some of the liquid or solid could be lost when it's separated from the reaction mixture.
Transferring solutions from one container to another often leaves behind traces on the containers.
There may be unexpected reactions happening resulting in extra products forming other than the ones you wanted.

4.1 $M_r(CuO) = 63.5 + 16 = 79.5$ ***[1 mark]***
So moles of CuO = 4.0 ÷ 79.5 = 0.0503... mol
From the equation, 1 mole of Cu is produced for every 1 mol of CuO that reacts. So 0.0503... mol CuO reacts to form 0.0503... mol Cu ***[1 mark]***.
So theoretical yield of Cu = number of moles × M_r
= 0.0503... × 63.5 = **3.2 g** (2 s.f.) ***[1 mark]***

4.2 Percentage yield = (2.8 ÷ 3.2) × 100 ***[1 mark]***
= **88%** (2 s.f.) ***[1 mark]***

Topic 4 — Chemical Changes

Page 93

Warm-Up Questions

1) 0-14
2) neutral
3) Strong acids ionise completely in water, so all of the acid particles dissociate to release H^+ ions.
Weak acids do not fully ionise, so only some of the acid particles dissociate to release H^+ ions.
4) Difference in pH = 5 – 2 = 3
Factor H^+ ion concentration changes by = 10^{-3} = **0.001**

You could give your answer here as '10^{-3}', '0.001' or 'it decreased by a factor of 1000' — they all mean the same thing.

5) Copper nitrate and water.

Exam Questions

1.1 H^+ ions/hydrogen ions ***[1 mark]***
1.2 neutralisation ***[1 mark]***
2.1 alkalis ***[1 mark]***
2.2 $2HNO_3 + Mg(OH)_2 \rightarrow \mathbf{Mg(NO_3)_2 + 2H_2O}$
[1 mark for formulas of both products correct, 1 mark for putting a 2 in front of H_2O to balance the equation]
3.1 E.g.the excess silver carbonate will sink to the bottom of the flask and stay there / the mixture will stop producing bubbles (of carbon dioxide) ***[1 mark]***.
3.2 filtration ***[1 mark]***
3.3 Heat the silver nitrate solution gently using a water bath/electric heater to evaporate some of the water ***[1 mark]***. Leave the solution to cool until crystals form ***[1 mark]***. Filter out the crystals and dry them ***[1 mark]***.
4 How to grade your answer:
Level 0: There is no relevant information. ***[No marks]***
Level 1: There is a brief explanation of how to carry out a titration, but some major details are missing. ***[1 to 2 marks]***
Level 2: There is a good explanation of how to carry out a titration. Some minor details may be missing. ***[3 to 4 marks]***
Level 3: There is a clear and detailed explanation of how to carry out a titration. ***[5 to 6 marks]***
Here are some points your answer may include:
Use a pipette (and pipette filler) to measure out 25.0 cm^3 of the alkali and put it into a conical flask.
Add two or three drops of an indicator / phenolphthalein / litmus / methyl orange to the alkali.
Use a funnel to fill a burette with the acid.
Record the initial volume of the acid in the burette.
Use the burette to add the acid to the alkali a bit at a time.
Swirl the flask regularly.
Go especially slowly as you approach the end-point / colour change.
The indicator will change colour when all the alkali has been neutralised/at the end-point.
Record the final volume of acid in the burette,

Page 99

Warm-Up Questions

1) How easily it loses electrons to form positive ions.
2) magnesium chloride
3) A metal hydroxide and hydrogen.
4) a) $Zn_{(s)} + Cu^{2+}_{(aq)} \rightarrow Zn^{2+}_{(aq)} + Cu_{(s)}$
b) zinc / Zn

Exam Questions

1.1 Any one from: potassium / sodium / calcium ***[1 mark]***.
1.2 copper ***[1 mark]***

Metals below hydrogen in the reactivity series don't react with acids.

1.3 E.g. iron and dilute acid would produce bubbles of hydrogen gas ***[1 mark]*** more slowly than zinc and dilute acid / very slowly ***[1 mark]***.

You don't need to use these exact words — the two points you need to cover are that it would still produce hydrogen gas, but at a slower rate than the zinc.

2.1 $2Fe_2O_3 + 3C \rightarrow 4Fe + 3CO_2$
[1 mark for formulas of all reactants and products correct, 1 mark for correctly balancing the equation]
2.2 Oxygen is being lost from iron oxide / the iron(III) ions in the iron oxide are gaining electrons ***[1 mark]***.
3 In A there will be no change ***[1 mark]***. In B the solution will have changed from blue to green ***[1 mark]*** and the grey metal will be coated in orange metal ***[1 mark]***. Iron is more reactive than copper ***[1 mark]***. So in A, the copper cannot displace the iron from iron sulfate ***[1 mark]***, but in B, the iron does displace the copper from copper sulfate ***[1 mark]***.

Page 104

Warm-Up Questions

1) the cathode / negative electrode
2) zinc and chlorine gas
3) reduced
4) hydrogen gas

Exam Questions

1.1 To lower the melting point of the aluminium oxide ***[1 mark]***.
1.2 $\mathbf{Al^{3+} + 3e^- \rightarrow Al}$ ***[1 mark]***
1.3 $2O^{2-} \rightarrow O_2 + 4e^-$ ***[1 mark]***
1.4 The oxygen made at the positive electrode reacts with the carbon to make carbon dioxide ***[1 mark]*** wearing the electrode away ***[1 mark]***.
2.1 hydrogen ***[1 mark]***
2.2 $2H^+ + 2e^- \rightarrow H_2$ ***[1 mark]***
2.3 chlorine ***[1 mark]***
2.4 $2Cl^- \rightarrow Cl_2 + 2e^-$ / $2Cl^- - 2e^- \rightarrow Cl_2$ ***[1 mark]***
2.5 Sodium is more reactive than hydrogen, so the sodium ions from the sodium chloride stay in solution ***[1 mark]***. Hydroxide ions are also left in solution when hydrogen is produced from water ***[1 mark]***.
2.6 Copper is less reactive than hydrogen ***[1 mark]***.

Topic 5 — Energy Changes

Page 110

Warm-Up Questions

1) An exothermic reaction is one which transfers energy to the surroundings.
2) An endothermic reaction is one which takes in energy from the surroundings.
3) When bonds are formed.
4) The activation energy is the minimum amount of energy the reactants need to collide with each other and react.

Exam Questions

1 B ***[1 mark]***
2.1 Exothermic, because the temperature of the mixture has increased ***[1 mark]***, therefore the particles have transferred energy to the reaction mixture ***[1 mark]***.

2.2 A lid was placed on the cup to reduce energy lost to the surroundings (by evaporation) ***[1 mark]***.

3.1 C – H and O = O ***[1 mark]***

3.2 (4 × 414) + (2 × 494) = 2644 kJ/mol

This is the amount of energy required to break the bonds in CH_4 and $2O_2$.

(2 × 800) + (4 × 459) = 3436 kJ/mol

This is the amount of energy released by forming the bonds in CO_2 and $2H_2O$.

2644 – 3436 = **–792 kJ/mole** ***[3 marks for correct answer, otherwise 1 mark for finding the energy needed to break the bonds in the reactants, 1 mark for finding the energy released by forming bonds in the products.]***

Page 115

Warm-Up Questions

1) A liquid that contains ions which react with the electrodes.
2) You connect them together in series.
3) A fuel cell is an electrical cell that's supplied with a fuel and oxygen (or air) and uses energy from the reaction between them to produce electrical energy.
4) E.g. potassium hydroxide/phosphoric acid

Exam Questions

1 voltmeter ***[1 mark]***

2.1 In the fuel cell, there is reduction at the cathode ***[1 mark]*** and oxidation at the anode ***[1 mark]***.

2.2 Reduction, because the reactant/oxygen is gaining electrons ***[1 mark]***.

2.3 $2H_2 + O_2 \rightarrow 2H_2O$ ***[1 mark]***

2.4 How to grade your answer:

Level 0: There is no relevant information. ***[No marks]***

Level 1: There is a brief explanation of some relevant advantages of using fuel cells rather than rechargeable batteries to power cars. ***[1 to 2 marks]***

Level 2: There is clear and detailed explanation of several advantages of using fuel cells rather than rechargeable batteries to power cars. ***[3 to 4 marks]***

Here are some points your answer may include:

Hydrogen-oxygen fuel cells are less polluting to dispose of than rechargeable batteries because rechargeable batteries are made from toxic metal compounds.

Fuel cells can be reused many times but there is a limit to how often batteries in electric cars can be recharged.

Fuel cells are less expensive than batteries to make.

Fuel cells store more energy than batteries and so batteries need to be recharged more often, which can take a long time.

3.1 Metal 2, Metal 3, Metal 1 ***[1 mark]***. Zinc is less reactive than all 3 metals, so the most reactive metal will make the cell with the greatest voltage, since there will be the biggest difference in reactivity ***[1 mark]***. So the order of reactivity is simply the order of the size of voltage ***[1 mark]***.

3.2 E.g. join together three cells in series ***[1 mark]***, using Metal 3 and zinc for the electrodes in each cell ***[1 mark]***. / Join together two cells in series, using Metal 2 and zinc as the electrodes ***[1 mark]*** and one cell using Metal 1 and zinc as the electrodes ***[1 mark]***.

Topic 6 — The Rate and Extent of Chemical Change

Pages 125–126

Warm-Up Questions

1) E.g. the rusting of iron is a reaction that happens very slowly. Explosions are very fast reactions.
2) They must collide with enough energy.
3) There will be more particles in the same volume, which will lead to more frequent collisions, so the rate of reaction will increase.
4) g/s
5) E.g. observe a mark through the solution in which the precipitate forms and measure how long it takes for it to disappear.

Exam Questions

1.1 stopwatch/stopclock/timer ***[1 mark]***

1.2 gas syringe ***[1 mark]***

1.3 E.g. place a conical flask on a mass balance ***[1 mark]*** and record the change in mass at regular intervals as the gas leaves the flask ***[1 mark]***.

2.1 Any two from: e.g. the concentration of sodium thiosulfate/ hydrochloric acid / the depth of liquid / the person judging when the black cross is obscured / the black cross used (size, darkness etc.). ***[2 marks — 1 mark for each correct answer.]***

Judging when a cross is completely obscured is quite subjective — two people might not agree on exactly when it happens. You can try to limit this problem by using the same person each time, but you can't remove the problem completely. The person might have changed their mind slightly by the time they do the next experiment — or be looking at it from a different angle, be a bit more bored, etc.

2.2

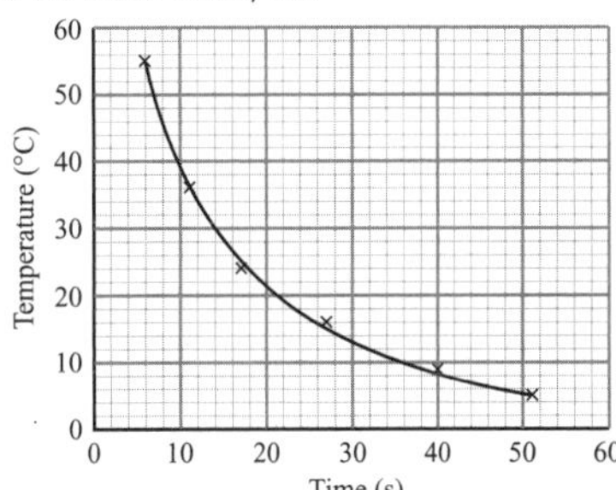

[1 mark for correctly drawn axes with a sensible scale, 2 marks for all points plotted correctly, otherwise 1 mark if 5 of 6 points plotted correctly, 1 mark for a suitable line of best fit]

2.3 As the temperature decreases the time decreases, meaning that the reaction is slower as the temperature decreases ***[1 mark]***.

2.4 At each temperature it would take longer for the reaction to complete ***[1 mark]***.

2.5 A lower concentration of reactant means there are fewer particles in the same volume of solution ***[1 mark]***. This makes collisions between the reactant particles less frequent ***[1 mark]***, and therefore decreases the rate of reaction ***[1 mark]***.

2.6 E.g. by repeating the experiment ***[1 mark]***. If the results gained are similar then the experiment is repeatable ***[1 mark]***.

3.1 A ***[1 mark]***

3.2

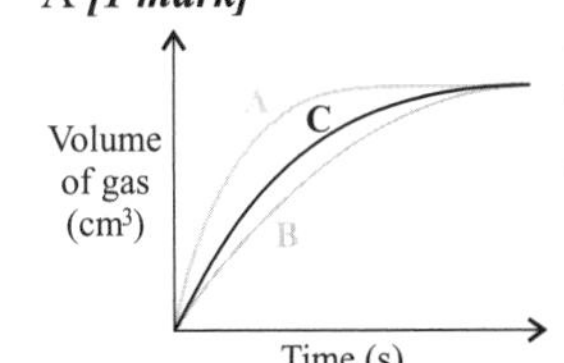

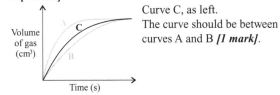

Curve C, as left. The curve should be between curves A and B ***[1 mark]***.

3.3 The variables are controlled so that you can tell if the variable you're changing/the independent variable is causing the results seen ***[1 mark]***.

3.4 E.g. yes, because the same volume of gas was produced in each experiment ***[1 mark]***.

This suggests that the same concentration and volume of acid (or an excess of acid) was used in both experiments.

3.5 Measuring how quickly the reaction loses mass ***[1 mark]***.

4.1 Mean rate = amount of product formed ÷ time = 40 ÷ 60 = **0.67 cm^3/s** ***[2 marks for correct answer, otherwise 1 mark for correct working]***

4.2 E.g.

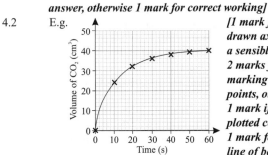

Volume of CO_2 (cm^3) 0 10 20 30 40 50; Time (s) 0 10 20 30 40 50 60

[1 mark for correctly drawn axes with a sensible scale, 2 marks for correctly marking on all 6 points, otherwise 1 mark if 5 of 6 points plotted correctly, 1 mark for a suitable line of best fit.]

4.3 E.g.

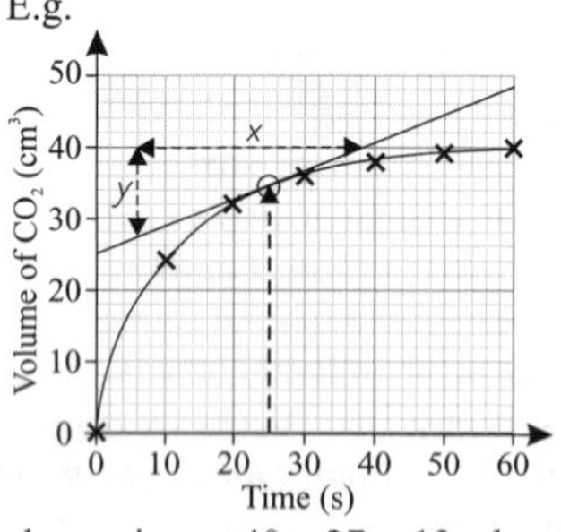

change in $y = 40 - 27 = 13$, change in $x = 38 - 6 = 32$,
rate = change in $y \div$ change in $x = 13 \div 32 =$ **0.4 cm^3/s**
[1 mark for drawing a tangent at 25 s, 1 mark for correctly calculating a change in y from the tangent, 1 mark for correctly calculating a change in x from the tangent and 1 mark for a rate between 0.3 cm/s and 0.5 cm/s]

5 How to grade your answer:
Level 0: No relevant information is given. ***[No marks]***
Level 1: One or two ways of increasing the rate are described and there's reference to collision theory. ***[1 to 2 marks]***
Level 2: At least two ways of increasing the rate are given with appropriate reference to collision theory. ***[3 to 4 marks]***
Level 3: There is a clear and detailed discussion of three ways by which the rate can be increased, which includes relevant references to collision theory. ***[5 to 6 marks]***

Here are some points your answer may include:
Collision theory says that the rate of reaction depends on how often and how hard the reacting particles collide with each other.
If the particles collide hard enough (with enough energy) they will react.
Increasing the temperature makes particles move faster, so they collide more often and with greater energy. This will increase the rate of reaction.
If the surface area of the nickel catalyst is increased then the particles around it will have more area to work on. This increases the frequency of successful collisions and will increase the rate of reaction.
Increasing the pressure of the hydrogen will mean the particles are more squashed up together. This will increase the frequency of the collisions and increase the rate of reaction.

Page 130

Warm-Up Questions

1) They are the same.
2) The equilibrium will move to the left/towards reactants and the amount of reactants will increase.
3) It would have no effect (because there are equal numbers of gas molecules on both sides).

Exam Questions

1.1 The reaction is reversible ***[1 mark]***.
1.2 Both (the forward and reverse) reactions are taking place at exactly the same rate ***[1 mark]***.
2.1 Less $NO_{2(g)}$ will be produced ***[1 mark]***, because there are more molecules of gas on the right hand side of the equation ***[1 mark]***.
2.2 It will have no effect ***[1 mark]***, because there are the same number of molecules of gas on both sides of the equation ***[1 mark]***.
3.1 It gives out energy ***[1 mark]***, because it's exothermic ***[1 mark]***.

As the forward reaction is endothermic you know that the reverse reaction is exothermic and therefore gives out energy.

3.2 A reversible reaction is always exothermic in one direction and endothermic in the other direction ***[1 mark]***, so a change in temperature will always favour one reaction more than the other ***[1 mark]***.
3.3 It would move to the right ***[1 mark]***.

Topic 7 — Organic Chemistry

Pages 136-137

Warm-Up Questions

1) methane, ethane, propane, butane
2) single covalent bonds
3) When burnt they release energy.
4) e.g. fuel for transport / feedstock for making new compounds
5) long-chain hydrocarbons

Exam Questions

1 Carbon-carbon single bonds and carbon-hydrogen single bonds ***[1 mark]***.
2.1 butane ***[1 mark]***
2.2 C_4H_{10} ***[1 mark]***
2.3 Any two from: e.g. polymers / solvents /lubricants / detergents ***[2 marks — 1 mark for each correct answer]***
3

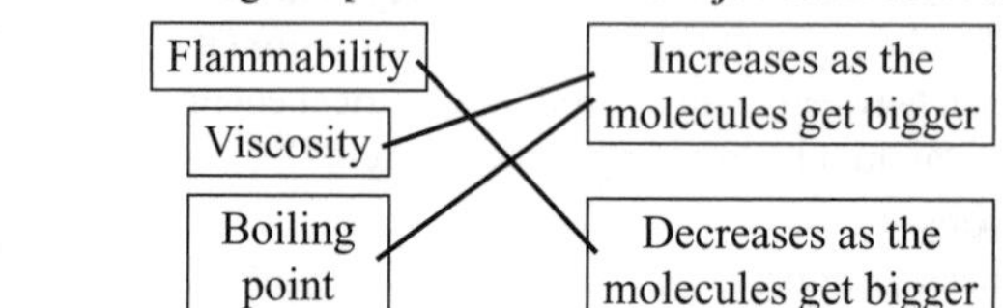

[1 mark for each correctly drawn line]

4.1 **F** ***[1 mark]***
4.2 **A** ***[1 mark]***
4.3 Inside the fractionating column there is a temperature gradient with the hottest part at the bottom and coolest at the top ***[1 mark]***. Crude oil that is gaseous moves up the column and when hydrocarbons in the gas reach a part of the column where the temperature is lower than their boiling points they condense and drain out of the column ***[1 mark]***. Hydrocarbons with longer chain lengths have higher boiling points so condense lower down the column whereas hydrocarbons with shorter chain lengths have lower boiling points so condense higher up the column ***[1 mark]***.
5.1 carbon dioxide/CO_2 ***[1 mark]*** and water/H_2O ***[1 mark]***
5.2 $C_3H_8 + 5O_2 \rightarrow 3CO_2 + 4H_2O$ ***[1 mark for correct reactants and products, 1 mark for correct balancing]***
5.3 oxidation ***[1 mark]***
6.1 Because shorter-chain hydrocarbons generated by cracking tend to be more useful than longer ones ***[1 mark]***.
6.2 $C_8H_{18} \rightarrow \mathbf{C_2H_6} + \mathbf{2}C_3H_6$ ***[1 mark for correct products, 1 mark for correct balancing]***
6.3 E.g. catalytic cracking is where long-chain hydrocarbons are vaporised ***[1 mark]*** and then passed over a hot powdered aluminium catalyst ***[1 mark]***. Or they can be cracked by steam cracking which is where long-chain hydrocarbons are vaporised and mixed with steam ***[1 mark]***. They are then heated to a very high temperature ***[1 mark]***.

You don't have to mention the names of the different cracking methods (catalytic and steam) but it might help structure your answer.

Page 143

Warm-Up Questions

1) Alkenes are hydrocarbons (compounds that only contain hydrogen and carbon atoms) and have the functional group C=C.
2) It contains two fewer hydrogen atoms than the alkane with the same number of carbon atoms.
3) The C=C bond opens up to make a single bond and the two carbon atoms can each bond to a halogen atom.
4) The bromine water will go colourless.

Exam Questions

1

Name of alkene	Formula	Displayed formula
ethene	C_2H_4	$H_2C=CH_2$ (H, H on each C)
propene	C_3H_6	$H_2C=CH-CH_3$ (H, H / C=C–C–H with H above and below)

[4 marks available — 1 mark for each correct name or formula]

The displayed formula for propene could also be drawn with the double bond between the second and third carbon atoms. As long as you've got the right number of hydrogen atoms attached to each carbon atom, it doesn't matter which two carbons the double bond is between.

2.1 $C_2H_{4(g)} + H_2O_{(g)} \rightarrow C_2H_5OH_{(g)}$
[1 mark for correct reactants and products, 1 mark for correct state symbols]

2.2 addition ***[1 mark]***

2.3 $C_2H_4 + H_2 \rightarrow C_2H_6$ ***[1 mark for correct reactants and products]***

2.4 Alkenes undergo incomplete combustion ***[1 mark]***.

3.1
```
  ( H   H  )
  ( |   |  )
 —( C — C )—
  ( |   |  )
  ( CH3 H  )n
```
[1 mark for all bonds drawn correctly]

3.2 poly(propene) ***[1 mark]***

Page 150

Warm-Up Questions

1) $C_nH_{2n+1}OH$
2) methanoic acid, ethanoic acid, propanoic acid, butanoic acid
3) -COO-
4) condensation polymerisation
5) amino acids

Exam Questions

1 **C *[1 mark]***

2.1
```
    H     O
    |   //
 H— C — C
    |    \
    H     O—H
```
[1 mark for showing the –COOH functional group correctly, 1 mark for showing three hydrogens surrounding the carbon atom that doesn't have the functional group attached]

2.2 ethyl ethanoate ***[1 mark]***

2.3 Carboxylic acids don't ionise completely in water/ not all acid molecules release their H^+ ions ***[1 mark]***. The have a higher pH than solutions of strong acids in the same concentration ***[1 mark]***.

3.1 propanol ***[1 mark]***

3.2 $C_2H_5OH + 3O_2 \rightarrow 2CO_2 + 3H_2O$
[1 mark for correctly balanced equation]

4 How to grade your answer:
Level 0: There is no relevant information. ***[0 marks]***
Level 1: There is a brief description of some differences between addition polymerisation and condensation polymerisation. ***[1 to 2 marks]***
Level 2: There is a detailed description of all the key differences between addition polymerisation and condensation polymerisation. ***[3 to 4 marks]***
Here are some points your answer may include:
Addition polymerisation only has one monomer type containing a carbon-carbon double bond whereas condensation polymerisation includes two monomer types or one type of monomer type with two different functional groups.
Condensation polymerisation produces two products, the polymer and a small molecule but addition polymerisation only produces one product.
Addition polymerisation involves only one functional group in the monomer whereas condensation polymerisation occurs between two functional groups on each monomer.

Topic 8 — Chemical Analysis

Page 156

Warm-Up Questions

1) A pure substance is a substance that only contains one compound or one element.
2) Impurities in a substance will increase the boiling point of the substance and may result in it boiling over a wider temperature range.
3) E.g. paint / drugs / cleaning products / fuels / cosmetics / fertilisers / metal alloys / food / drink.
4) Bubble a gas through (or shake it with) an aqueous solution of calcium hydroxide/limewater. If carbon dioxide is present, the solution will turn cloudy.
5) Chromatography separates out the substances in a mixture.

Exam Questions

1.1 Light the gas with a lighted splint ***[1 mark]***. Hydrogen will burn with a squeaky pop ***[1 mark]***.

1.2 Place a glowing spint in the sample of gas ***[1 mark]***. Oxygen will relight it ***[1 mark]***.

2.1 D ***[1 mark]***

2.2 A ***[1 mark]***

2.3 C ***[1 mark]***

2.4 R_f = distance travelled by substance in sunrise yellow ÷ distance travelled by solvent
$R_f = 9.0 \div 12.0 =$ **0.75**
[2 marks for correct answer, otherwise 1 mark for using the correct formula to calculate R_f]

Page 161

Warm-Up Questions

1) potassium (ions)
2) a) blue
   b) $Cu^{2+}_{(aq)} + 2OH^-_{(aq)} \rightarrow Cu(OH)_{2(s)}$
3) The concentration of the ion in the solution.

Exam Questions

1

Flame colour	Metal ion
green	**Ba^{2+}**
crimson	Li^+
yellow	**Na^+**
orange-red	Ca^{2+}

[1 mark for each correct answer.]

2.1 A sample of the element is put into a flame ***[1 mark]***. The sample gives out light when heated / Heating the sample excites the electrons in the atoms. When the electrons drop back to their original energy levels, they release energy as light ***[1 mark]***. This light is passed through a spectroscope, which produces the line spectrum ***[1 mark]***.

2.2 Caesium and rubidium have different arrangements of electrons ***[1 mark]***, so emit different wavelengths of light when heated ***[1 mark]***.

3.1 iron(II) iodide ***[1 mark for iron(II), 1 mark for iodide]***

3.2 No, the student is not correct. The sample produced a white precipitate with hydrochloric acid and barium chloride solution, so he can be sure that it contains sulfate ions ***[1 mark]***. Magnesium does produce a white precipitate with sodium hydroxide ***[1 mark]***, but so do other metal ions / aluminium and calcium, so he cannot be sure that it contains magnesium ions ***[1 mark]***.

Topic 9 — Chemistry of the Atmosphere

Page 170

Warm-Up Questions

1) volcanic activity
2) E.g. limestone / coal
3) E.g. coal / crude oil / natural gas
4) E.g. walk, cycle, or take public transport instead of using their car for a journey / reduce their use of air travel / turn down their heating at home.
5) Any two from: e.g. carbon monoxide / soot / unburned hydrocarbons / particulates.

Exam Questions

1 Once green plants had evolved, they thrived in Earth's early atmosphere which was rich in **carbon dioxide** ***[1 mark]***. These plants also produced **oxygen** ***[1 mark]*** by the process of **photosynthesis** ***[1 mark]***. Some of the **carbon** ***[1 mark]*** from dead plants eventually became 'locked up' in fossil fuels.

2.1 When sulfur dioxide gas reacts with water in clouds ***[1 mark]*** it forms dilute sulfuric acid, which falls as rain ***[1 mark]***.

2.2 E.g. it kills plants / damages buildings and statues / makes metals corrode ***[1 mark]***.

2.3 Incomplete combustion of fossil fuels produces particulates ***[1 mark]***. Particulates (and the clouds they help to produce) reflect sunlight back into space ***[1 mark]*** so less light reaches the Earth ***[1 mark]***.

2.4 Short wavelength radiation from the sun passes through the Earth's atmosphere ***[1 mark]***. It hits the Earth's surface and is reflected back as long wavelength radiation ***[1 mark]***. This long wavelength radiation is absorbed by greenhouse gases in the atmosphere ***[1 mark]***. They re-radiate it in all directions, including back towards the Earth, warming its surface ***[1 mark]***.

3.1 A measure of the amount of carbon dioxide and other greenhouse gases released over the full life-cycle of a product, service or event. ***[1 mark]***.

3.2 How to grade your answer:

Level 0: There is no relevant information. ***[No marks]***

Level 1: There is a brief description of at least one potential effect of an increase in global temperature. Some attempt is made to explain what impact this effect could have on humans. ***[1 to 2 marks]***

Level 2: There is some description of at least two potential effects of an increase in global temperature and an explanation of the potential impact of these effects on humans. ***[3 to 4 marks]***

Level 3: There is a clear and detailed explanation of at least three effects of an increase in global temperature and a clear explanation of the potential impact of these effects on humans. ***[5 to 6 marks]***

Here are some points your answer may include:

Effect: An increase in global temperature could lead to polar ice caps melting. This could cause a rise in sea levels.

Impacts: Rising seas levels could cause flooding in coastal areas and cause coastal erosion.

Flooding and coastal erosion could make people who live near the coast homeless / cause costly damage to buildings and roads / pose a threat to human life.

Effect: Climate change could lead to changes in rainfall patterns.

Impacts: Some regions may have too much or too little water. Having too little water may affect the ability of certain regions to produce enough food for the human population. This could lead to famine or even starvation.

Too much water could lead to flooding.

Flooding could make people homeless / cause costly damage to buildings and roads / pose a threat to human life.

Effect: Climate change may lead to an increase in both the frequency and severity of storms.

Impacts: Storms can cause damage to buildings and roads and can pose a direct threat to human life.

Effect: Rising temperatures can directly affect the human and wildlife populations in an area.

Impacts: An increase in temperature may affect what food plants can be grown in a certain region.

This could lead to famine or even starvation.

Topic 10 — Using Resources

Page 177

Warm-Up Questions

1) Non-metal solids with high melting points that aren't made from carbon-based compounds.

2) They contain cross-links between polymer chains that hold the chains together in a solid structure.

3) copper and zinc

4) iron + oxygen + water → hydrated iron (iii) oxide

5) It coats the iron with a layer of different metal that won't be corroded.

Exam Questions

1.1 Their properties make them more suitable for the application ***[1 mark]***.

1.2 E.g. high carbon steel is strong/brittle whereas low carbon steel is softer/more easily shaped ***[1 mark]***.

1.3 E.g. blades for cutting tools / bridges ***[1 mark]***

1.4 stainless steel ***[1 mark]***.

2.1 A ***[1 mark]***

2.2 The nail in tube D would not rust whereas the nail in tube B would ***[1 mark]***. The paint covering the nail in tube D creates a barrier that keeps out water and oxygen and so prevents rust from forming ***[1 mark]***.

2.3 Zinc is a more reactive metal than iron ***[1 mark]***, so water and oxygen will react with this before iron ***[1 mark]***. Also, the coating of zinc acts as a barrier that keeps out water and oxygen ***[1 mark]***.

Page 183

Warm-Up Questions

1) E.g. natural rubber

2) Development that takes account of the needs of present society while not damaging the lives of future generations.

3) E.g. saves energy needed to extract metals from the earth / conserves limited supplies of metals from the earth / cuts down on the amount of waste going to landfill.

4) E.g. getting the raw materials / manufacturing and packaging / using the product / product disposal.

Exam Questions

1.1 A natural resource that doesn't renew itself quick enough to be considered replaceable ***[1 mark]***.

1.2 Any two from: e.g. reuse the cans / recycle the aluminium / use fewer cans ***[2 marks]***.

2.1 E.g. because the supply of copper-rich ores is limited ***[1 mark]***.

2.2 Bacteria convert copper compounds in the ore into soluble copper compounds which separate out from the ore ***[1 mark]***. This produces a (leachate) solution containing copper ions, from which copper can be extracted by electrolysis/ displacement with a more reactive metal e.g. scrap iron ***[1 mark]***.

2.3 E.g. these processes have a smaller environmental impact ***[1 mark]***.

2.4 E.g. these processes are slower ***[1 mark]***.

3.1 The effect of some pollutants produced in the life cycle of a product can be difficult to quantify ***[1 mark]***.

3.2 E.g. the person carrying out the company's LCA may have purposefully not included details of the disposal of the product ***[1 mark]***, because the company wanted to use the LCA to support that their product was environmentally friendly for advertising reasons ***[1 mark]***.

Page 187

Warm-Up Questions

1) Any two from: e.g. the levels of dissolved salts in the water aren't too high / the water has a pH of between 6.5 and 8.5 / there aren't any bacteria or other microbes in the water.

2) filtration

3) harmful chemicals

4) effluent and sludge

Exam Questions

1.1 Potable water contains dissolved substances whereas pure water only contains water molecules ***[1 mark]***.

1.2 First, large bits of material are removed by a wire mesh ***[1 mark]***, then sand and gravel beds filter out any other solid bits ***[1 mark]***.

1.3 E.g. chlorine / ozone / ultraviolet light ***[1 mark for each correct answer up to a maximum of three marks.]***

2.1 screening ***[1 mark]***

2.2 The effluent undergoes aerobic digestion ***[1 mark]***. During aerobic digestion air is pumped through the effluent which encourages aerobic bacteria to break down any organic matter in the water ***[1 mark]***.

2.3 E.g. processes used to treat waste and sea water use more energy than treating fresh water and are therefore more expensive ***[1 mark]***. People may not like the idea of drinking treated waste water ***[1 mark]***.

3.1 E.g. sodium ions can be tested for by carrying out a flame test using a sample of water ***[1 mark]***. If sodium ions are present the flame will turn yellow ***[1 mark]***. To test for chloride ions a few drops of dilute nitric acid should be added to a sample of water, followed by a few drops of silver nitrate solution ***[1 mark]***. If chloride ions are present, a white precipitate will form ***[1 mark]***.

3.2 E.g. the pH of the water would need to be tested to see if it is between 6.5 and 8.5 ***[1 mark]***.

Page 192

Warm-Up Questions

1) hydrogen and nitrogen
2) true
3) iron
4) Fertilisers that are formulations containing salts of nitrogen, phosphorus and potassium.
5) potassium chloride / potassium sulfate

Exam Questions

1.1 phosphorus ***[1 mark]*** / potassium ***[1 mark]***

1.2 E.g. fertilisers provide/replace missing elements in the soil that are essential for plant growth ***[1 mark]***. This helps to increase crop yield ***[1 mark]***.

2.1 It contains nitrogen from two sources — the ammonia and the nitric acid ***[1 mark]***

2.2 Phosphate salts in the rock are insoluble so plants can't directly absorb them and use them as nutrients ***[1 mark]***.

2.3 calcium sulfate and calcium phosphate ***[1 mark]***

3.1 **A**: natural gas ***[1 mark]*** **B**: air ***[1 mark]***

3.2 $3H_{2(g)} + N_{2(g)} \rightleftharpoons \mathbf{2NH_{3(g)}}$ ***[1 mark]***

3.3 It would increase ***[1 mark]***. Lowering the temperature moves the position of equilibrium towards the products ***[1 mark]***.

3.4 If the pressure was higher the process may become too dangerous/expensive to build and maintain ***[1 mark]***.

Practice Paper 1

Pages 200-218

1.1 +1 ***[1 mark]***

1.2 NaF ***[1 mark]***

1.3 Any two from: e.g. potassium fizzes more than sodium, which fizzes more than lithium / potassium moves more quickly than sodium, which moves more quickly than lithium / potassium melts and a flame is seen, sodium melts but no flame is seen and lithium does not melt ***[1 mark for each correct observation]***.

1.4 As you go down Group 1 the outer electron gets further from the nucleus ***[1 mark]***. This means that the outer electron is more easily lost because it feels less attraction from the nucleus ***[1 mark]***. So reactivity increases as you go down Group 1/from lithium to sodium to potassium ***[1 mark]***.

1.5 $2Li + 2H_2O \rightarrow \mathbf{2LiOH + H_2}$
[1 mark for formulas of both products correct, 1 mark for correct balancing]

1.6 It contains hydroxide ions ***[1 mark]***.

2.1 James Chadwick ***[1 mark]***

2.2

Particle	Relative mass	Relative charge
Proton	1	+1
Neutron	1	0
Electron	Very small	–1

[2 marks for all correct, 1 mark for at least two correct]

2.3 C and D ***[1 mark]***. They are the only pair to have the same number of protons but a different number of neutrons ***[1 mark]***.

2.4 Protons = 30 ***[1 mark]***
Neutrons = 64 – 30 = 34 ***[1 mark]***
Electrons = 30 ***[1 mark]***

Remember: number of protons = atomic number, number of neutrons = mass number – atomic number, number of electrons (in a neutral atom) = number of protons.

2.5 A compound is a substance made of two or more different elements ***[1 mark]*** that are present in fixed proportions and chemically bonded together ***[1 mark]***.

2.6 $M_r = 56 + 32 + (4 \times 16) = \mathbf{152}$
[2 marks for correct answer, otherwise 1 mark for writing an expression that could be used to calculate the relative formula mass of iron(II) sulfate.]

3.1

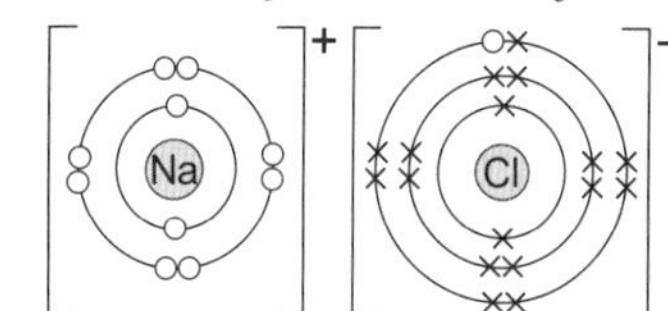

[1 mark for adding seven crosses and one dot to outer shell of Cl^- ion, 1 mark for correct charge on both ions.]

3.2 Sodium chloride has a giant lattice structure ***[1 mark]*** held together by strong electrostatic forces of attraction/bonds between positive and negative ions ***[1 mark]***.

3.3 The molecules in substance A are only attracted to each other by weak intermolecular forces ***[1 mark]*** which don't need much energy to break/overcome ***[1 mark]***.

3.4 Substance C ***[1 mark]***. It has a high melting and boiling point and can conduct electricity when molten or dissolved but not when solid ***[1 mark]***.

3.5 How to grade your answer:

Level 0: There is no relevant information. ***[No marks]***

Level 1: A brief attempt is made to explain one or two of these properties in terms of structure and bonding. ***[1 to 2 marks]***

Level 2: Some explanation of three or four of the properties, in terms of their structure and bonding, is given. ***[3 to 4 marks]***

Level 3: Clear and detailed explanation of five or all of the properties, in terms of their structure and bonding, is given. ***[5 to 6 marks]***

Here are some points your answer may include:

Diamond

Each carbon atom in diamond forms four covalent bonds in a rigid giant covalent structure/tetrahedral lattice, making it very hard.
Because it is made up of lots of covalent bonds, which take a lot of energy to break, diamond has a very high melting point.
There are no free/delocalised electrons in the structure of diamond, so it can't conduct electricity.

Graphite

Each carbon atom in graphite forms three covalent bonds, creating sheets of carbon atoms that can slide over each other.
The carbon layers are only held together weakly, which is what makes graphite soft and slippery.
The covalent bonds between the carbon atoms take a lot of energy to break, giving graphite a very high melting point.
Only three out of each carbon's four outer electrons are used in bonds, so graphite has lots of free/delocalised electrons and can conduct electricity.

4.1 Any two from: e.g. have a lid on the beaker / use insulation (e.g. cotton wool) around the beaker / use a polystyrene beaker ***[1 mark for each correct answer]***.

4.2 1 °C ***[1 mark]***

4.3 37.5 °C ***[1 mark]***

4.4 Temperature change = final temperature – initial temperature
= 37.5 – 17 = **20.5 °C**
[2 marks for correct answer, otherwise 1 mark for writing a correct expression for calculating the temperature change.]

If your answer to 4.3 was wrong, you can still have both marks for correctly subtracting 17 from it to find the temperature change.

4.5 The reaction was exothermic as the temperature of the surroundings increased during the reaction ***[1 mark]***.

5.1 hydrogen/H_2 ***[1 mark]***

5.2 Reaction D ***[1 mark]***. The most reactive metal will react fastest with the acid ***[1 mark]***. In reaction D the largest volume of gas has been collected in the syringe / the most bubbles are being given off ***[1 mark]***.

5.3 e.g. zinc ***[1 mark]***

You would get this mark if you named any metal between magnesium and iron in the reactivity series (for example, aluminium would be fine here too).

5.4 Sodium is very reactive and would react explosively with acid ***[1 mark]***.

5.5 Gap for reaction of copper sulfate with iron: yes ***[1 mark]***
Gap for reaction of iron sulfate with magnesium: yes ***[1 mark]***

5.6 $Mg_{(s)} + Cu^{2+}_{(aq)} \rightarrow Mg^{2+}_{(aq)} + Cu_{(s)}$
[1 mark for the correct reactants, 1 mark for the correct products]

5.7 Mg / magnesium ***[1 mark]***

Remember, when you're talking about oxidation and reduction in terms of electrons, oxidation is the loss of electrons. In this reaction, the magnesium went from being neutral atoms to positively charged ions — so they must have lost electrons.

6.1 E.g. the solid will stop dissolving and some solid will still remain at the bottom of the beaker ***[1 mark]***.

6.2 water ***[1 mark]***

6.3 $ZnCl_2$ ***[1 mark]***

6.4 Filter the solution to remove the excess solid zinc oxide ***[1 mark]***, then heat the solution and allow some of the water to evaporate ***[1 mark]***. Leave the solution to cool until crystals form ***[1 mark]***. Filter the crystals from the remaining solution and allow them to dry ***[1 mark]***.

6.5 sodium chloride ***[1 mark]***

6.6 carbon dioxide ***[1 mark]***

6.7 E.g. test a sample of the solution using universal indicator paper / test the solution using a pH probe. If the paper turns green / the probe shows a pH of seven, then the solution is neutral. ***[1 mark for any sensible method you could use to test the pH of the solution, 1 mark for a correct, matching observation of what that method would show if the solution was neutral.]***

7.1 Oxygen is being removed from the metal oxide / Fe^{3+} ions (iron(III) ions) are gaining electrons ***[1 mark]***.

7.2 M_r of $Fe_2O_3 = (2 \times 56) + (3 \times 16) = 160$
Moles of Fe_2O_3 = mass $\div M_r = 40 \div 160 = 0.25$
From the balanced equation you know that 2 moles of Fe_2O_3 produce 4 moles of Fe. So 0.25 moles of Fe_2O_3 will produce $0.25 \times 2 = 0.50$ moles of Fe
Mass of Fe = moles $\times A_r = 0.50 \times 56 =$ **28 g**
[4 marks for correct answer, otherwise 1 mark for M_r of Fe_2O_3, 1 mark for number of moles of Fe_2O_3, 1 mark for number of moles of Fe.]

7.3 Aluminium is more reactive than carbon/above carbon in the reactivity series ***[1 mark]***. Reduction with carbon can only be used to extract metals that are less reactive than carbon/below carbon in the reactivity series ***[1 mark]***.

7.4 In pure iron, the atoms are arranged in layers ***[1 mark]***. These layers can slide over each other freely ***[1 mark]***. Steel also contains carbon atoms, which are a different size to the iron atoms ***[1 mark]***. The carbon atoms distort the layers, so they can't slide over one another so easily ***[1 mark]***.

7.5 Any three from: e.g. the rods should be the same size / the rods should be the same distance away from the heat / the same amount of petroleum jelly should be used on each rod / the wooden splints should be the same distance from the end of each rod / the wooden splints should be the same size ***[1 mark for each correct answer]***.

8.1 In a solution of copper sulfate, the ions are free to move and carry an electric charge ***[1 mark]***. When copper sulfate is solid, the ions are held in fixed positions and can't move ***[1 mark]***.

8.2 Cu^{2+}, SO_4^{2-}, H^+, OH^- ***[1 mark]***

The Cu^{2+} and SO_4^{2-} ions come from the ionic compound. The H^+ and OH^- ions come from the water.

8.3 $4OH^- \rightarrow O_2 + 2H_2O + 4e^-$ ***[1 mark]***

8.4 Mean change of mass for negative electrode = **2.54** ***[1 mark]***

8.5 At the negative electrode, copper ions/Cu^{2+} ions are being reduced to copper metal/Cu ***[1 mark]***. The half equation for this reaction is: $Cu^{2+} + 2e^- \rightarrow Cu$ ***[1 mark]***. So a layer of copper metal forms on the negative electrode during the electrolysis ***[1 mark]***.

8.6 The negative electrode would change in mass by twice as much ***[1 mark]***.

9.1 How to grade your answer:

Level 0: There is no relevant information. ***[No marks]***

Level 1: A brief attempt is made to explain one or two advantages of using high atom economy reactions in industry. ***[1 to 2 marks]***

Level 2: A good explanation of at least one environmental benefit and one economic benefit of using high atom economy reactions in industry is given. ***[3 to 4 marks]***

Level 3: A clear and detailed explanation of at least two environmental benefits and two economic benefits of using high atom economy reactions in industry is given. ***[5 to 6 marks]***

Here are some points your answer may include:

Environmental benefits

Reactions with high atom economies use less of their raw materials/use up raw materials more slowly than reactions with low atom economies.
Less raw materials will need to be extracted for a high atom economy reaction, reducing the amount of energy used and the pollution caused by extracting the raw material.
Reactions with high atom economies produce less waste than reactions with low atom economies.
This means that using a reaction with a high atom economy sends less waste to landfill and creates less pollution.
Using a high atom economy reaction also reduces the amount of energy used to treat waste made by the reaction.
Reactions with high atom economies are more sustainable than reactions with low atom economies.

Economic benefits

Since high atom economy reactions use less of their raw materials/use up raw materials more slowly, it is usually cheaper to buy the raw materials needed for a high atom economy reaction.
As high atom economy reactions produce less waste, a company using a high atom economy reaction will spend less money on treating waste.
A company using a high atom economy reaction will spend less money on energy bills, since they will use less energy extracting raw materials/treating waste.

9.2 M_r of desired product $= 2 \times M_r$ of LiCl $= 2 \times 42.5 = 85$
M_r of all reactants $= M_r$ of $Li_2CO_3 + (2 \times M_r$ of HCl)
$= 74 + (2 \times 36.5) = 147$

$$\text{Atom economy} = \frac{M_r \text{ of desired products}}{M_r \text{ of all reactants}} \times 100$$
$= (85 \div 147) \times 100 = 57.8... =$ **58%** (2 s.f.)

[4 marks for the correct answer, but deduct 1 mark if answer is not given to 2 s.f.. Otherwise 1 mark for finding the M_r of the desired product, 1 mark for finding the M_r of all reactants and 1 mark for substituting these values correctly into the atom economy formula.]

9.3 100% ***[1 mark]***

If a reaction only has one product, its atom economy must be 100%.

9.4 percentage yield = (actual yield ÷ theoretical yield) × 100
$= (170 \div 212.5) \times 100 =$ **80%** ***[1 mark]***

Practice Paper 2

Pages 219-238

1.1 *[1 mark for all four points correctly plotted, 1 mark for a smooth curve that passes through all the points]*

1.2 36 °C *[1 mark for any answer in the range 34-38 °C]*

1.3 The larger the alkane molecule, the higher its boiling point / the smaller the alkane, the lower its boiling point *[1 mark]*.

1.4 C_nH_{2n+2} *[1 mark]*

1.5
```
    H  H  H
    |  |  |
 H—C—C—C—H
    |  |  |
    H  H  H
```
[1 mark]

1.6 Add a few drops of bromine water to both compounds and shake *[1 mark]*. With propane, nothing will happen *[1 mark]*. Propene will decolourise the bromine water / turn the solution from orange to colourless *[1 mark]*.

1.7 $C_3H_8 + 5O_2 \rightarrow 3CO_2 + \mathbf{4H_2O}$
[1 mark for correctly giving H_2O as the missing product, 1 mark for correct number of moles of H_2O.]

1.8 Decane has a higher boiling point than pentane *[1 mark]*

2.1 As the level of nitrogen dioxide increases, the severity of asthma attacks increases / there is a positive correlation between nitrogen dioxide levels and severity of asthma attacks *[1 mark]*

2.2 The sample of people that the scientists collected data from was too big *[1 mark]*.

2.3 Oxides of nitrogen are formed when nitrogen and oxygen from the air react at high temperatures *[1 mark]*. This happens in car engines (which operate at high enough temperatures for nitrogen oxides to form) *[1 mark]*. In cities, there are lots of cars, so the levels of nitrogen oxides tend to be higher *[1 mark]*.

2.4 nitric acid *[1 mark]*

2.5 e.g. sulfur dioxide *[1 mark]*

2.6 Carbon monoxide binds with haemoglobin in the blood *[1 mark]*. This lowers the amount of oxygen that can be transported around the body *[1 mark]*.

3.1 E.g. they don't have enough fresh water / groundwater / surface water to treat and use as drinking water *[1 mark]*.

3.2 Desalination *[1 mark]*

3.3 How to grade your answer:

Level 0: There is no relevant information *[No marks]*

Level 1: There is a brief description of the method to distil seawater, but many details are missing or incorrect. *[1 to 2 marks]*

Level 2: There is some explanation of the method used to distil sea water. Most of the information is correct, but a few small details may be incorrect or missing. *[3 to 4 marks]*

Level 3: There is a clear, detailed and fully correct explanation of the method used to distil sea water in the lab. *[5 to 6 marks]*

Here are some points your answer may include:
Pour the sea water into a (round bottom) flask.
Attach the flask to a condenser and secure both with clamps.
Connect a supply of cold water to the condenser.
Place a beaker under the condenser to collect the fresh water.
Place a Bunsen burner under the round bottom flask and heat the sea water slowly.
The water will boil and form steam.
The steam will condense back to pure liquid water in the condenser and be collected in the beaker as it runs out.
Continue to heat the (round bottom) flask until you have collected a reasonable amount of pure water.

3.4 Large bits of material are removed from the sewage *[1 mark]*.

3.5 The effluent undergoes aerobic biological treatment *[1 mark]*. Air is added to encourage aerobic bacteria to break down any organic matter that is present *[1 mark]*.

4.1 –COOH *[1 mark]*

4.2 $(0.39 + 0.55 + 0.45) \div 3 =$ **0.46 g** (2 s.f.)
[2 marks for correct answer, otherwise 1 mark for writing a correct expression that could be used to find the mean]

4.3 The loss of mass for each run is less than 1 g *[1 mark]*.

4.4 The gas is carbon dioxide *[1 mark]*. The student should bubble the gas through/shake the gas with an aqueous solution of calcium hydroxide/lime water *[1 mark]*. If carbon dioxide is present the solution will turn cloudy *[1 mark]*.

4.5 The student could perform a flame test *[1 mark]*. If sodium ions are present the substance would burn with a yellow flame *[1 mark]*.

5.1 Any two from: e.g. PVC is an electrical insulator/does not conduct electricity, so it makes the wires safe to touch / PVC is very flexible, so wires that are coated in it can bend / PVC is light, so it won't weigh down the wires / PVC has high corrosion resistance, so it protects the wires inside from corrosion/exposure *[1 mark for each correct reason]*.

5.2 ceramic *[1 mark]*

5.3 Aluminium alloys are far less dense than steel (3 g/cm^3 compared to 7.8 g/cm^3), so the parts are much lighter *[1 mark]*. They are also more resistant to corrosion *[1 mark]*.

5.4 Plants are grown in soil containing copper compounds *[1 mark]*. The plants don't use much of the copper so it accumulates in the leaves *[1 mark]*. The plants are then dried and burnt, leaving ash containing soluble copper compounds *[1 mark]*. The copper can then be extracted by electrolysis / displacement using scrap iron *[1 mark]*.

5.5 How to grade your answer:

Level 0: There is no relevant information *[No marks]*

Level 1: A sensible material has been chosen. An attempt has been made to justify the choice by referring to one of the properties in the table. *[1 to 2 marks]*

Level 2: A sensible material has been chosen, and there is a good justification of the choice that refers to two or three of the properties shown in the table. *[3 to 4 marks]*

Level 3: A sensible material has been chosen, and there is a clear and detailed justification of the choice that refers to all four of the properties given in the table. *[5 to 6 marks]*

Here are some points your answer may include:
The golf club shaft needs to be strong to withstand forces applied when the ball is hit.
Carbon fibre and steel have the greatest strength (4100 MPa and 780 MPa, respectively). Neither iron (200 MPa) or lead (12 MPa) would be strong enough.
The shaft needs to be as lightweight as possible, since the club will be both carried around and lifted into the air.
Carbon fibre has the lowest density of all the materials (1.5 g/cm^3), so would make the lightest club.
Stainless steel and iron are both denser than carbon fibre, but with a density of 7.8 g/cm^3 they would make lighter clubs than lead.

The club will be used outdoors in all weather, so it should be corrosion resistant.
Carbon fibre, stainless steel and lead are all corrosion resistant, but iron is not, so would not be suitable.
The cost of carbon fibre is very high. Any of the other materials shown would be cheaper.
E.g. though carbon fibre is expensive, because the golf club is a professional golf club, the cost of the club can be high, so carbon fibre would be the best choice.

Make sure that you remember to say what you think is the best choice of material in questions like this. It would have been OK to pick stainless steel here too (as long as you justified your choice clearly), because it would still be a suitable material. It wouldn't be OK to pick lead or iron because they are unsuitable (neither of them is strong enough, iron is not corrosion resistant and lead is very dense).

6.1 carbon dioxide ***[1 mark]***

6.2 nitrogen ***[1 mark]***

6.3 Algae and plants evolved to carry out photosynthesis ***[1 mark]***. During photosynthesis carbon dioxide is absorbed, so the amount of carbon dioxide in the atmosphere decreased ***[1 mark]***. Photosynthesis also produces oxygen, so the amount of oxygen in the atmosphere increased ***[1 mark]***.

6.4 Deforestation — e.g. plants take carbon dioxide out of the atmosphere, so removing trees causes carbon dioxide levels to rise ***[1 mark]***.
Increased energy consumption — e.g. burning fossil fuels releases carbon dioxide into the atmosphere ***[1 mark]***.
Increasing population — any one from: e.g. more people will be respiring and releasing carbon dioxide / more energy will be used (and carbon dioxide is released when fuels are burnt) / more land will be needed, so trees will have to be cut down, and less carbon dioxide will be removed from the air ***[1 mark]***.

6.5 E.g. carbon dioxide is a greenhouse gas ***[1 mark]***, so an increase in carbon dioxide in the atmosphere could lead to an increase in global temperatures/global warming ***[1 mark]***. As temperatures rise, global food production could decrease due to more extreme weather events / loss of farming land available / water shortages ***[1 mark]***.

7.1 Any two from: e.g. volume of acid / size of solid particles / temperature ***[1 mark for each correct answer]***

7.2 The rate of the reaction decreases as the reaction proceeds ***[1 mark]***.

7.3 Draw a tangent to the graph at 50 s, e.g.:

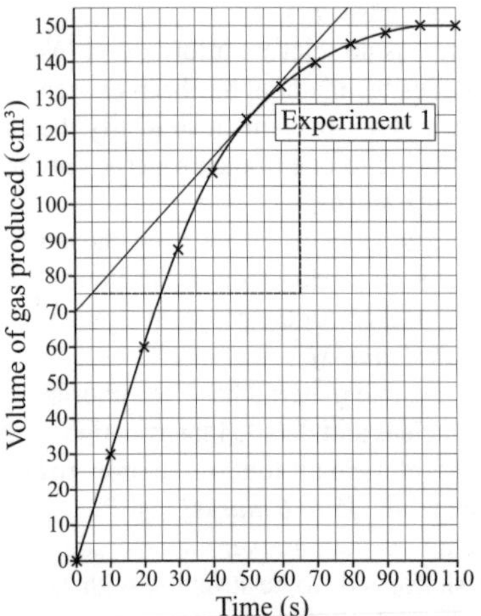

Then find the gradient of your tangent, e.g.:

$$\text{Gradient of tangent} = \frac{\text{change in } y}{\text{change in } x} = \frac{(140 - 75)}{(65 - 5)}$$

$= 65 \div 60 = 1.08...$

units = $cm^3 \div s = cm^3/s$

rate of reaction at 50 s = **1.1 cm^3/s** (2 s.f.)

[4 marks in total — 1 mark for a correctly drawn tangent, 1 mark for gradient calculation, 1 mark for correct answer in the range 1 - 1.2, 1 mark for correct units]

7.4 75 cm^3 ***[1 mark]***

You can see from the graph that the first experiment produced 150 cm^3 of gas in total. Calcium carbonate must be the limiting reactant (as the acid is in excess), so if you use half the mass, you'll produce half the amount of gas.

7.5 The gradient of the graph at the start of the reaction is steeper for Experiment 2 ***[1 mark]***.

7.6 Increasing the concentration of a reactant increases the number of particles of that substance in the reaction mixture ***[1 mark]***. This increases the frequency of collisions between particles of the reactants, increasing the rate of the reaction ***[1 mark]***.

8.1 $H_2C{=}CH_2$ (H and H on each carbon, C=C) ***[1 mark]***

8.2 alkene ***[1 mark]***

8.3 The polymer chains in thermosetting polymers have cross-links between them ***[1 mark]***. The polymer chains in thermosoftening polymers are separate, and only held together by weak (intermolecular) forces ***[1 mark]***. When thermosetting polymers are heated the cross-links stop the polymer from melting ***[1 mark]***. When thermosoftening polymers are heated, the weak intermolecular forces holding the chain together are broken and the polymer melts ***[1 mark]***.

8.4 Condensation polymerisation ***[1 mark]***. You can tell this is condensation polymerisation because water/a small molecule is lost as the polymer is formed ***[1 mark]***.

8.5 amino acid ***[1 mark]***

8.6 Any one from: e.g. DNA / starch / cellulose / protein ***[1 mark]***

9.1 The forward and reverse reactions are happening at exactly the same rate ***[1 mark]***.

9.2 Le Chatelier's principle states that if you change the conditions of a reversible reaction at equilibrium the system will try to counteract that change ***[1 mark]***.

9.3 The yield of sulfur trioxide will increase ***[1 mark]***. When pressure is increased the position of equilibrium moves towards the side with fewer gas molecules ***[1 mark]***. There are two molecules of gas on the right, compared to three on the left, so the position of equilibrium will move to the right ***[1 mark]***.

9.4 The yield of sulfur dioxide will decrease ***[1 mark]***. Increasing the temperature will move the position of equilibrium towards the endothermic reaction ***[1 mark]***. As the reaction to form sulfur trioxide / the forward reaction is exothermic, the position of equilibrium will move to the left to counteract the increase in temperature ***[1 mark]***.

9.5 catalyst ***[1 mark]***

9.6 E.g. 15% of 7500 kg = 7500 × 0.15 = 1125 kg
new mass made in a day = 7500 + 1125 = 8625 kg
new mass made in 5 days = 8625 × 5 = **43 125 kg**
[3 marks for correct answer, otherwise 1 mark for finding the new mass made in a day and 1 mark for multiplying it by five to find the new mass made in five days.]

Index

Index

CAS45, CHS45